COUNTRYSIDE PLANNING

Should rural Britain continue to be dominated by farmed landscapes, or should these be transformed into managed countryside or woodland, or even into low-density exurbia on the American model? These are the issues that face an increasingly post-agricultural and post-industrial society such as Britain.

Countryside Planning addresses these concerns and provides an in-depth study of the rural debate. Beginning with the key concepts and issues, the author sets out the context in which planning operates and how society has constructed its own images of the countryside. Using three theoretical perspectives, the book describes the evolution of the current planning system and provides a basis for further discussion about the possible future for the countryside.

In the wake of the recent Rural White Paper, the book includes the major issues that affect contemporary rural Britain including the current reforms of the Common Agricultural Policy, the role of farmers as land managers and the hypocrisy of sustainable development and green tourism. Using boxed policy summaries throughout the text, as well as key question-and-answer sections in every chapter, the author treats policy and trends across the whole spectrum of countryside planning.

Countryside Planning is an in-depth and authoritative analysis of rural policy, and makes an important contribution to the countryside planning debate and the future of rural Britain.

Andrew Gilg is a Reader in the Department of Geography, University of Exeter.

DEDICATION

Saturday 17 March 1990: Murrayfield, Edinburgh
SCOTLAND 13 ENGLAND 7
The Grand Slam decider for both teams.

Scotland team
Hastings, Stanger, Hastings, Lineen, Tukalo, Chalmers, Armstrong, Sole, Milne, Burnell, Gray, Cronin, Jeffrey, White, Calder, and Turnbull substitute for White.

*

Saturday 18 February 1995: Parc des Princes, Paris
FRANCE 21 SCOTLAND 23
Scotland's first win in Paris since I was a student in 1969.

Scotland team
Hastings, Joiner, Townsend, Jardine, Logan, Chalmers, Redpath, Hilton, Milne, Wright, Cronin, Campbell, Wainwright, Peters, Morrison, and Weir for Cronin.

COUNTRYSIDE PLANNING

The First Half Century

Second Edition

Andrew W. Gilg

London and New York

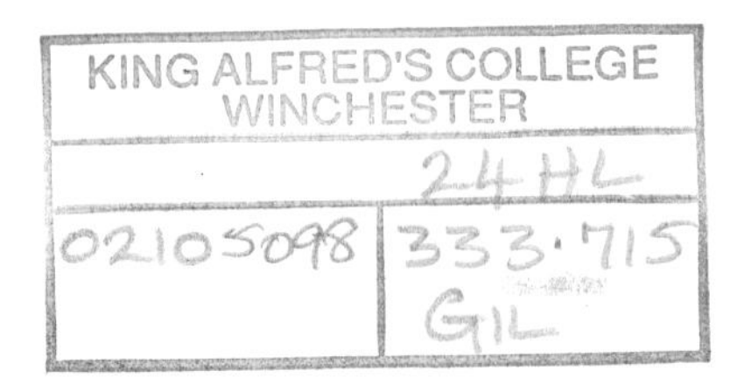

First published 1996
by Routledge
11 New Fetter Lane, London EC4P 4EE

Simultaneously published in the USA and Canada
by Routledge
29 West 35th Street, New York, NY 10001

Typeset in Garamond by LaserScript, Mitcham, Surrey
Printed and bound in Great Britain by
TJ Press (Padstow) Ltd, Padstow, Cornwall

British Library Cataloguing in Publication Data
A catalogue record for this book is available from the British Library

Library of Congress Cataloging in Publication Data
A catalogue record for this book has been requested

ISBN 0–415–05489–3 (hbk)
ISBN 0–415–05490–7 (pbk)

CONTENTS

ILLUSTRATIONS

FIGURES

TABLES

PREFACE

This book is a totally new edition of the first, which was published in 1978. Since then there have been three major changes. First, the thrust of policy has been markedly altered by 17 years of Conservative Government between 1979 and 1996. Second, agriculture has lost its primacy in rural planning as a result of European food surpluses, and this has presented other land users with an opportunity to fill the partial void. Third, there has been an explosion of literature on rural planning topics, notably from academia and the pressure groups. In addition to these changes, 1995 witnessed the publication of the first major 'White Paper' on rural areas since the Scott Report of 1942. The preparation of this new edition provided a perfect opportunity for a more critical commentary than that provided by the bland syrup of the 'White Paper'.

There was thus a need totally to reconceive the concept for the book used in 1978. The one central concept that has survived is that of incremental and evolutionary change as the main explanatory factor. The two books can thus be usefully read in parallel, since the first edition provides not only more detail on the 1945–75 period, but also a contrast, with themes and topics that were important then, but have since fallen by the wayside. In particular, the first edition included much material on planning techniques, but the literature has largely fallen silent on this topic since then, in favour of debates over research methodologies. Indeed, while empirical analyses dominated the first edition, this edition divides research material by the methodological and theoretical approach adopted by the researchers. The analysis is thus more discursive and hopefully more thought provoking than before.

In more detail, three major forms of analysis are used in the text:

1 Traditional meta-narrative empirical analyses based on the assumption that there is a policy effect, and that the impact of policies can be measured quantitatively in a cause-and-effect relationship, with, for example, winners and losers being identified usually by statistical or econometric methods.

2 Structural analyses which focus around a number of themes: political economy studies which examine the interplay between key agencies in formulating and implementing policies, and structural or regulation theory studies which often assume the primacy of capital in this interplay, or other structural factors which predetermine to a certain extent who will gain from policy changes. Both these approaches have the central concept that life is a struggle between people usually acting as organised groups.
3 A human agency or behaviouralist approach, however, argues that human irrationality can undermine these structural forces, and that actors can act in different ways dependent on the issue at stake; thus farmers may pass up the chance to drain ponds and grow more crops because they are keen bird-watchers. Similar approaches identify the importance of chance events, and factors such as chance have encouraged the development of a post-modern approach which rejects both the meta-narrative and structural approaches as arrogant simplifications of a chaotic complex universe.

These three groupings are very loose and could be easily reordered, but they have been used to provide the reader with some guidance about the perspective adopted by researchers and thus a stance with which to evaluate their findings. It is important to remember that public policy and research into policy is deeply value laden and judgemental, even though policy, notably town and country planning policy, is presented in terms of some common good, usually in the form of words like the 'public good' or 'amenity', as if everybody agreed what should happen, and that planners had been able to define what the public wants.

Instead, the reality of the situation is that planners have to act on imperfect information, about both the physical and economic environment and the socio-political culture they live within, and that public policy emerges and develops, not as a rational response to carefully researched issues, but in a process of crisis-driven and pragmatic incrementalism.

The nature of our imperfect world also leads to a number of other key assumptions about the evolution of countryside planning policies:

a) Policies not only evolve via crisis-driven incrementalism, but are added to in a process of accretionary incrementalism in which the policy matrix, like Topsy, just keeps growing, since policies are only rarely junked.

b) Policies are affected by unexpected responses by those affected, and by unforeseen events.

c) Policies, for example on education and housing, can be affected by policy developments in seemingly unrelated areas, notably general social policies, which can have major impacts.

d) Policies are created by individual departments or ministries and implemented in turn by discrete organisations. This leads to a fragmentation of policy and poor coordination between policies, as highlighted in c) above.

e) Policies are underresearched in spite of the attempt since the mid 1980s to force public bodies to set out performance targets and develop systems of policy evaluation, in a 'bottom-line' accountancy style of policy assessment used by the National Audit Office in their scathing attacks on the deemed inefficiency of many public policies. Research by academia and by pressure groups, in contrast, is often ad hoc, self-indulgent and focused on 'flavour of the month' topics, in spite of attempts by bodies awarding research grants to set the research agenda. Thus, though there has an explosion in the number of research papers and of research journals since the mid 1970s, this research output has focused on a few topics, for example, ESAs, and also contains a lot of replication by the relatively small group of active researchers in the field.

Thus the book adopts a broadly historical approach based on the key concept that people, and the context within which they find themselves, make history. Therefore, the best way to understand the present is to explain the route by which we arrived, even though, like the Irishman when asked the way to Donegal, we wouldn't start from here. In other words, public policy never starts with a blank sheet, or from an ideal place. We are prisoners of our past, but every day we are released into the future and the need to make our own history as we see fit.

This book also adopts an organisational and sectoral approach, because policy is made by organisations which have been based on sectoral land uses, for example, agriculture, forestry, nature conservation, recreation and so on. For this reason, the book is based on three key approaches: historical evolution; organisational structures; and sectoral land uses.

Indeed the book considers the physical countryside only. Thus socio-economic factors are only considered as they impact on the physical environment of the countryside, for example, especially, how people react to this environment, and how they try to mould it for their own purposes. It is not a book per se about rural people and, indeed, the title of the book, *Countryside Planning* rather than *Rural Planning*, has purposely been chosen to reflect the emphasis placed on the physical fabric of both the natural and built environments of rural areas.

The book has been structured to reflect the key assumptions and approaches outlined above, and apart from the first and last chapters, the remaining chapters are structured in a broadly similar way into the following sections:

Introductory / key concepts and questions / issues
Evolutionary events
The situation in 1995 / the 1990s
Analysis by three broad themes of: empirical meta-narrative; structural analysis; and human agency / post-modernism
Commentary and some thoughts on the future.

In more detail, each section is based on different literatures and is intended for different audiences. The introductory section A is based on the wider conceptual literature, my own reading of what issues are important as derived from a continuing project designed to produce a general model for *Regulating Rural Environments*, and is intended to be read by all readers, except those with a good grounding in theoretical rural geography. Section B on evolutionary events provides the essential history of planning legislation without which the rest of each chapter could not be understood, and represents the bedrock of the book. It is lightly referenced since it has been derived almost entirely from the 15 annual reviews of countryside planning I have written for the various versions of the *Countryside Planning Yearbook*, the *International Yearbook of Rural Planning* and *Progress in Rural Policy and Planning* between 1980 and 1995. Those with a good background in the histories of any of the planning sectors can, however, use these sections as an *aide-mémoire*. Section C on the Situation in the 1990s/1995 provides a brief summary of the situation when the book was being written, during the long hot summer of 1995. The Analysis section D provides a selective review of the literature, which because of space constraints cannot be comprehensive, and has thus been focused around the three broad methodological and theoretical perspectives outlined above. However, the book is not meant to be a text on rural geography or rural studies, and only discusses trends in these areas as they relate back to the key topic of the book, countryside planning. In addition, the research that has been carried out has only focused on certain policies. Readers are reminded that the analysis is partial, since there is incomplete evidence, and students in particular should thus find in these research lacunae inspiration for filling research gaps with their own projects, dissertations, or even PhD theses. In spite of these gaps, the section is intended to provide students with the material to answer the sting in the tail of exam and essay titles which typically ask: *Outline the development of forestry policy in the UK* (topic covered in sections B and C) *and assess its impact on the environment* (covered in section D). The literature has also been chosen to reflect the longer-term evolution of events, not only because the past helps to explain the present, but because policies which have been discarded or modified in the past may once again be relevant for the future. The Commentary section E thus looks at the future through the lens of the past, and attempts to answer the key questions posed by a checklist derived in Chapter 1.

The book is mainly intended to be read by students seeking a holistic understanding of countryside planning, and is thus targeted at students reading for degrees in geography, environmental studies, and land management. However, it will also be useful for students in agriculture, forestry, town and country planning, conservation, and recreation, who will be able to supplement their understanding of their own topic by an appreciation of the wider issues involved, and how different policies interact with each other.

The main aims of the book are therefore threefold. First, to provide an authoritative account and explanation of the evolution of countryside planning powers; second, to examine different interpretations of the impact of planning policies on the rural environment; and third to take a long-term view and assess how successful countryside planning has been so far in achieving its diverse aims.

Finally, many thanks are due to all those who helped me write the book. In particular the students at Exeter University who over the years since 1970 have listened to my increasingly post-modern discourses on rugby internationals, the meaning of life, and occasionally countryside planning. I also need to thank over a dozen post graduates who have produced countryside planning theses and helped to fill some of the research lacunae referred to above. Most of them find a place in the bibliography, but two of the most recent deserve a special mention, Martin Battershill and Mike Kelly. In addition, my colleagues have continued to support me with study leaves and advice, most notably, my erstwhile colleague, Mark Blacksell, and my new colleague, Jo Little. Further afield, I owe a huge debt of gratitude to the editors and contributors of the 15-volume series of countryside planning books referred to above, which should have mutated further into *Perspectives on British Rural Policy and Planning* by the time this book is published. In particular, I would like to thank Owen Furuseth, Robert Dilley, Geoff McDonald, Clive Potter and Philip Lowe. Finally, I need to thank my family for various secretarial tasks, and for helping me to renovate our rambling *Thatch in urbe* and in particular the 'East Wing' which contains my study, which was indeed gutted and rebuilt during the period in which this book was written, necessitating a move to another, even hotter, room *pro tem*. Therefore, Joyce, Julie and Alastair are thanked for all their help, notably for driving me back from the Alps in January 1995 with a dislocated and smashed shoulder.

The Alps, and in particular the slopes of Crans-Montana in Switzerland, also need to be thanked for providing the inspirational landscape which drives us forward in our endeavours. Returning to the first paragraph of this preface my progress from failed skier in 1971 (knee cartilege) to over-confident good intermediate skier in 1995 (dislocated shoulder) is a fourth major change between the two editions of this book.

The biggest and most important change, however, is that unlike the situation in 1978, when Scotland's Rugby Union team had failed to win a Five Nations championship in my lifetime, they have since then won everything on offer twice, with two Grand Slams in 1984 and 1990, notably in 1990, with an Exeter agricultural economics graduate, David Sole, as captain.

This book, like all my other sole-authored books, is thus dedicated to the

Scottish Rugby Union, in the hope that the potentially disastrous change to professionalism announced in August 1995, while Chapter 4 was being written, will not spoil the camaraderie and atmosphere of Five Nations Rugby weekends in Edinburgh, Dublin, Paris, Cardiff and London. Thus this preface closes with a dedication to the players who gave me and my family two perfect days.

Andrew W. Gilg
Moorview Lodge
Exeter
December 1995

ACKNOWLEDGEMENTS

I gratefully acknowledge permission to reproduce sometimes with modifications the following material: CAB International for Figures 2.1, 3.1 and 3.2; Addison, Wesley Longman for Figures 4.1, 4.3 and 6.2; HMSO, the Council for the Protection of Rural England and *Ecos* for Figure 4.2; *Ecos* for Table 3.1; *The Architects Journal* for Figure 5.1; The Town and Country Planning Association for Figures 5.2 and 7.2; *Town Planning Review* for Figures 5.3 and 7.1; HMSO for Figure 6.1 and Tables 2.1, 3.2 and 4.1; The Countryside Commission for Figure 6.3; and UCL Press for Table 6.1.

1

COUNTRYSIDE PLANNING

An introductory model based on elementary assumptions and some first principles

INTRODUCTION

In the second half of the twentieth century countryside planning has become one of the most vibrant and lively issues in current affairs, and it looks set to become one of the key tasks facing those who would try to shape the destiny of the new twenty-first century. For example, the debate over the Wildlife and Countryside Act 1981 prompted more letters to *The Times* newspaper than it has received on any other similar occasion in its history; more people are members of conservation groups like the National Trust and the Royal Society for the Protection of Birds than watch sport at the weekend; and programmes on wildlife and countryside access command prime time on TV. The Government published a major 'Rural White Paper' in 1995 which reported a Countryside Commission survey that found that 93 per cent of people considered that the countryside is valuable, and 91 per cent believed that society has a moral duty to protect the countryside for future generations (Cm 3016, 1995). What is it about the countryside and its evolution that causes such passion? There are two basic answers.

First, the countryside resonates with meanings relating to our most basic needs for food, shelter and procreation. A walk in the countryside will thus reveal to us: fields full of food; woods or hedgerows as places of shelter or abundant materials for building shelters; and wildlife rushing about in the throes of mating, building nests or rearing young. The countryside is thus a powerful metaphor for our own mortality as animals rather than sophisticated people cocooned in the artificial world of cyberspace.

Second, since 52 per cent of the countryside is owned by well under 1 per cent of the population, and a massive 75 per cent by only 3 per cent of the population, most of us feel powerless to control countryside change directly (Norton-Taylor, 1982). These feelings of frustration are compounded by the many perceived and real threats to the countryside which are constantly reported in the media, fuelled by specialised pressure groups like the Council for the Protection of Rural England. The only sensible option is thus indirect control via the political process and influencing the mechanisms of the various planning systems that have arisen in response to such pressures.

Although each planning system has arisen in its own peculiar way and has its own unique procedures, as the rest of this book will show all of them exhibit the key features of the cyclical model based on the classical sequence of actions first outlined by the famous polymath planner Patrick Geddes in the first years of this century. In essence, as the accompanying model shows (Groome, 1993), it involves a cyclical process in which the decision to make a planning response is triggered by a crisis, in for example, the economy or the environment. Once the process is triggered, a sequence of actions is then taken which leads eventually to a review of whether the initial crisis that triggered the activity has been remedied or not.

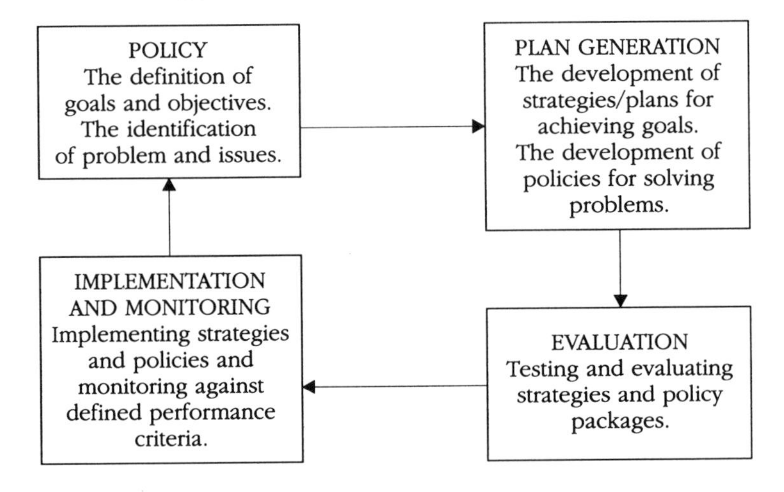

As the rest of this chapter and book will show, the model, although a useful conceptual device of the core activities in planning, is complicated by the chaotic events that happen in the real world, and in particular by three factors. First, there is an overlap over time as new crises impose themselves on previous crises. In addition, the causes of the initial crisis will change over time, but the planning response will often remain focused on the initial crisis. A classic example of this is provided by agricultural planning which even in the food surplus years of the 1970s–1990s remained rooted in the policies derived in the food shortage years of the 1940s and 1950s. Second, the planning response to crises in other areas of the economy or the environment may affect the efficacy of existing policies. For example, general changes in taxation policy can impact heavily on forestry. Third, not everyone will agree with the original diagnosis of the causes of the crisis, or with the policies and procedures set in train to remedy it. But the model assumes a consensus, whereas in reality society is a sea of conflict, although one of planning's main claims to legitimacy is that it can pose as an arbitration or neutral professional service.

There is therefore a need to derive a more complex model of the planning process which can take these complicating factors into account and also provide a framework for this introductory chapter. A simplified version of a much more complex model which is being developed by the author for another purpose (Gilg, 1994) is thus shown in Figure 1.1. This model expands the vision which portrays planning merely as a technical exercise in problem solving, by adding four core concepts: value systems; society; power; knowledge; and also a discussion of how the subject area, in this case the countryside, is conceived. The rest of this chapter outlines the key features of the model in order to provide a conceptual framework for the subject-specific chapters which follow.

VALUE SYSTEMS

The countryside and planning are both social constructs based on individual value systems. The key point to be made here is that traditional value systems based on a common religion and moral philosophy and the legal system based on Judaeo-Christian ethics have both become increasingly challenged during this century, as religious belief has fallen away. The main movement has thus been one away from absolute beliefs, as typified by the Ten Commandments, to one of relative beliefs which has been labelled 'permissiveness' by its critics and 'political liberalism' by its advocates (Rawls, 1993). Planners are thus faced with an increasing variety of opinions about how we should control our collective actions, and thus how we should try to manage countryside change and to what purpose. Traditionally, planners have taken the side of ruling elites and focused their policies through the lenses of landowners, although research has shown considerable disparities between the classes on these issues (Pattie *et al.*, 1991). Since the 1960s, however, an increasingly vocal electorate, in the form of pressure groups, has demanded more participation in decision making and that their value systems be considered, so that an 'old' and 'new' politics has emerged which to some extent cuts across the traditional spectrum of left to right wing politics (Abercrombie *et al.*, 1988). In particular, environmental groups ranging from the traditional to the radical have entered the scene, and while most are happy to employ peaceful forms of action, a radical fringe is willing to embark on 'direct' action, notably where animal welfare in agriculture is involved.

Planners can either celebrate this diversity of viewpoint or, more traditionally, they can try to seek consensus. To do this their favoured technique is 'public participation', by which the views of the public are sought at various stages of the planning process. Typically, therefore the cyclical planning model can be modified as shown by Turoff (1975) and de Loe (1995):

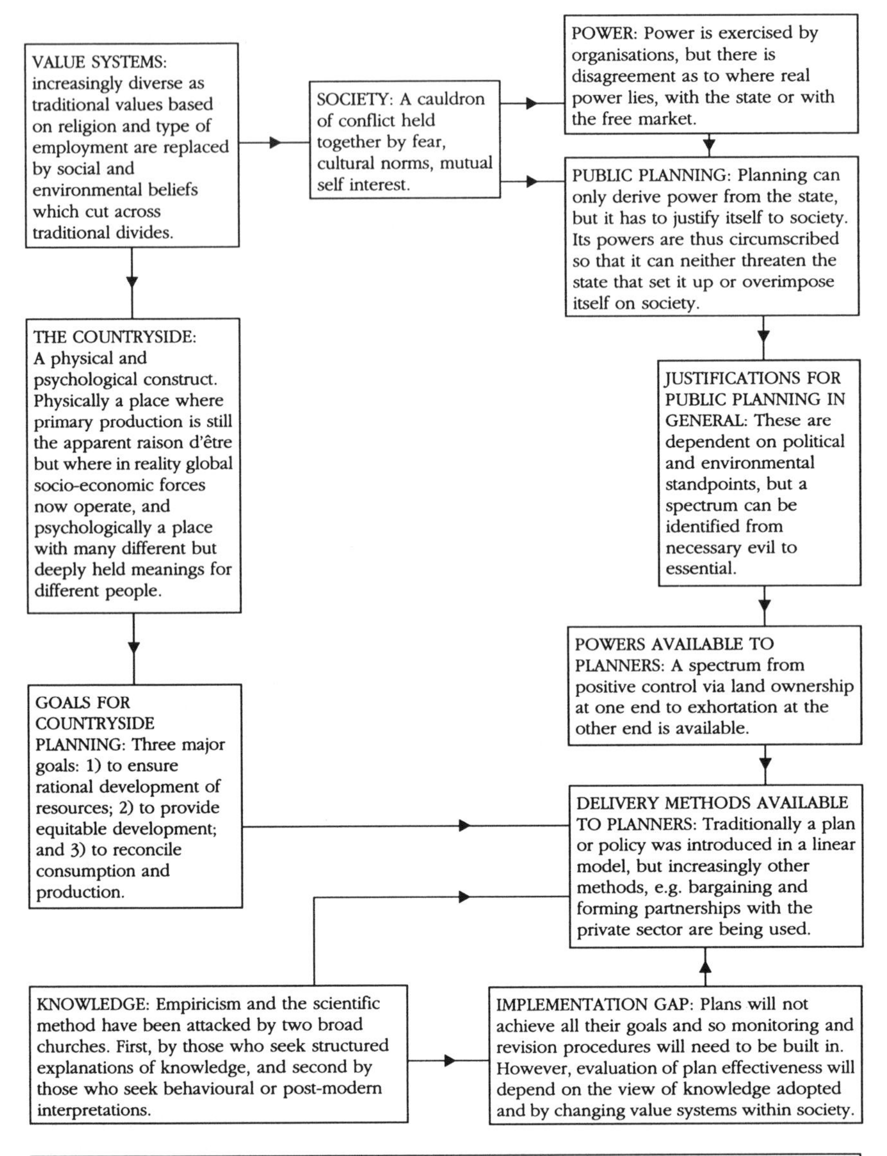

Figure 1.1 Simplified model of the relationships between society, power and countryside planning

Source: Author

1 Formulation of the issues. What is the issue that really should be under consideration?
2 Given the issue, what are the policy options available?
3 Determining initial positions on the issues. Which are the ones everyone already agrees upon, and which are the ones exhibiting disagreement among the respondents?
4 Exploring and obtaining reasons for the disagreements. What underlying assumptions, views or facts are being used by the individuals/groups to support their respective positions?
5 Evaluating the underlying reasons. How do individuals/groups view the separate arguments used to defend various positions and how do they compare to one another on a relative basis?
6 Re-evaluating the options. Re-evaluation is based upon the views of the underlying 'evidence' and the assessment of its relevance to each position taken.

SOCIETY

Society is a word that has many meanings, but three key ones stand out: first, *the customs and organisation of a civilised nation*; second, *a social community in which no society can retain members who flout its principles*; and third, *an association of persons united by a common aim or interest or principle*. Society thus implies a state of semi-equilibrium where people have foregone individual freedoms and desires in favour of a greater common good upon which all can agree as proposed by the influential philosopher John Rawls (1993). Crucially in doing so they have surrendered many powers over their own destiny to organisations who can impose their own norms and regulations. Thus, although a society may be in reality a cauldron of conflict, the fear of punishment by the state or retribution from others prevents most of us from departing from both legal and cultural norms, and long-term self-interest encourages us not to act in short-term anti-social ways. A prime example of how this process works is provided by the example of motor cars. Under the left-liberalism of John Rawls it is easy to defend the right to drive a car anywhere, anytime, but hard to support the collective goods of clean air, quiet and calm that excessive driving jeopardises. In response to this defect the increasingly influential school of 'communitarianism' has developed from both empirical evidence (Putnam *et al.*, 1993) and ideological first principles (Etzioni, 1993). Its key point is that the performance of institutions is shaped by the social context in which they operate, and that beneficial contexts are provided by a high degree of mutual trust, social cooperation and a well-developed sense of civic duty. Unfortunately, 'modernity' the creation of a technologically advanced industrial, commercial and urban society since the eighteenth century, has led to the separation of spheres of activity, secularisation, and the weakening of social ties. Thus Selznick (1992) in his book *The Moral Commonwealth: Social Theory and the Promise of Community* has advocated a hybrid, retaining features of liberalism but integrating more community-based values and systems, and combining it with elements of pragmatism, to produce a model which he calls 'communitarian liberalism'.

The concept of society is thus crucial to understanding planning, since planning has to assume eventually some common value systems around which to base its policies, and the degree of compliance or deviation from these policies that it can expect, depending on how far these policies accord with the views of all of society or only certain sections of it. In order to gain acceptance of these policies planners will also need to gauge the correct mix between self-interest incentives and retributive sanctions that it will employ. Put simply, society controls itself by sticks and carrots and planners similarly have this basic choice in attempting to mould the evolution of society and its economy and environment.

POWER

The previous section assumed a benevolent view of power in which individuals had willingly surrendered freedom to the state and its agency the Government, who govern by consent. This, however, represents one end of a spectrum which terminates at dictatorship. In between the most realistic models explain power being exercised in either a consensus society or a conflict-ridden society. Lukes (1974) sees power as an essentially contested concept which can be analysed at three levels. The first level, the classical pluralist approach, looks for the exercise of power in observable overt conflicts of interest, as defined by the participants to the conflict. The second level includes covert as well as overt conflicts. This occurs when self-defined conflicts of interest are recognised by participants, but are not actually expressed in the political arena. Power is thus exercised by suppressing conflicts and preventing their overt expression. The third level involves the acceptance that people or groups do not always accurately define their own interests. Through ignorance or manipulation, they wrongly perceive their own interests and therefore may not recognise a conflict with another group, whether overt or covert. Power is then exercised through the manipulation of people's perceptions of their interests, that is, in the arena of ideology.

In addition to these overall concepts there are also several other differences in how the exercise of power can be explained or executed, namely: the concept of the state; spatial considerations; top-down/bottom-up systems; democratic systems; the desirable degree of intervention; and the power of personality and chance events. These are now considered in turn.

The concept of the state. The traditional view of the state in Britain has been based on the 'pluralist' model in which each individual or group has equal access and influence to those in power. More realistically, an 'elitist' model argues that some individuals or groups have favoured access and unequal influence, for example farmers compared to ramblers. A variant of the 'elitist' view, the 'corporatist' model, contends that organised groups have the greatest influence, notably those groups that are thought to form the 'establishment', for example, Oxbridge-educated civil servants or professionals. A more radical view is based on Marxist rather than orthodox economics and argues that only the interests of capital have any influence, and that even the state is subservient to the imperative to accumulate capital. Examples of each view will be provided by the material discussed in this book.

Spatial considerations. We live in an increasingly global world, and power is now exercised extra-nationally at the global, continental and international levels. Nation states including Britain have thus had to surrender certain powers to these extra-national organisations. Examples include global agreements on atmospheric pollution, continental agreements on bird protection, and, most famously of all, international agreement on the Common Agricultural Policy.

Top-down/bottom-up systems. Within the nation state there are increasing calls for both federal/regional systems of government, for more decisions to be made at the local level', and for planning to take an integrating role in coordinating decisions at these different spatial scales. This is controversial in a country like the United Kingdom, which has traditionally employed a top-down system of power in which legislative decisions are taken at the centre and then executed with little or no discretion locally. The principle of subsidiarity is crucial here. Its core idea is that the overall framework in which decisions are made should be set centrally, but that the detailed execution should be decided locally. In essence this is encapsulated in the slogan: 'think globally, act locally.'

Democratic systems. Power is apparently exercised by politicians, but their accountability varies between different planning systems and at different spatial scales. The central principle in Britain is, however, one of 'first past the post'. This means that most people in power have been elected on less than 50 per cent of the vote, thus disenfranchising the majority. Nationally, voting is done on party lines, but in local, especially rural, areas, voting is often done more independently by both the electorate and the elected. There is virtually no precedent for either proportional representation or participation by referendum, and without these planning loses much of its legitimacy.

The desirable degree of intervention. The decision by the state or the government to intervene depends as much on the ideology involved as it does on the issue. Ideologically left-of-centre or centre parties like the Labour Party or the Liberal Democratic Party are instinctive intervenors, while right-of-centre parties like the Conservative Party are instinctive non-intervenors. Within this framework, however, all politicians have four basic options: first, do nothing or pretend to do something by setting up a committee and hope that the issue will go away or divert the blame elsewhere, for example, Europe; second, do the bare minimum to deflect public criticism; third, appropriate the policies of the opposition and steal their thunder; and fourth, make a virtue of a problem and actively espouse intervention as a good thing, thus turning defence into attack. In actual practice the most common option taken is a fifth one: modify existing policies by gradual incremental change, usually only in response to a crisis, so that policy making over time can be seen as a process of crisis-driven accretionary incrementalism, which thus produces a chaotic set of increasingly complex and contradictory policies.

The power of personality and chance events. Even with apparently robust systems powerful personalities can still change the course of events against not only the opposition of opponents, but even supposed allies. Thus, for example, Mrs Thatcher was able to not only move the goalposts but even the playing field, even though she never commanded more than just over 40 per cent of the votes cast, and only 30 per cent of the electorate overall in her term as Prime Minister between 1979 and 1990. Powerful personalities can also have considerable influence elsewhere so that the apparent direction of events can from time to time be quite markedly shifted.

Similarly, chance events can have a great influence. For example, in the 1980s the Countryside Commission had been under attack for several years, until the gale of October of 1987, which devastated large parts of south-east England. Faced with a baying media, the Secretary of State for the Environment suddenly found that the Countryside Commission could offer him a lifeline with a plan to repair the damage to the landscape caused by the gale, with its Task Force Trees programme. The then Director of the Commission, Adrian Phillips, has recounted (1993) that this event helped the Commission to raise its profile and to contribute to the general greening of government policy that took place in the late 1980s.

In conclusion, the exercise of power over any length of time can thus be seen as a somewhat random sequence of events shaped by personalities and pragmatism rather than conspiratorial or dictatorial powers, with incrementalism rather than revolution ruling the day.

PUBLIC PLANNING

All forms of public planning, as opposed to planning by individuals or private-sector organisations, imply the intervention of a third party into current activities. It is thus axiomatic that planners have to tread a delicate path between alienating the state by taking over too many of its powers or alienating the public who have sought their intervention. In addition, planning powers cannot be so onerous that they force legitimate reasonable activities to close, or be so lenient that they may as well not exist. In another case planners may have been asked to intervene or to continue to intervene in private-sector businesses, the most notable example being provided by farm and forest production subsidies. Whatever the form of public planning, however, the planners will find it hard to win since they are trying to impose the concepts of a common good on a diverse and pluralist society. Thus to survive, planners must continually justify their activities.

JUSTIFICATIONS FOR PUBLIC PLANNING IN GENERAL

Public planning is very much a creature of the twentieth century and has thus had to justify its birth, growth and survival into maturity. It has done so by advancing four groups of arguments across a broad spectrum so that as many sections of society will be seduced by their appeal as possible. The four groups can be labelled: necessary evil, reactive, proactive, and essential, and broadly correspond to the groupings to be found in the political spectrum (O'Riordan, 1985a), which range from the right wing (free marketeeers) to the left wing (socialists), although, as Abercrombie *et al.* (1988) have observed, the 'new politics' with its emphasis on environmental issues, specific groups and consumption does not always fit neatly with the 'old politics' based on the division between capital and labour.

Necessary evil. These are the arguments used to convince right wingers and are centred around negative actions preventing things. Key examples are: stopping or reducing damage; preventing profiteering; preventing chaos; preserving property rights/values; protecting the poor and the weak from exploitation; and preventing poor landscapes from being created.

Reactive. These are the arguments employed to convince people on the left wing of right-wing parties and the right wing of left-wing parties and are focused around making the market more efficient or correcting market failures. More left-wing arguments centre on the market's failure to take environmental goods and losses (pollution) into account, namely the 'externalities' of any activity. In the British context arguments include the fact that Britain is a small crowded island with a high and growing degree of conflict between land uses, most notably between urban and rural interests, and between food producers and consumers. In the wider ecological context, arguments range across different shades of 'greenness', varying from a need for conservation ranging from self interest, for example, conserving wildlife as a gene pool, to the more prosaic argument that wildlife has a right to exist equal to that of mankind. Typical arguments include: seeking consensus; resolving existing conflicts; preventing possible conflicts; balancing economic growth with the preservation of the environment; balancing efficiency with equity; managing change; coordinating development; and seeking the greatest good for the greatest number (utilitarianism). Ecologically, key arguments include the precautionary principle, the polluter pays principle, the concept of the best practicable environmental option, and keeping as many options as possible open, which is one of the core messages of conservation.

Proactive. These are the arguments put forward by the socialist wing of the Labour Party and include: putting the community in control; making things happen that wouldn't happen otherwise and to make things better (utopianism), which was one of the arguments of the founding fathers of town and country planning.

Essential. These are the arguments spelt out by communists and deep green ecologists. Communists argue that planning should promote social justice, preferably by transferring productive activity to communal ownership, which would axiomatically reduce the role of intervention third-party planning to whatever activities were left in the private sector, but would at the same time cause an enormous growth in central/state planning activities and apparatus. In contrast deep green ecologists predict that planning of a draconian sort is needed to prevent eco-catastrophe.

As with all spectrums, each section is not exclusive and at various times and for various activities any combination of arguments may be employed. Whatever part(s) of the rainbow are used there should be some combination of arguments to make a convincing enough case, and plenty of examples of how different combinations have been used are provided in Chapters 2 to 6.

POWERS AVAILABLE TO PLANNERS

Working independently from each other, Gilg (1991a) and Selman (1988a) have produced a spectrum of options which may be made available to planners if politicians so wish. As before, these range across the political rainbow and can be combined, but this time the spectrum is set out from the left wing to the right wing, although it could be recast along positive to negative lines.

The Gilg/Selman spectrum

1 Public ownership or management of land via long-term leases;
2 Regulatory controls, mainly negative, for example, planning permission;
3 Monetary disincentives to discourage production and/or undesirable uses;
4 Financial incentives to encourage production and/or desirable uses; and
5 Voluntary methods based on exhortation, advice, and demonstration, but often backed up with the threat or promise of one of the four methods outlined above.

If the spectrum were to be recast along positive/negative lines, examples of positive planning powers would include public ownership and financial incentives, while negative planning powers would include regulatory controls and monetary disincentives, for example, quotas and licences. In this spectrum, exhortation, advice and demonstration would be seen as broadly neutral.

In addition to these powers there are three other options: first, set up some form of administrative body; second, set up special areas in which the powers are modified, for example, National Parks; and third, use market methods to sell the right to produce or pollute, as for example widely practised in the United States, where the US Constitution forbids many of the powers available to British planners under the so called 'taking' away of individuals right to use their property as they see fit.

In Britain, the traditional process has been to start with the weakest power available and then to proceed up the regulatory spectrum as and when needed. The only exception was the Labour Government of 1945–51 which introduced the entire spectrum in a series of Acts which are outlined in the rest of the book.

DELIVERY METHODS AVAILABLE TO PLANNERS

It has already been demonstrated that the favoured system of policy delivery has been the so-called rational comprehensive linear systems approach, in which planners portray themselves as scientifically perfect, value free and neutral technicians who present to decision makers a series of calculated options using the range of powers set out in the previous section. Such systems were very popular in the 1960s. The underlying assumption is the 'utilitarian' one that the best option will be the one that will produce the greatest good for most people and the least harm for the smallest number of people. The main vehicle for delivery was an end-state plan representing the utopia to which the plan aspired. Since 1970, however, a number of other methods have been advocated. In particular, public participation became more important, and the degree to which planners could be seen as value-free technicians began to be widely queried. Planners have reacted by putting forward more alternatives and making much clearer both the technical and judgemental basis on which their proposals were founded, in an espousal of post-modern uncertainty.

A more self-confident stream has focused on planning as a decision-centred exercise within wider groups in society, with the planner being seen as the centre of a very large grouping of corporate bodies. Even more self-confident planners have advocated a value-laden agenda based on normative ethics in which certain standards are set and then attempted to be imposed. A more pragmatic group of planners has espoused the method of irrational idealism and have sought to make advances as opportunities arise.

In essence three broad delivery systems can be discerned in practice. First, variations on the rational cyclical process system model; second; an incremental crisis response model limited by imperfect resources of time, knowledge, and political realities; and third, a muddling through managing change model.

A more complex typology has been developed by McDonald (1989) based on the classic work of Friedmann (1973). This forms a matrix divided between the distribution of power from the centre to the periphery, and the method of delivery invoked. In more detail, the matrix shows four styles of allocative planning: *command planning*, which sets compulsory targets; *policies planning*, which structures decision environments; *corporate planning*, which uses processes of bargaining and negotiation; and *participant planning*, which relies on voluntary compliance with preferences reached in group deliberation. In a critique of how these four styles may be delivered in practice, McDonald rejects the *rational planning approach* as being too simplistic for a complex world of conflict, except where narrow sectoral plans are involved. McDonald also rejects the *conflict resolution role* because there is no concensus among policy makers. Thus McDonald advocates the role of planning as *bargaining*.

This is, in fact, the main delivery system that has emerged since the 1970s, based on the plan as a framework only, which is then employed only as a starting-point in negotiations between interested parties. The main activities are seen as bargaining with one party, acting as an arbitration service between parties, and thus attempting to resolve or prevent conflicts. The essential process is thus one of disjointed incrementalism set within only loose overall guidance.

An ideal system can nonetheless be predicated and indeed it would be in the nature of planning to strive for such a system. Such a system would set out the desired goals, the negative factors preventing these goals being achieved and the positive factors aiding these goals being achieved, the potential for eliminating negative factors and for enhancing positive goals, and finally the percentage of failure that would be acceptable before a policy review would be needed. The system should also set out the policies in a statement and/or maps and diagrams demonstrating how the policies will work, who is affected and how, and what they will need to do, if anything.

An ideal system would also note the problems likely to be encountered. First, the problem of overlapping time horizons between past and present policies, differential priorities between different goals, and overlapping spatial scales. These problems are further exacerbated unless the policy cycle and the assessment cycle are synchronised.

The system would also need to take into account the workability, efficiency and affordability of their policies, the degree to which they are undermined by partial knowledge of the present and uncertain predictions about the future, and thus the likelihood that the policies will be blown off course by unforeseen or chance events in the future.

IMPLEMENTATION GAP

Given the problems encountered by even an ideal system, it is not surprising that many planners encounter an implementation gap between what they desire to achieve and what they actually achieve. Cloke (1987) has pointed out that there are a number of difficulties with the apparently simple link between policy development and implementation, and argues that in reality the link between cause and effect is very blurred. In more detail Ham and Hill (1984), Pearce (1992) and Gilg and Kelly (1996a) have listed eight difficulties in assessing the link between policy and impact on the ground.

The first difficulty is that there is no general agreement about how to define policy and thus to provide a definitive list of policies to be evaluated. Second, it is hard to identify particular occasions when policy was made. Third, policies may be set out in a variety of forms, for example, policy documents, speeches and legal decisions. Fourth, policies may not be prioritised and contradictory policies may exist not only between policy areas, but even within policy areas. Fifth, policies are subject to change, leading to a time lag and overlapping of possibly contradictory policies. Sixth, decision makers will be affected by the policy and will alter their behaviour by either 'displacing' it elsewhere or not doing it at all, in what may be called the 'AIDS' and 'birth-control' effects (Gilg and Kelly, 1996a). Seventh, policies may be ignored by decision makers either through ignorance or wilfully. And eighth, a series of ad hoc decisions can subsequently be rationalised and developed into policy in order to legitimise past decisions.

There are thus a host of difficulties involved in analysing the impact of planning policies and the implementation gap between policy and reality. Depressingly there are also many other issues that could be discussed, for example, the limited policy agenda employed by policy makers (Cloke, 1987), and the many distributional side-effects that planning decisions set in train, for example, on land values or social mix.

Thus planners operate under many constraints which can be graphically depicted in the six-sided 'planning constraints coffin' shown in Figure 1.2. It is not surprising therefore that Cherry (1979) has concluded that 'Planning legislation is the story of the incremental adoption of measures imperfectly conceived in respect of problems only partially understood' (316). Or, to paraphrase Samuel Butler's aphorism that life is the art of drawing sufficient conclusions from insufficient premises, we can propose that, planning is the art of drawing sufficient policies from insufficient data.

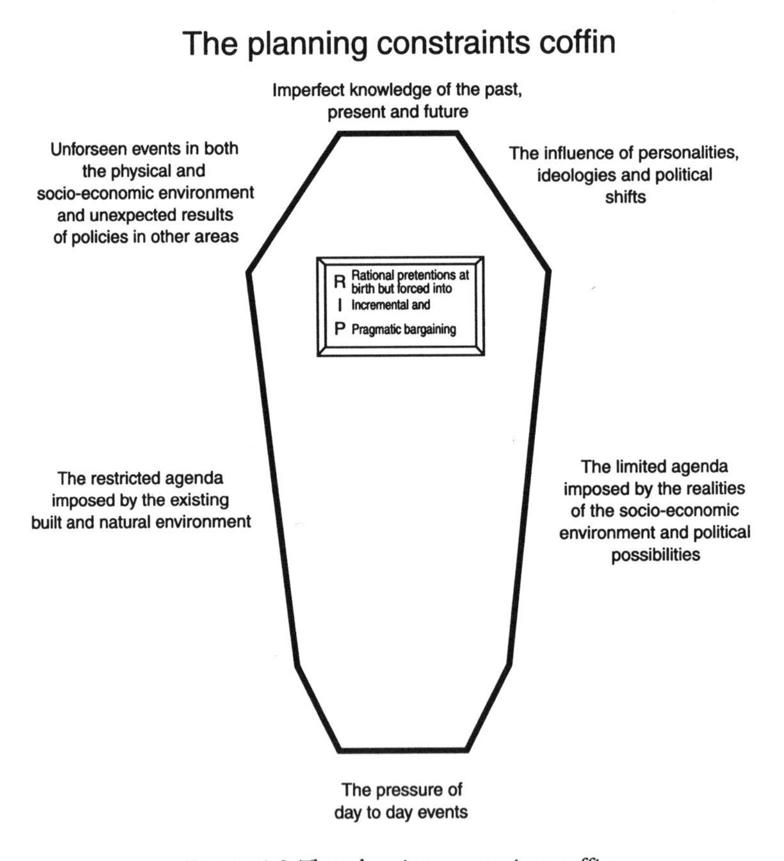

Figure 1.2 The planning constraints coffin

Source: Author

When attempting to assess the impact of a public policy a number of other factors need to be taken into account, for example, unforeseen events and the unintended effects of other policies, such as taxation. Models for impact assessment and causality have been provided by Schrader (1994) and also by Hill and Young (1989). These demonstrate how various factors interact, and either reinforce the intended effect, for example multiplier effects, or reduce it, for example many producers react to policy signals by doing the opposite or running faster to stand still. When milk prices were cut in the 1970s and 1980s, in an attempt to reduce production too many producers tried to offset the fall in income by increasing production, in a so-called 'perverse response'. Another problem is that the world is increasingly interconnected and pollution, for example, is no respector of international boundaries.

Tarrant (1992) has proposed three approaches for assessing the impact of government policies on agriculture and these can be adapted as follows. First, by selecting areas with similar socio-economic and natural environments, but different policies. Second, by simulating what might happen under different policies (the approach favoured by agricultural economists) and comparing this with what actually happened. Third, by examining the production or other patterns that developed and comparing them with the policy predictions. However, at this stage it is best to remember the aphorism

To plan is human
To implement is divine

Any assessment of the implementation gap will depend on the way in which knowledge is used and this can be crucial, as the next section will demonstrate.

KNOWLEDGE

For over two centuries, since the age of enlightenment, the scientific view of the world based on empiricism and observation has held sway. The related concept of modernism with its central belief in progress and the basic goodness of human beings has also been a powerful concept. It was felt that linking the scientific method with modernism would inevitably lead to utopia once the scientific laws linking human happiness to the physical world could be unearthed. For most of this century planners were drawn into this beguiling equation and produced so-called master plans depicting scientifically perfect cities reaching towards the sky.

Since the 1970s, however, although the scientific method still remains immensely powerful and continues to alter our lives by never-ending improvements to our productive and communication capacities, in the more complex world of human beings and how they use the finite and fragile resources of the planet earth, serious doubts have been cast on the efficacy and neutrality of the scientific method. In particular three levels of uncertainty have been identified (O'Riordan, 1995): first, data shortages; second, deficiencies in the models used; and third, the discovery that some natural systems appear to act randomly and chaotically, thus taking us beyond the knowable into a truly 'black hole'. Three main systems of formulating and evaluating knowledge have thus been debated about and used since the 1970s. As already described in the Preface, these are, first, traditional meta-narrative empirical analyses based on the assumption that there is a policy effect, and that the impact of policies can be measured quantitatively in a cause-and-effect relationship, with, for example, winners and losers being identified usually by statistical or econometric methods. Second, structural analyses which focus around a number of themes including: a) political economy studies which examine the interplay between key agencies in formulating and implementing policies, and b) structural studies which often assume the primacy of capital in this interplay, or other structural factors which predetermine to a certain extent who will gain from policy changes. Both these approaches have the central concept that life is a struggle between people usually acting as organised groups. But a third, human agency or behaviouralist, approach, however, argues that human irrationality can undermine these structural forces, and that actors can act in different ways dependent on the issue at stake. Thus farmers may pass up the chance to drain ponds and grow more crops because they are keen bird-watchers. Similar approaches identify the importance of chance events, and factors such as chance have encouraged the development of a post-modern approach which rejects both the meta-narrative and structural approaches as arrogant simplifications of a chaotic complex universe.

These approaches are not mutually exclusive, and although some of their proponents have erroneously claimed that only their perspective is valid, the great characteristic of recent years has been a diversity of approaches as acknowledged by Gardner and Hay (1992) in their review of British geography in the early 1990s. Accordingly, these approaches are now considered in turn.

Traditional meta-narratives based on empirical observation. Until the 1960s empiricism was the main method employed by planners. They observed the real world and 'induced' general laws from these observations. Encouraged by the so-called Vienna school of logical positivists set up in the 1930s, planners and other social scientists revised the empirical method by trying to seek general laws from first principles or *a priori* deduction. According to O'Riordan (1995), the resulting *scientific method* claims that science evolves by *theory building, theory testing* and *normative evaluation*. The basic theories themselves are examined for their *correctness* in terms of their internal logicality, and for their *consistency*, that is, their inherent plausibility. These theories in turn are converted into hypotheses or propositions whose truth or applicability is subjected to analysis. Normally that analysis relies upon *observations* and meticulous recording; *experimentation*, also with meticulous recording; or *modelling*, through which representations of 'reality' are created to provide a more manageable basis for examination and prediction. Where there is a historical record, the model can be calibrated against measured output to test its *robustness* and *accuracy*. Where there is no historical record, or where the model is essentially designed to depict the future, the only test for reliability is *peer group criticism* of the model's assumptions, interactions and sensitivities to relationships between cause and outcome.

This approach was aided and abetted by the parallel development of systems theory and computers which promised ways of testing general laws about human behaviour in previously impossible ways, and was enthusiastically espoused by planners. Thus the 1960s and 1970s witnessed ever more complex attempts at modelling the human environment, and explaining and predicting human behaviour. Eventually, these multi-variate methods became so complex that it became difficult to interpret what the eigen values and the other statistical measures were actually showing. More problematically, when the interpretation could be made, the results were not very exciting and often confirmed common-sense observations.

A simpler variant of the meta-narrative was developed by civil servants in response to one of Mrs Thatcher's few good innovations, the so-called 'bottom-line' or accountancy-driven approach to policy evaluation. This involved the introduction of much modified annual reports to Parliament in the late 1980s. These take the form of looking both forward and backward. The forward-looking sections set out the Government's expenditure plans for the next three years, while the backward-looking sections provide information on the past performance of policies, current activity, research and development programmes, and the work of executive agencies. The reports also provide 'performance indicators' usually based on the achievement of certain financial targets (hence the 'bottom-line' epithet). However, this approach has its flaws, which are even admitted by those who have to use it. For example, the 1995 MAFF report (Cm 2803) argued:

> There are risks associated with using indicators as measures of policy effectiveness. The outcomes reflected by an indicator may be the result of many factors, including some completely outside MAFF's control, and it can be difficult to measure the exact influence of MAFF actions alone. Even where MAFF actions do have an unequivocal influence on events, it is not always possible to define the 'optimum' outcome.
>
> (6)

In spite of these difficulties the performance-target approach based on financial measures has been the one adopted by research in the public sector, either in the form of contracted research, or in the form of policy assessments by House of Commons and House of Lords Committees. Its most noble proponent, however, has been the feared National Audit Office, as Phillips (1993) recounts in chapter 6.

Structural analyses focused on political economy and regulation theory. As the previous sub-section has shown, academic researchers became dissatisfied with the restricted results produced by the scientific method. Moreover, some academics argued that the empirical and logical-positivist systems had also failed to live up to their promise because they took a neutral view of the world or, in the case of classical/orthodox economics, they assumed perfect knowledge by all participating players. In a reaction to these simplistic views social scientists and planners turned to the century-old writings of Marx and found there a set of theories which appeared to have major relevance to the modern world. In essence a set of workers reworked Marx's writings and developed the so-called political economy approach, which focuses on the primary need for capital to accumulate and thus on the ensuing struggle between capital and other factors. A bewildering number of sub-themes were also developed based on the underlying structures within society caused by the imperatives of capital accummulation, most notably, Habermas's 'critical theory', and Althusser's 'structuralist' theory. In a connected but not synonymous development Gidden's theory of structuration attempted to find a theory for the links between structure and agency, two of the cornerstones of Marxist analysis.

In the 1990s, a variation of the theme emerged under the guise of regulation theory (Jessop, 1990), whose key notion according to Goodwin, Cloke and Millbourne (1995) is:

> that the expanded social reproduction of capitalism is never guaranteed, but has continually to be secured through a range of social norms, mechanisms and institutions which help to stabilise the system's inherent contradiction temporarily around a particular regime of accumulation.
>
> (1247)

The regulation school thus insists, according to Harvey (1989), that we look at the total package of relations and arrangements that contribute to the stabilisation of output growth and aggregate distribution of income and consumption in a particular historical period and place, and thus helps to explain why the character of capitalism varies over time and space. With regard to rural areas Lowe *et al.* (1993) have pointed out that uneven development in rural areas will be the result of moving from one regulatory regime, namely agricultural production, to a post-productivist model, and research by Goodwin, Cloke and Millbourne (1995) in rural Wales has confirmed that regulation does not just happen, but has to be continuously imposed and culturally negotiated.

Human agency, behaviouralist and post-modern perspectives. At the same time that the political economists were reformulating Marx's general laws, another group who were also dissatisfied with the failures of the scientific method developed the idea of behaviouralism and its related field of humanism. This group argued that human beings were not essentially logical and that, especially collectively, they could act in unpredictable and illogical ways. For example, the 1995 Nobel Prize winner for Economics, Lucas (Lucas and Sargent, 1981) has argued that people are too clever to be outwitted by government policies, and that they will react to policy in perverse ways, thus producing the opposite result to that intended. Such arguments led some academics to conclude that any endeavour involving human behaviour would need to take this into account and place humanity at the centre of any such study.

In parallel with this group, people working independently of each other produced a number of observations which have since been labelled as post-modernism. Post-modernism, as the name implies, rejects the modernist project and celebrates diversity and difference, rejects the idea of meta-narratives, and points out that words used by planners to convey consensus and confidence are in fact a misuse of words to imply a scientific precision that can never exist. A key concept in both schools is that what really matters is not how the world actually is, but how it is *perceived to be.*

Another key concept is that of deconstruction, as formulated by Derrida (1995), which draws attention to the dominant polarities of western metaphysics and contraries such as culture/nature, spirit/matter, writing/speech, positive/negative, which always privilege the former term at the expense of the latter. To *deconstruct* such contraries is to discover that each opposition is founded on the desire for presence or the full availability of the privileged term, whereas the very opposition indicates a necessary absence which can never be fully represented. The anarchic utopianism which is at the heart of post-modernism is the antithesis of Marx's scientific socialism based on theories of class structure and conflict, ideology and the state, and instead relies on the imported Heideggerian theme of spirit. However, post-modernism offers nothing on the structuring of injustice and justice in our changing institutions with their mixture of individualism and authoritarianism.

Whichever methodological approach is adopted, research resources are scarce, and research has been very selective in its choice of topics. Often research has not addressed policy issues directly, or only those that are topical or in crisis. There has been too much emphasis on policy failures and not enough on policy successes and why they succeeded.

Knowledge is thus not the apparently certain and value-free commodity that it appears to be, and planners should make it clear at every stage of planning process what knowledge system and assumptions they are adopting. Most notably, this should be done at the information-gathering/analysis stage and at the monitoring/review stage. In this book the first three sections of each chapter are presented from the viewpoint of the empirical method, and the author makes the claim that as far as possible this is his white, middle-class, middle-aged and male observation of the events that have taken place. In section D of each chapter, however, the three analytical perspectives outlined above are used in order to reflect the current heterogeneity in the social sciences (Gardner and Hay, 1992), which has been welcomed by some (for example, Cosgrove (1990)) as giving space for a variety of paradigms to exist, while others, notably Harvey (1989), have criticised the 're-assertion of the primacy of empirical research' and the emergence of 'pallid, depoloticised and ultimately vacuous versions of the theory of structure and agency'.

THE COUNTRYSIDE

Our knowledge of the countryside is immensely coloured by the knowledge system we approach it with, and in particular whether we wish to use it as a productive device or as consumers of either its produce or its environment. Planners also instinctively view the countryside through the lens of land use. This section thus examines land uses in the physical and socio-economic environments. Significantly, the physical environment is usually referred to as the 'countryside', while a consideration of the socio-economic environment usually refers to 'rural areas'.

The physical environment of the countryside. Physically the countryside can be defined as the place where extensive land uses take place. Land uses can then be sub-defined by their primary purpose or economic activity, for example, agriculture or forestry, and a system of related land uses and demands on rural land can be developed as for example by Clout (1972). Some land uses, however, may coexist alongside the primary land use, for example, leisure and nature conservation, but more typically the environmentally unfriendly nature of modern farming has caused conflicts between farmers and other groups. Traditionally, agriculture and forestry have held sway over all other users, but in recent years the crisis of European overproduction has forced farming to retreat from its once pre-eminent position and accept that not only may other uses coexist within it, but that these other uses may even become the primary use, for example a golf course or nature reserve. This process of change can cause conflict and together with the continuing capacity of modern agriculture to cause environmental damage are thus prime reasons for intervention by planners.

The socio-economic environment of rural areas. Traditionally the socio-economic environment was dominated by agriculture. In the last 50 years, however, agricultural employment has been drastically reduced and new houses and employment centres have been built so that the countryside is now dominated by people who do not derive their living from the land, but place great store on living and working in a rural environment. With modern telecommunications many more jobs and houses could be transferred to rural areas. If this trend is not to be self defeating and turn the current exurbia into future suburbia, then planners will need to balance the desire for rural settlements against the need to protect the countryside as a natural environment. To a large extent they are aided and abetted in this process by the new incomers to the countryside, the so-called service class (Cloke and Thrift, 1987) who have a 'last bastard in' or 'drawbridge' mentality to any more development once they have secured their place in the sun.

Representations of the countryside. Countryside change is far more complex than the above introductory comments imply and so a number of summarising concepts have been established. The first relies on the traditional geographical concept of landscape. Using this approach two core and two transitional/fringe landscapes can be defined (Coleman, 1976) in a continuum working out from urban areas as follows: *the urban fringe* (the zone of transition between town and country with clashes of interest between house building and landscape preservation, and excessive recreational use); *the farmscape* (the area still dominated by lowland farming but where agriculture has adversely affected the environment); *the marginal fringe* (the zone of transition between lowland and upland farming, dominated by reclamation/reversion, afforestation and recreation changes); and *the uplands* (the area dominated by semi-natural vegetation which is controlled by grazing and burning). This approach has also been fundamental to many planning policies which have taken a landscape-specific approach, for example, National Parks in the 1950s, Country Parks in the 1960s, Environmentally Sensitive Areas in the 1980s, and English Nature's 'Natural Areas' and the Countryside Commission's 'New Landscapes of England' project in the 1990s.

The second approach relies on socio-economic classifications of the social and economic fabric of the rural areas. Traditionally these employed census data and statistical procedures to identify rural areas. Examples have been provided for the 1971 and 1981 censuses by Cloke (1977), and Cloke and Edwards (1985), and for the 1991 census by the Office of Population Censuses and Surveys (1995a, 1995b), who statistically clustered 37 census-data variables to produce a map of types of area, including a rural area category. In recent years the approach has moved on from these rather static snap-shot approaches to one based on a conceptualisation of change. This has been the approach adopted by the team involved in the Economic and Social Research Council's (ESRC) Countryside Change Research Inititative in the late 1980s and early 1990s. This team (Marsden, Murdoch, Lowe, Munton and Flynn, 1993) used four analytical concepts to examine: the ıanging relationships between *production* and *consumption*; the *commoditisation* of social and economic processes; the *representation* of rural issues by various actors; and the integration of *property relations* as a key structuring mechanism. Stressing the role of key actors in these processes, the team then divided the countryside into four areas for the future: the *preserved countryside* (the countryside within the major commuter catchments or with attractive landscapes where a strong anti-development ethos dominates); the *contested countryside* (the countryside where more traditional farming interests still dominate, but where incomers are attempting to change the ethos to preservation); the *paternalistic countryside* (the countryside where large private estates and big farms still significantly shape events); and the *clientilist countryside* (the countryside where direct agricultural support supplies most of the income in the disadvantaged uplands or in remote areas).

The various possible approaches to understanding our concepts of the countryside have recently been summarised and discussed in *Writing the Rural: Five Cultural Geographies* (Cloke *et al.*, 1994). In their introduction Cloke and Thrift identify four phases in attempts to identify rurality. The first phase equated rurality with particular spaces and functions, notably extensive land uses, with small settlements, and with a special way of life. The second phase was to replace these functional definitions with more pragmatic concepts suggested by the use of political economic approaches. But since these stressed the role of wider economic and political structures, a third phase stressed the existence of not one but several rural spaces. This post-modern phase sees rurality as a social construct reflecting a world of social, moral and cultural values. To accept the rural as a social and cultural construct thus allowed the rural to be rescued as an important research category, as the way in which the meanings of rurality are constructed, negotiated and experienced, and interconnect with the agencies and structures being played out in the space concerned. The fourth phase is centred around postructuralist deconstruction, for example, Halfacree (1993); drawing on the work of Baudrillard, has argued that the sign (= rurality) is becoming increasingly detached from the signification (= meanings of rurality) and in turn that sign and signification are both becoming divorced from the referent (= the rural locale). If symbols are becoming ever more detached from their referential moorings, then socially constructed rural space becomes increasingly detached from geographically functional rural space, and there is an increasing danger that myths lose even greater touch with reality.

In a development of these ideas Halfacree (1995) has argued that there are two ways for defining rural. First, insights gained from critical realism; which involves attempting to define the rural as a distinct locality; and second, social representation by ordinary people. Halfacree has used this second approach and he interviewed or obtained questionnaire responses from 267 individuals. The answers concentrated on the physical rather than the social environment, and showed an inherent belief in the rural idyll which was not overt at first but was revealed as respondents revealed themselves. In another study of real people, not of academic theory, Jones (1995) from a survey from a small village has shown that lay discourses of the rural are both spatially and conceptually complex and incoherent and cannot be easily be assimilated into academic definitions. Notably, those definitions by academics that exclude villages close to large towns and cities, but which are seen as rural by real people. Such research thus throws doubt on whether countryside myths as portrayed by academics, but believed by the public at large, are then myths at all.

Jones (1995) thus argues for a new approach to rural areas which takes the views of ordinary or 'lay' people more into account, but also in the changing way people interact across space, as the following quote shows:

> These emerging interpretations of how to approach the idea of rural space and place can be tied into Massey's (1994), more general retheorization of place in which she argues strongly for the continued concern and recognition of the importance and uniqueness of places, but within a new theoretical framework which has three main characteristics: that places should not be seen as static entities but rather as fluid processes; that they should not be seen as bounded in the traditional sense but rather as interesections of social relations and understandings which trace out into global networks; and finally, and most importantly in the context of this work, they should not be seen as homogenous in construction and interpretation but rather as a melee of conflicting interpretations, but still retaining a distinct uniqueness.
>
> (47–8)

Nonetheless, one of the most enduring and dangerous myths is likely to endure, namely, that the countryside is *different* and that people can act differently there. In fact they do; for example, perfect strangers acknowledge each other on country walks, whereas they would not dream of doing so in other environments. The concept of countryside as a sort of fantasy world does not, however, give individuals the right to use the countryside as they see fit as some sort of extra-urban space in which to do what they want or construct images of ideal retreats. Most importantly, policy makers and planners will have their own image of the countryside and these prejudices will be crucially important in shaping their understanding of the issues involved, and the possible goals for which they will be aiming. For much of the twentieth century the twin ideas of the countryside as an essential food producer, which needed to be protected from urban growth, have dominated policy making, notably for those policy makers who lived through the severe food shortages between 1939 and 1954 (the year in which food rationing ended). Only now are policy makers coming to power who have less painful memories of food rationing or of a period when some foods like chicken were a treat. I was born in 1946, and for most of my childhood, chicken, most fresh fruits and chocolate were once a week or even annual treats. Now chicken is throw-away food, and deserves to be, since it is tasteless or even foul, rather than fowl. Once again we are prisoners of our past, and although I should be a fan of food production, other baggage, like being brought up on a semi-feudal estate where my parents were exploited to some extent, but also favourably treated in comparison to the farm labourers (Gilg, 1992b), and then marrying into an extended farm family where non-farmers were treated with suspicion, if not contempt, have perhaps made me more anti-farming than I care to admit. Readers are thus advised to treat my analyses in Chapter 2 with this in mind, since our early conceptions of the world are hard to ignore.

The countryside is thus an arena of conflict between different land uses, the way in which land is used by each activity, and different perceptions of what the countryside should be and who should be allowed to live in it or use it. There are thus not one but many countrysides and countryside crises for planners to solve.

GOALS FOR COUNTRYSIDE PLANNING

There is an almost limitless number of goals that could be formulated, and one recent book (Gilg, 1991a) found over 1500 different documents setting out policy prescriptions for the countryside. Other attempts have been made by Hill and Young (1989) who defined four main goals: first, providing food security; second, improving the welfare of rural residents; third, protecting the rural environment; and fourth, promoting rural recreation. Other visions and objectives can be derived from actual practice, as shown by Errington (1994) in his analysis of the 1994 Berkshire *Rural Community Strategy*. If these attempts and attempts by Orwin (1944), Thorns (1970), Wibberley (1972), Lassey (1977), the Royal Town Planning Institute (1976) and McDonald (1989) are analysed, three major goals or aims for countryside planning stand out. First, to ensure the rational development of resources on a sustainable basis; second, to provide equitable development between different activities and groups within society; and third, to reconcile consumption and production internally, and externally to reconcile both of them with preserving the heritage. As the rest of the book will show only too clearly, existing policies because they have grown up incrementally and are based on short-term sectoral interests often put forward by professional groups guarding their own territories of expertise, often manifestly fail to set consensus goals, let alone achieve them in practice. This is not to say that an ideal system could not be predicated and that we should not strive towards it as set out next.

AN IDEAL COUNTRYSIDE PLANNING SYSTEM?

In a parallel piece of work to this book the author has been formulating a general theory of countryside planning which is known as the Regulating Rural Environments project. Work is still proceeding on the general theory, but a much simplified version appears in Figure 1.1. Part of the project has also been involved with producing an ideal system of countryside planning from first principles using a hermeneutic approach. The preliminary system which is still at the development stage is shown in Figure 1.3. Readers are invited to use this system to evaluate the efficacy, equity and effectiveness of the policies discussed in this book, and also to use the following checklist when evaluating each policy area that is discussed in the book:

Will the problem go away?
Is there anything we can do?
If so, is it politically acceptable?
Is it economically affordable? and
Is it technically feasible?

CODA ON THE BRITISH SYSTEM OF GOVERNMENT, FOR READERS UNFAMILIAR WITH IT OR SEEKING A REFRESHER

The British system of government is based on precedent and has no constitution. This is one reason why countryside planning policies have grown incrementally. The British system is also based on a Prime Minister, who is first among equals, and derives power from the ability to appoint or dismiss members of the Cabinet. Members of the Cabinet are drawn from either house of Parliament and most of them represent departments of state or ministries which are drawn up on sectoral grounds. This is one reason why policies are often in conflict with each other, because ministers regard their own sector as their own domain to be defended and promoted at all costs. To some extent the bi-weekly morning meetings of the Cabinet are supposed to provide coordination, but time is strictly limited at these meetings, and so a series of Cabinet sub-committees seek to provide coordination between sectoral interests. Significantly there never has been a separate Countryside Minister, although there have been junior ministers with responsibilities for the countryside in both the Ministry of Agriculture and the Department of the Environment. In addition there have been hardly any government statements on overall countryside policy. A rare example was provided by the wartime Scott report, and the report by the Countryside Review Committee (1976) of senior civil servants, which stated that present policies were geared towards single goals and that these could conflict. The Committee therefore concluded that the sectional approach was insufficient

1 Planners should **think** about the problems and opportunities represented by the current situation by carrying out **surveys** and subjecting the findings to careful **contemplation** based on a **classificatory** analysis which **groups** and **ranks** the issues. The public should be aware however that planners are only human and may not always examine every aspect of the issues, through either laziness or prejudice, and that even when they use their abilities and techniques to their best advantage these may still not be good enough for the complexity of the issues involved.
2 These issues should then be extrapolated into the future derived from not only how planners **expect** the future to develop based on a number of **assumptions** but on how other groups may forecast the future depending on their **aspirations or concerns**. These different **aspirations or concerns** will cause **conflicts** over who gets what and **planners** may need to **resolve** these by a series of **resource allocation decisions**. Such **decisions** may also be avoided by **anticipating** future developments and if possible **preventing problems** by employing **the precautionary principle**.
3 One of the goals of planning is thus a reactive approach to problem-solving which is rooted in the Enlightenment idea of progress in general, and in progress via the scientific method in particular. Planning is not however just about problem-solving and reactive policies can be matched by pro-active proposals which make good things happen that wouldn't otherwise. Both types of policy are more likely to be **acceptable** if they are based on a **vision** of how things may be **better or worse** if the policies are not employed in the future by setting goals used on certain **beliefs** and seeking a **balance** between different groups on the 'greatest good of the greatest number' principle or put simply **common sense**. **Beliefs** can best be **balanced** if they are **ranked** into an order of priority. Planner's **explicit motives** can thus also be **ranked**.
4 Proposals for possible methods for achieving these **explicit motives** which include **resolving and preventing conflicts** may involve **exhortation via mediation**, **arbitration**, **and reconciliation**, but some degree of **intervention** will probably be necessary and **plans** may often be best implemented by using a **regulatory approach**. When assessing possible **regulatory** methods for achieving stated **goals** planners can use their '**superior' judgemental skills**, but they should also consult widely. The principle of consultation can also extend to reaching agreements which will be most effective if they are legally binding agreements. All planning processes should also identify resources that need **protection** and methods for **preventing** damage to these resources may include **site protection** and employing the core principle of **conservation/preservation**, good **husbandry**. The environment may best be protected however by reducing consumption of resources by exercising **thrift**, or 'making do and mend' this however is a concept which planners may find hard to get across in a society dominated by built in obsolescence. In assessing which **methods** to employ planners should be **cautious** and set **realistic targets** based on what is **achievable** given their role in society, which may in essence be one of **coordination** only.
5 Whichever methods are chosen an organisation will need to be set up which can institute procedures for implementing the chosen methods which will need to be set out in a series of **plans, schemes or designs**, following a process of plan-building based on consultation, in which a spectrum of planning powers from exhortation to coercion have been outlined. This spectrum should contain three sub-elements and three main mechanisms as shown below:

Indirect Influence	(**leadership**, **adjustment**, **guidance**, **codes**)
	Exhortation
	Financial Incentives and Disincentives
Direct Influence	(**government**, **rules**, **directions**, **precepts**, **decrees**)
	Regulations backed up by sanctions
Direct Control	(**conducting**, **orders**, **commands**)
	Direct control or ownership of resources

These mechanisms need to contain measures to ensure that they are not **exploited** or ignored and a set of sanctions and enforcement procedures need to be set up to police them.
6 The performance of these **plans** should be **evaluated** to assess who or what is **winning and losing**, and **planners** should set up a series of **performance indicators** by which their own work and their **plans** can be appraised. This **appraisal** should accept that there will be an implementation gap and should thus set acceptable levels of effectiveness in **achieving** the **goals** of the plan. In both cases the appraisal should attempt to find out why any **deviation** between the **plan** and reality has occurred, and propose **amendments** to the plan. This process will probably need to go on for ever and **planners** will have to be **eternally vigilant** about new threats to the environment.

Figure 1.3 An ideal planning system?

Source: Author

and so called for an early reasessment of priorities. Given the speed of government decision making this assessment was delayed till 1995, when a Rural White Paper was published, as discussed in Chapter 7.

The sectoral system also leads to a decision-making process which is often confrontational rather than consensus seeking. Likewise, the opposition parties sit opposite the ruling party in the House of Commons, and this serves to engender confrontation rather than a search for policies for the good of the nation as a whole. Real power thus resides with the Prime Minister and the Cabinet, leaving backbenchers with the role of querying policies by formal questions or through the system of committeees set up to examine issues of public concern. Similarly, the House of Lords does not act as a powerful second chamber but often as the conscience of the government and as an informed area of criticism through its own system of committees.

Traditionally government ministries and departments were subject to little scrutiny, but since the 1980s one of Mrs Thatcher's best reforms has seen an accountancy 'bottom line' approach which has swept through Whitehall. In the 1990s each ministry or department, and most of the sub-agencies, have to have a corporate/business/management plan and annual reports setting out their priorities for the medium term future, how they intend to achieve them and what cost. These plans and annual reports provide extremely useful information and have become a prime source for students of countryside planning as section C in each of the sectoral chapters reveals. These plans are subject to scrutiny by the much feared National Audit Office, the Comptroller General, and the Public Accounts Committee. The work of government organisations is very much performance-vetted these days.

The system is very much top-down, with key strategic policy decisions being made in the Cabinet, leaving the detail to be filled in at departmental level. Policies are then executed by the executive agencies which since the 1980s have increasingly been agencies acting on contract and not formally part of the government machine. The Government also sets up its own advisory agencies, for example, the Countryside Commission in 1968, not only to give it advice, but even to act as an *agent provocateur* and criticise government policies in its own area of responsibility. Real opposition or support comes, however, from the myriad pressure groups set up by activists to alter government policies, for example the National Farmers Union. The electorate is conspicuously absent from the process, since they only have one vote, and then only once every five years, or more frequently if a government cannot survive its potential maximum term of five years. In addition Britain runs a first-past-the-post system which has meant that all postwar governments bar one have been elected with only between 40 and 50 per cent of the votes cast. This means that up to 60 per cent of the electorate may be in disagreement with the policies being operated.

The system thus runs as a series of meetings between the various groups involved. In theory this leads to a consensus being achieved, and in fairness the system makes every attempt to let every group have their say, but in reality some groups are more influential and have greater access to key personalities than others. Those most likely to be in this preferential position are those which are in tune with the ideologies of the ruling party, or which have something to trade in return for a policy concession, or whose campaign is in tune with current public opinion. In a review based on 40 years' experience of advising ministers Delafons (1995), a former civil servant, has argued that there are six main forces driving policy change: political parties; ministers; parliament; pressure groups or lobbyists; international bodies; and academics or the research community. Delafons concludes that the process is messy and complex, and that much work is spent on putting the best practicable face on things and accounting for policy failures. From his new perspective as a visiting professor, Delafons calls for a greater coordination between research and policy making, in common with Tewdwr-Jones, (1995) an academic, who bemoans the chasm between planning practice and theory, as does Phillips (1993), another former official (Director of the Countryside Commission in the 1980s), who has also become a visiting professor.

Delafons points out in addition that bureaucrats rarely propose policy changes, since they are far too busy implementing current policies, or clearing up old ones. Furthermore new policies mean more work. Therefore, officials subject policy proposals to the following formidable checklist, which can usefully be compared with the theoretical list provided earlier.

1 What is the problem?
2 What do we know/need to know about it?
3 What is the objective?
4 What is the proposal?
5 Is it relevant?
6 Is it practical?
7 Will it do any good?
8 How much will it cost?
9 Is it fair?
10 Is it legal?
11 Are there EU implications?
12 Who else is involved?
13 How do we present it?
14 How much work will it involve?
15 What alternatives are there – if any? Zero option: do nothing?
16 Go or no go?
17 Forget it or start again? An iterative process.

Major policy changes have to be made by an Act of Parliament. All Acts of Parliament start as Bills, most of which are sponsored by a department or ministry in response to various pressures from the party in government, the economy, from interest groups, and day-to-day events. Some short Bills are put forward each year by backbenchers (not members of the Government but may be members of the ruling party). These short Bills often involve controversial countryside issues, and favourites have included Bills to abolish field sports, notably fox and stag hunting, and the protection of hedgerows. A government Bill in any one sector will often have a long gestation period, having been the subject of various Green and White consultative papers as the Government formulates a policy that will gain consent by both Parliament and society. No minister will put forward a Bill that will not be widely accepted, since the state has few mechanisms to govern without consent. This period of gestation may take five or even ten years, and only one major piece of legislation is granted to a minister in every five-year parliament, although a minister will be granted several smaller Bills by the cabinet meetings which decide the priorities for legislation in a crowded programme. This explains why the process of policy change is so slow. In addition, it takes a whole year for the Bill to proceed through all its parliamentary stages and become an Act.

An Act is often only a broad framework for subsequent action, and very little of an Act will make its way into direct law. More often the Act gives a minister discretion to implement certain sections by virtue of a Statutory Instrument or Order which puts flesh on the bones. For example, the day-to-day operation of development control in the town and country planning system is set out in a series of General Development Orders. Advice on how to operate and interpret this detailed legislation is provided in a series of circulars, advice notes, books, leaflets, and by speeches from ministers. An important safeguard on abuse of the system is provided by the fact that the legal operation of all Acts is not controlled by the government but by the independent legal system, thus providing a check.

Most legislation is predicated on a number of possible actions by either public organisations or by private-sector organisations and individuals. First, things that have to be done, namely, *compulsory powers*, for example, the production of a Structure Plan by a County Council. Second, things that can or may be done, namely, *permissive powers*, for example, creating a long distance public footpath. In addition to the powers given to public bodies, legislation may place *obligations* on organisations or individuals, or *prevent* them from doing things, but may also give them certain *rights*.

Crucially, the nature of policy evolution by infrequent major strategic legislation backed up by considerable discretionary powers of frequent detailed legislative change by Statutory Instrument is one of the key reasons why British public policy is so incremental in nature. It is so much easier to tack on either a modification by Statutory Instrument or by a short Act of Parliament than to scrap the lot and start again. Like my firewood saw, which I have had for 20 years and which has had 14 new blades and three new handles!

The system also extends extra-nationally, and here the key factor is membership of the European Union (EU, until 1992 the European Community or EC), which dates from the early 1970s, although there were and are many other links with international organisations, most notably the General Agreement on Tariffs and Trade (GATT) in agriculture. In the 1970s and 1980s the links with Europe were mainly limited to trading issues and the Common Agricultural Policy. In the 1990s, however, there has been a concerted attempt by some groups within Europe towards a more integrated system. The first step on this road was provided by the Maastricht agreement of 1991, which laid down the route for convergence towards a common currency and other common policies by the end of the century if not before, but at the same time enshrined the principle of subsidiarity in which decisions as far as possible are made at the lowest local level possible. The two principles are combined in the slogan, diversity within unity.

Britain was not in favour of all aspects of Maastricht and indeed opted out of one of its key elements, the social chapter on working conditions. In general the British Government is concerned about a loss of sovereignity. This is largely because at the moment the unelected European Commission, composed of 17 appointed Commissioners who control 13,000 civil servants, is seen to have too much power over the legislative process, and in the medium-term future they fear that this power will transfer to the formerly somewhat toothless European Parliament of 518 members. Under the Maastricht agreement, however, the European Parliament was elevated to a position of equality with the Commission at the second stage of the policy-making procedure, and joined the Council of Ministers in a process of 'co-decision' from then onwards. However, the Commission remained the only body able to propose legislation.

In more detail, there are three principal sources of European Union law: a) Treaty provisions; b) primary legislation adopted by the Council and Commission acting under treaty provisions; and c) secondary legislation adopted by the institutions of the European Union, normally the Commission, to implement primary legislation (Cole and Cole, 1993). The first two sources of EU law, treaty provisions and primary legislation deriving from treaty provisions, are applied directly by the Commission. Secondary legislation is a more complex and dynamic area, with EU legal acts divided up into different categories: regulations; directives; decisions; and recommendations:

Regulations are laws which are binding and directly applicable in all Member States without any implementing national legislation. Both Council and Commission can adopt regulations, used in many areas, such as the regulation of agricultural markets.

Directives are laws binding on Member States as regards the results to be

achieved, but are implemented into national legislation in the form each Member State sees appropriate.
Decisions are acts binding entirely on those to whom they are addressed, with no national implementing legislation being needed.
Recommendations have no binding effect and can be adopted by either Council or Commission.

These more day-to-day decisions are overseen by the Council of Ministers, which normally meets monthly, while the strategic decisions are taken by the European Council of national leaders who meet every six months in the country which has the Presidency for that six months. In common with the UK, the legal system – in this case the European Court of Justice – determines the legality of community laws and the actions of institutions. A European Court of Auditors ensures that Community institutions spend money legally and properly, but there is still much concern that fraud is widespread, notably in the CAP.

Internally, the government devolves some powers to Scotland and Wales, and indeed, Scotland has many separate rural planning organisations, but all primary legislation and even secondary legislation has to be referred to Westminster. In Wales many of the organisations are sub-sets of the English organisation, but in recent years more truly Welsh organisations have been set up. There is no formal regional tier of government in England, but most organisations are divided into regions or provinces. The most effective form of locally devolved power is represented by elected County and District Councils, but they have to operate within strict financial and legal limits imposed by the national Government and many people query whether local democracy is really effective. However, as Chapter 5 will show real power does exist in the form of decisions over whether to allow new houses or other developments in the countryside. Many other forms of devolved government are carried out by non-elected quangos, for example, the Forestry Commission, peopled by the 'great and the good'.

CONCLUDING SUMMARY

Britain is a semi-democratic country in the sense that the electorate has limited power over both the legislature and the executive, and that many documents and decision-making procedures are unavailable to the public. Effective power is wielded by the inner circle of government organisations and those groups which have real access to them. This is not to deny that there is a vigorous and public debate about rural planning issues and that the processes of planning policy development are not fascinating in their own right. Most planning powers, organisations, and the professional organisations and techniques that operate the system have grown incrementally by a process of pragmatic trial and error, rather than by planned design, and have been accretionary.

This means that the only sensible way to address countryside planning is by an historical narrative, and then to try and impose some of the rigour offered by the conceptual constructs outlined in this chapter, notably the concepts of: *a spectrum of powers* (set out on page 11); *the typology of delivery systems* (set out on pages 11–13); *the aims* (set out on page 24); *and most of all the model of a possibly ideal planning system* (set out in Figure 1.3). The book will also evaluate each policy area from the perspectives of the checklist shown below.

Policy evaluation checklist derived from Chapter 1.

1. *Were the problems understood and which value system dominated?*
2. *Which goals were to be achieved and were the policy objectives clearly spelt out?*
3. *Were the policy responses accretionary, pragmatic and incremental?*
4. *Which powers in the Gilg/Selman spectrum were mainly used?*
5. *Were planners given sufficient powers/resources and sufficient flexibility for implementation?*
6. *Were performance indicators built into the policies?*
7. *Have the policies been underresearched?*
8. *Have the policies been effective overall?*
9. *Have there been uneven policy impacts, by group or by area?*
10. *Have the policies been cost-effective and are they environmentally sustainable?*
11. *What have been the side-effects?*
12. *Were the policies modified by unforeseen events/policy changes elsewhere?*
13. *Were the policies overtly sectoral and were they compatible with other policy areas?*
14. *What have been the effects of 16 years of Tory rule?*
15. *Should the policies be modified?*
16. *What alternatives are there?*
17. *Overall verdict.*

Before we move onto the sectoral chapters it is useful to recap on the main themes and questions that have been covered so far in both the Preface and this chapter.

The main themes so far in terms of public policy in general. First, policy developments are characterised by accretionary, pragmatic incrementalism. Second, policies are often blown off-course by unexpected events and reactions by those affected by the policy. Third, policies are often undermined by policy changes in other areas with completely separate aims. Fourth, this is exacerbated by the division of public policy into sectoral organisations. Fifth, public policy is underresearched, and the impacts of many policies can only be estimated. Sixth, public policy has been markedly affected by the 'bottom-line' performance target approach of the Conservative Government between 1979 and 1995.

Main themes so far in relation to countryside planning

1 Planning is needed since a minority of people own land or have proprietorial rights over it which are reinforced by the legal system. The only way for the majority to exercise some sort of control is by a planning system.
2 In order to justify itself to both camps, planning attempts to increase productive capacity but at the same time safeguard resources, the heritage and the long-term sustainability of the activity.
3 It does this *either* by attempting to resolve and if possible prevent conflicts over land use arising from different attitudes, *or*,
4 by attempting to manage change by providing an information-gathering and arbitration service to both public and land users. Accordingly,
5 planning is controversial, complex and multi-faceted, but one theme is constant, the desire to create a better tomorrow by the application of systematic methods and thus providing order out of chaos.
6 Countryside planning should thus: a) conserve and prevent waste and pollution of our natural and environmental resources; b) increase production of food and timber; and c) provide opportunities for physical, intellectual and emotional enrichment. Our essentially urban culture does not always lead to spontaneous action to achieve this and hence some conscious planning and management of the countryside is necessary (Royal Town Plannning Institute, 1976).

Finally some awkward questions:

1 Should there be a Countryside Planning Ministry?
2 Countryside planning should resolve and prevent conflicts between different land uses and users with the ultimate aim of managing the resource on a long-term multiple basis. Should this be done as a technical exercise with others being left to sort out priorities?
3 Planning is concerned with providing the right site at the right time for the right people in the right place. Discuss.

The only way to answer these questions is to read on.

2

PLANNING FOR AGRICULTURE AS A PRODUCTIVE INDUSTRY

A: INTRODUCTORY CONCEPTS

There are three concepts that underlie all thinking about planning for agriculture as a productive industry: the world food system; reasons for state intervention into agriculture; and mechanisms for state intervention into agriculture.

The world food system. There are two key trends here: first, the globalisation of production and distribution; and second, the integration of agriculture into the wider economy, and thus the globalisation of agriculture. Bowler (1992a) has termed this the 'industrialisation' of agriculture and a 'third agricultural revolution' which has resulted in a 'food supply system' which has the food chain as its spine. The food chain begins with agricultural inputs, which lead into farm production, followed by product processing, food distribution and consumption. External factors which influence the food chain are: state farm policies (planning); international trade in food; the physical environment; and the financial markets. Together, the food chain and these external factors make the food supply system. The development of a food supply system does not mean that agriculture itself has yet been fully industrialised, but that it has been integrated into wider industrial processes so that 'production on the farm' now forms only a part in the process. This revolution has been charted in detail by Goodman and Redclift (1991), who claim that it implies a major change in the natural environment since key decisions are now effectively made by capitalists in places that may be continents away from where farmers activate these decisions as pawns in a global game. These globalisation trends have also been brilliantly analysed by Le Heron (1993).

Nonetheless, Pierce (1990) reminds us that agriculture still depends very heavily on the natural environment, and is still characterised by very large numbers of relatively small producers. The degree to which agriculture has resisted the wider processes of industrialisation and globalisation has thus prompted attempts, notably by Whatmore, Munton, Marsden and Little

(1987b), to develop a conceptual typology of farm businesses, differentiated on the basis of external relations (for example, the market, technology and credit) and internal relations (ownership of capital, land use rights, the labour process and managerial practices), which depict the degree of penetration of capitalist social relations in production in terms of 'commodification' and 'subsumption'.

Reasons for state intervention. From first principles and from empirical observation five obvious reasons for state intervention suggest themselves. First, to even out the effects of the weather; second, to even out peaks of supply and demand; third, to make agriculture more efficient; fourth, to use the principle of comparative advantage; and fifth, to make it fairer by ensuring reasonable incomes for farmers, and cheap food for consumers. Bowler (1979) has developed these basic reasons into the more detailed set of goals shown in Figure 2.1, which has since been modified and simplified by Tarrant (1992).

McInerney (1986) from the perspective of an agricultural economist has also provided five reasons, namely: stability; growth; efficiency; sustainability; and equity. From a similar perspective de Gorter and Swinnen (1994) have provided a list of nine broad justifications based on 'market failure' which when subdivided include: the inherent instability of markets; imperfect and costly information; imperfect capital markets; yield variability; the difficulty of assessing the public goods or nuisances associated with agricultural production; monopolistic suppliers of inputs and purchasers of outputs; chronically low farm incomes; imperfect labour mobility; asset fixity; the technological treadmill; inelastic supply/demand for food; wrongly valued exchange rates; soil conservation; subsidies for new technologies; the need to provide R & D for an industry characterised by small units incapable of sustaining their own research: coordinating production; and moral imperatives to help poorer people with food aid.

Basically, in an uncertain economic and natural environment, small producers need the security offered by state intervention if they are to invest with confidence. In the British context an extra dimension is provided by the fact that Britain is only about 70 per cent self-sufficient in food, imposing a major drag on the balance of payments, which in 1992 amounted to £ 5.5 billion. By contrast, the European Union as a whole and the United States are net food exporters.

Mechanisms for state intervention. Overall mechanisms have already been discussed in Chapter 1, notably the spectrum of powers developed by Gilg and Selman. There is not, however, any general agreement as to how to systematise the mechanisms used to deliver agricultural policies, and various classifications have been used. For example, Bowler (1979) used a fourfold classification divided into measures designed to influence:

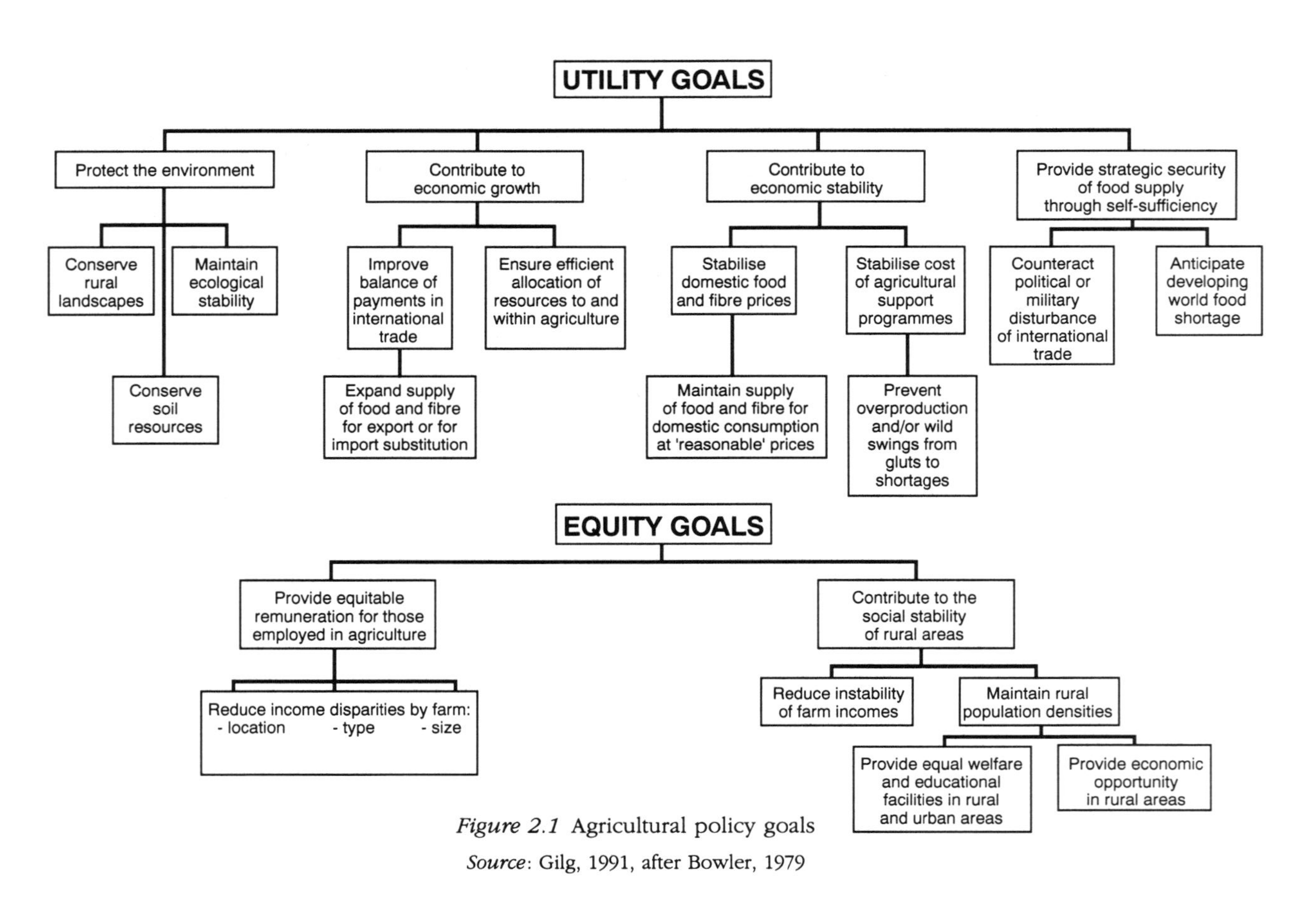

Figure 2.1 Agricultural policy goals

Source: Gilg, 1991, after Bowler, 1979

The growth and development of agriculture, for example, subsidies on inputs, preferential taxes and interest rates, and financial assistance for research and education;

the terms and levels of compensation for production, for example, demand management by advertising and grading schemes; supply management by land retirement, quotas, import levies and tariffs, target prices with support buying, buffer stock schemes, export subsidies, and bi-multilateral commodity agreements; and direct payments like deficiency payments, subsidies, production grants and direct income support payments;

the economic structure of agriculture, for example, financial assistance for land consolidation schemes, farm amalgamation grants, land-reform schemes, retraining or retirement schemes, financial assistance for production or marketing, inheritance laws, taxation, and farm wage determination; and

the environmental quality of rural areas, for example, by pollution controls and land use planning regulations.

In another attempt, Tarrant (1992) has used a threefold classification. First, *increase demand* by, for example, domestic food subsidies, intervention buying, export subsidy, and food aid. Second, *reduce supply* by quotas, co-responsibility levies, and set-aside. Third, affect *farm incomes*, either by *reducing costs*, for example, by input subsidies, providing grants for restructuring, for retirement, and by conducting and disseminating research, or *supplementing costs* by the use of deficiency payments, headage payments, and by direct income support.

In a third attempt, Pierce (1990) has provided not only a fourfold typology but also an accompanying explanation. First, *price/income* measures such as deficiency payments, target prices and crop insurance, to ensure a minimum return for farmers based upon previously agreed prices for products and base acreages and yields. Second, *input* measures such as energy/chemical /transport/water subsidies, credit, and tax relief to influence the cost of inputs, their relative allocation, and the competitiveness of operations, which in turn will probably lead to increased intensification and a narrowing of the range of products. Third, *market* measures such as supply management, quality control and tariffs to influence the quantity, quality and price of food marketed, which in turn will raise domestic prices above the international price. Fourth, *research* measures intended to lower the costs of production and increase returns per unit of output, which will limit the choice of production technologies and resource strategies.

Clearly, the same measures are to be found in each typology, but often under a separate heading. The next section of this chapter will add to the heterogeneity of classification when it will be seen that the Ministry of

Agriculture has constantly changed the headings under which its many and diverse policies have operated. For the meantime it is best to simplify matters by referring to Wilson's (1994) concept of an agricultural policy triangle which at any one time is a working compromise between the three dimensions of *state regulation*, the *free market, and social welfare* or the agricultural economist's key concepts of *supply and demand measures.*

B: EVOLUTIONARY EVENTS

The first period: Reversing 100 years of decline with 30 years of unquestioned growth

Agricultural planning in the United Kingdom has been strongly influenced by the experiences in the hundred years between 1840 and 1940. In the first 90 years of this period the repeal of the protectionist corn laws in the 1840s and the general policy of free trade – exporting industrial goods in return for Empire food – allowed UK agriculture to decline. The net result was that between 1860 and 1933, the area of tillage fell from 5,587 to 3,522 million hectares, and the proportion of food imported rose to a massive 70 per cent, which prompted an initial revaluation of the policy and the introduction of some support measures in the mid 1930s.

The advent of the Second World War in 1939 brought a major revaluation of the policy, as three problems emerged with the previous policy. First, the wartime blockade highlighted the vulnerability of such a dependence on imports; second, it was realised that imported food supplies would be costly and insecure after the war; and third, the depressed state of British agriculture was seen as a waste of resources.

So the wartime Scott Report (Cmd 6378, 1942) on *Land Utilisiation in Rural Areas* made four observations for postwar agricultural policy:

1 It is essential for the Government to formulate and adopt a long-term policy for agriculture;
2 Agricultural land must be properly farmed and maintained;
3 Measures should be taken to secure, as far as possible, stability of conditions; and
4 Agriculture requires a considerable amount of new capital to enable it to produce more economically and efficiently

The immediate response was a 'ploughing up campaign' to increase food production and rationing, which made the general public aware of the need for a vigorous postwar food policy. The long-term response came in the Agriculture Act 1947, which provided a massive momentum for expansion based on the resurgence of 'agricultural fundamentalism' under which agriculture became once again a key feature of overall government policy.

In more detail the 1947 Act called in its preamble for:

> a stable and efficient agriculture capable of producing such part of the nation's food as in the national interest, it is desirable to produce in the UK, and of producing it at minimum prices, consistent with proper remuneration and living conditions for farmers and workers in agriculture, and an adequate return on capital invested in the industry.

The Act was put into operation throughout the 1950s and 1960s by the Ministry of Agriculture Fisheries and Food, and the related Department of Agriculture for Scotland and the Welsh Office for Agricultural Development. These organisations evolved a set of policies based on the Act by a series of 'Annual Reviews' which followed a five-stage process of: preparation of economic and statistical data; commodity considerations; general policy considerations; consultation with farmers' representatives; and decisions to change/modify policies (Bowler, 1979). Over the years other pressure groups, notably the food industry, consumer associations and environmental groups, joined the process. The main policy weapons were price manipulation, production grants and subsidies, and increases in productivity to be achieved by the application of science and restructuring the industry.

Market manipulation was based on the guaranteed price/deficiency payments system under which a guaranteed price of, say, £65 was set. If the market price was only £55, however, the farmer was then paid a deficiency payment of £10, to make up the difference. The main advantages of this system were that: if surpluses did occur they led to lower food prices; payment was limited to goods sold and thus poor quality produce (that is, bad farmers) were excluded; and government liability was limited to free market sales, and also by the introduction of standard quotas in the mid 1950s. Stability was also introduced in the late 1950s by a band which restricted the amount by which guaranteed prices could move up or down. The market was also manipulated by the use of quotas of production, by the use of marketing boards, and by the fixing of some prices, notably for milk, which was set at different levels according to the season, location, and market sold into (Gilg, 1986b). These policies gave the Government considerable power over the type of production that was encouraged by giving farmers very clear signals.

Production grants and subsidies, which only accounted for 25 per cent of government expenditure in 1955, grew steadily from then, so that by 1970 they accounted for 40 per cent of spending, the same proportion as deficiency payments, which had declined from 70 per cent in 1955 (Bowler, 1979). In addition special grants and subsidies were gradually introduced for special types of farming or farming in disadvantaged areas.

Increases in productivity by the application of science and restructuring the industry were achieved by a number of measures. First, pure and applied research was carried out by ministry scientists, the results were tested on experimental farms, and then when the new productive system was seen to work, officials of the National Agricultural Advisory Service (in the 1960s which became today's Agricultural and Advisory Service) went out to farmers and suggested they try the new techniques, crops or whatever. Second, structural measures included schemes for early retirement, farm amalgamation, and training workers for the new industrialised agriculture. Third, there were generous tax and fiscal measures which encouraged farmers to invest in long-term productivity gains by improving machinery, buildings, and land by, for example, drainage and hedgerow removal.

The net result was a massive economic momentum (Korbey, 1984) which encouraged farm production to grow with no questions asked. This mood of expansion continued into the 1970s with two white papers. First, *Food from Our Own Resources* (Cmd 6020, 1975), which advocated a 2.5 per cent per annum expansion of production, and second, *Farming and the Nation* (Cmd 7458, 1979), which noted that the continued expansion (10–20 per cent over 5 years) of agricultural net product would be in the national interest. By the 1970s, however, Britain was no longer master of its own farm policy because it had joined the European Community in 1972 and notably would become increasingly subject to the vagaries of the Common Agricultural Policy.

The second period: transition to the CAP

Initial development of the CAP

The Common Agricultural Policy (CAP) was set up by the original European Economic Community (EEC) of six nations in the 1960s as an essential first stage to European integration. The genesis of both the EEC and the CAP, like the genesis of agricultural policy in the United Kingdom, is to be found in the need to rebuild Europe after the Second World War ended in 1945. In particular, there was a need to create a trading and political bloc that could compete with the giants of the United States and the then Soviet Union, and also be a major player in the development of new postwar international organisations, like the GATT and the Organization for Economic Co-operation and Development (OECD). Accordingly in 1957 the original six countries (West Germany, France, Italy, Belgium, Luxembourg, and the Netherlands) signed the Treaty of Rome and began to set up community institutions. Britain decided to stay out, largely because of its strong links with the Empire, which was rapidly becoming the Commonwealth, and the United States. In an attempt to defuse the power of the EEC, Britain helped to form an alternative bloc, European Free Trade Association (EFTA), with mainly the Scandinavian countries.

The nascent EEC soon turned its attention to agriculture for a number of reasons. First, the memory of war shortages was still very strong; second, because agriculture contributed then over 30 per cent to average living costs; and third, because of a percieved need to forge a link between the two old adversaries of France and Germany by helping to coordinate France's strength in farming with Germany's strength in industry. Fourth, and perhaps most significant, 25 per cent of voters were at the time engaged in agriculture.

After 45 meetings, 137 hours of talks and 582,000 pages of documentation, six basic principles for the CAP emerged, namely that there should be: 1) a single market area; 2) free internal movement of farm goods; 3) a uniform external tarrif; 4) common prices within the market area; 5) community preference in agriculture trade; and 6) the financial burden should be shared (Bowler, 1985). None of these principles have, however, been more than partly achieved in practice.

Following from these principles, five objectives emerged, namely: 1) to increase agricultural productivity by promoting technical progress, by ensuring the rational development of agricultural production, and the optimal utilisation of the factors of production, in particular labour; 2) thus to ensure a fair standard of living for the agricultural community, in particular by increasing the individual earnings of persons engaged in agriculture; 3) to stabilise markets; 4) to assure the availability of food supplies; and 5) to ensure that supplies reach consumers at reasonable prices (Cole and Cole, 1993). Not all of these objectives are compatible and to some extent the inherent conflict they contain have continued to plague any rational evolution of the CAP.

Another problem plaguing the rational development of the CAP is that the logical development from principles to objectives was not followed by a related set of mechanisms. In contrast a series of ad hoc and increasingly incremental mechanisms emerged, notably: 1) variable import levies and export subsidies; 2) a system of intervention purchases; 3) production grants and subsidies; 4) measures to improve the structure of farming; 5) measures to aid Less Favoured Areas; and 6) in response to a growing divergence rather than a convergence of currencies, a system of 'green' currencies.

Another reason accounting for the rather haphazard evolution of mechanisms is the system of decision making which encourages a process of bargaining between parties, and notably between member states, especially during the transition period when everybody had to agree in order for a new policy to emerge. The net result is a series of trade-offs (Petit *et al.* (1987) and a set of less-than-optimal 'satisficing' rather than 'optimising' solutions (Fearne, 1989).

The British accession

For Britain, accession to the EEC and transition to the CAP meant a combination of trends from no change to revolution. First, a number of policies remained entirely in the UK domain, for example, those relating to potatoes. Second, a number of mechanisms already employed incorporated similar methods to those within the CAP, and thus only minor modifications were necessary; these mechanisms included the system of production grants and subsidies, structural measures, and measures for disadvantaged areas or types of farming. Third, in the system of market manipulation, however, a revolution had to take place in both the system of internal market support, and also over import levies and export subsidies. This third trend was very important, since it is the area where most CAP expenditure is incurred and because it ran most counter to Britain's traditional status as a major free market trader in agricultural produce.

Taking the method of price support first, the CAP was using a system of intervention purchases, which worked as follows:

1 A guide/basic/target price was set each year which would give an average farmer a reasonable profit;
2 Somewhat below this an intervention price was set which provided a floor to the market; and at which
3 the EEC had to buy up all produce offered to it at the intervention price, assuming it met certain basic standards.

In theory the market should have balanced out with one year's glut being sold off in next year's shortage. In practice, however, the target price was set too high, and without any quotas on sales being imposed, or the rigour of competition from cheaper imports, farmers were given *carte blanche* to overproduce, as they were given a virtually guaranteed price above production costs. The result was a growing surplus of produce as the financial incentives given by the CAP were harnessed to rapidly growing productivity, against a background of inelastic demand for food from a static population.

One way to get rid of the unwanted surpluses was to sell them onto the world market, but by and large EEC prices were way above world food prices, and so a third element to the system was created, a system of export refunds to bring EEC prices down to a level where they could be competitive.

The net result was high prices for farmers, low prices for foreign producers, an increased cost of living, and higher taxes. All of these were regressive – economically, socially and morally – for everybody outside the narrow confines of the EEC's farming community. Even within the farming community some farmers did better than others, notably large arable farmers in lowland areas of northern Europe.

A complicating factor also soon arose in the form of green currencies and the resulting system of Monetary Compensation Amounts (MCAs) set up in 1969 to deal with the problem (Le Heron, 1993). In essence the problem was that agricultural support prices were set infrequently on the assumption that exchange rates would be stable, but in practice some currencies, notably the German Mark, proved far stronger than others, and so the MCA system attempted to bridge the gap between exchange rates in the real world, and exchange rates set in the unreal world of agricultural support prices. This gave national governments the power to either raise or reduce support prices by delaying the rate at which their 'green' or agricultural support exchange rate was adjusted upwards or downwards to reflect its real value.

Evolution of CAP from 1970 to 1980

The first surpluses began to occur as soon as 1968. The first attempt to deal with the problem was the Mansholt Plan (Potter 1990) which in 1972, after much watering down, projected a reduction in the cultivable area for the EEC of the time from 70 to 65 million hectares, via a combination of lower prices and retirement of farmers.

Within the year, however, matters were complicated by the expansion of the EEC to nine countries with the accession of the United Kingdom, Ireland and Denmark, which brought new tensions. Even though there were several attempts to bring prices into line and to shift the emphasis from the producer to the consumer, they all failed as a result of national interest trade-offs, and the problems of cost and surpluses continued to grow.

By 1980 the following had become big issues for the CAP:

1 the cost both absolutely and relatively with the CAP consuming around two-thirds of the EEC budget and the intervention support system consuming over 90 per cent of the CAP budget (Bowler, 1985);
2 widening gaps in exchange rates and thus increasing MCAs;
3 CAP prices were above world prices often by as much as a factor of 1.5 or 2.0;
4 there were widening regional divisions notably between the north and the south;
5 big farmers were gaining over small farmers and arable farmers were gaining over livestock farmers;
6 there were also big differences between countries with regard to being either net beneficiaries or contributors, for example France was a big beneficiary, while Germany and the United Kingdom were major net contributors (Bowler, 1985); and
7 there was a growing body of public opinion which viewed surpluses as not only being expensive but also immoral.

The third period: incremental reform of the CAP from 1980 to 1995

Not surprisingly, this period saw a series of ever more fevered attempts at reform, most of which were postponed until a crisis point had been reached. The two main thrusts for reform came from external and internal forces.

External forces

The main external force for reform came from the need to update the General Agreement on Tariffs and Trade (GATT) to allow the continued expansion of genuine free trade around the globe (Greenaway, 1991). The restrictive practices operated by the CAP were seen to be a major stumbling block to such trade, notably the distortions caused by import levies and export subsidies, especially by the Cairns Group of agricultural exporting countries – which includes Canada, Australia and Argentina (Robinson, 1993). These countries can produce at much lower prices than the EEC, mainly by farming extensively rather than intensively. For example, in the United Kingdom one hectare could produce 7.9 tonnes of wheat and a gross income of £876 at a break-even price of £79 per tonne in 1992, whereas in the United States the respective figures were 2.7 tonnes, £201 and £64 per tonne. The nub of the problem was thus severe competition between two trading blocs operating very different agricultural systems against a background of surplus world stocks. A few more figures highlight the differences. For example, in the EEC farming accounted for about 8 per cent of the workforce compared to 3 per cent in the United States, the EEC had a farm trade deficit averaging $18.1 billion annually between 1986 and 1988 (Cole and Cole, 1993) compared to a US farm trade *surplus* of $18 billion in 1988, and the United States had embarked on major measures to cut back on farm support and production via the 1985 Farm Bill, leading to a loss of confidence and bankruptcies in mid-West farming states, in contrast to the protracted debate in Europe.

The initial Cairns Group demand in 1987 was for a 75 per cent cut in production support and a 90 per cent cut in export subsidies. The EEC reaction was that this was wholly unacceptable and they offered a 30 per cent overall cut, since according to Guyomard *et al.* (1993) the only people to gain would be consumers! After several years of crisis talks, and prolonged impasses, agreement was provisionally reached at Blair House in Washington in 1992 and finally sealed in Brussels in 1993. Under the so called Uruguay round of GATT therefore the EEC (by now the 15-state European Union (EU) agreed to phase in three cuts in agricultural support between 1995 and 2000, namely: a 20 per cent cut in farm subsidies for production (this still allowed the EU to subsidise farmers, as will be seen later with the 1992 CAP reforms); a 36 per cent cut in import tariffs and export subsidies; and a 21 per cent cut in the volume of subsidised exports based on 1991 levels but spread differently across product sectors (Cramon-Taubadel, 1993).

Internal forces 1980–8

A number of modifications, some quite dramatic, but mostly uncoordinated, and either incremental or complete U-turns were made in this period. The most dramatic was probably the introduction of milk quotas in 1984, reversing policy signals encouraging dairy farmers to expand production which had been announced only a few days before. Unfortunately, the quotas were set above consumption, and subsidised disposals were still needed for some 20 per cent of production. In turn quota became a saleable asset, and dairy farmers have in general adjusted well given the freedom once again to expand or reduce production by buying or selling quota.

Price freezes were also proposed by the Commission during this period, but these were generally adjusted upwards by politicians (Bowler, 1985) and were then further offset by general productivity rises of an average 3 per cent a year. In addition price freezes encourage farmers to run faster in order to stand still, in the view of a former senior planner in MAFF (Raymond, 1985). Accordingly, an extra weapon was added in 1987 with the introduction of a scheme for set-aside and extensification under Regulation 1760/87 which was mandatory for each state, but voluntary for each farmer. The target was a 20 per cent reduction in the arable area.

Under the main scheme, farmers who set aside some or all of their land from production were paid rates of around £200 per hectare as compensation for the production foregone (Potter, 1987). The scheme was not very successful and was severely criticised in that it contained an essentially bad principle, paying people to do nothing. Other weaknesses were: the short-term (1988–93) nature of the scheme; the fact that the best land for conservation gain was not targetted (Potter *et al.*, 1987); the voluntary nature of the scheme; very different rates for different countries and different types of set-aside (Ilbery, 1992a); very uneven take-up rates (Briggs and Kerrell, 1992); the potential for fraud; and perhaps most problematically, the American experience since the 1930s that the poorest land is set aside, while the best land is farmed more intensively, thus offsetting the gain (Ervin, 1988).

Other changes in the period included: Regulation 797/85, which added to measures from the 1970s, for example retirement schemes deriving from the Mansholt plan, produced new or larger Less Favoured Areas, aid for farm diversification, provisions for national schemes for Environmentally Sensitive Areas (ESAs), and the new concept of pluriactivity whereby farmers and their households take on many other gainful activities in addition to farming. In the United Kingdom provisions to allow aid for farm diversification were included in the Farm Land and Rural Development Act 1988, which allowed the Farm Diversification Grant Scheme to be set up in 1988. This was aimed at stimulating additional investment in the development of alternative enterprises on farms and the effective marketing of the products and services of diversified enterprises. In the early 1990s some elements of this scheme, notably grants to aid farm tourism, were abolished (Cm 1903, 1992), and the remaining elements were transferred to the Farm and Conservation Grant Scheme which had been set up in 1989 and which is discussed in more detail in Chapter 3.

The EEC agricultural budget, however, came under renewed strain from the accession of Spain, Portugal and Greece in the 1980s with their poorly structured, inefficient farms and much bigger percentages of the workforce in agriculture (Cole and Cole, 1993). So by 1988 the EEC had reached a crisis in its evolution in spite of several proposals for reform, for example, green papers in 1985 ((7872)8480/85) and 1987.

Internal forces, 1988–92

Matters came to a head at the December 1987 Council and a crisis meeting was called in Brussels in February 1988. This meeting agreed that the only way to move the EU forward from the budgetary stranglehold so long imposed by the CAP was: first, to limit the amount spent on CAP Guarantee spending; second, to double the CAP Guidance Fund alongside the Social and Regional Development Funds; and third, to expand the budget by around 15 per cent between 1988 and 1993.

The second of these changes was made in the Official Journals of L185 and L374 in 1988, notably the creation of Objective 1 (lagging) regions where prosperity is less than 25 per cent of the EU average, and Objective 5a and 5b regions with agricultural/rural problems. Both of these measures were an attempt to wrest back resources from the Guarantee section of the CAP and to divert it to long-term strategic measures based on structures.

In the short term, however, the Guarantee Fund still dwarfed the Guidance and other funds, and so in 1992 the so-called Delors I and II packages (named after the then President of the Commission) proposed to raise overall spending from 51 billion ECU in 1987 to 66 billion in 1992 and 83 billion in 1997. The CAP's budget would rise from 33 to 40 billion ECU, but the Structural Funds budget would rise much faster from 9 billion to 29 billion ECU between 1987 and 1997 (Bandarra, 1993). These packages were of course a classic example of the truism that, when cutting a cake differently, it is a big help if the cake gets bigger.

However, the CAP's problems would not go away by employing the 30-year-old solution of increasing expenditure, albeit at a slower pro rata rate, and in 1992 the CAP still accounted for 53.4 per cent of spending (Cole and Cole, 1993). Indeed, the same old problems – excessive self-sufficiency in temperate foods, food mountains and lakes, costs of all types, but ironically declining numbers of farmers – were still around in 1992 in spite of the attempts at reform made in 1988, which in agriculture had included specific reforms to the CAP, notably the implementation of a new policy measure based on the related concepts of thresholds/stabilisers/co-responsibility levies).

Under these measures thresholds were set for certain types of production. For example, a threshold of 160 million tonnes was set for cereals. If this total was exceeded a 3 per cent co-responsibility levy would be levied on all producers, accompanied by a 3 per cent cut in the intervention price. Unfortunately, the market was then only 135 million tonnes and productivity increases of 3 per cent were enough to shield farmers from the worst effects, so the plan was another example of a half measure. Nonetheless, farmers could be in no doubt, after over 20 years of various attempts to curb production by various measures, that the long postwar period of expansion would have eventually to come to an end. For some commentators this was achieved by major reforms to the CAP in May 1992.

The May 1992 reforms

In common with many previous reforms, those of May 1992 had a long and complex gestation and were intertwined with the negotiations over the Uruguay round of GATT. The process began in the winter of 1991 with proposals by the then Agriculture Commissioner, the Irishman, Ray MacSharry (COM(91)100). The key feature of these proposals was a modulation of support to small farmers. In more detail, the proposals rested on 10 objectives which included: maintaining the maximum number of farmers; recognising the dual role of farmers as producers and custodians of the countryside; helping rural areas develop their economic activity; avoiding the build-up of intervention stocks by controlling production; encouraging extensification and environmentally-friendly farming; making sure that the budget went to those in greatest need; and introducing measures whereby quotas and other restraints are increased progressively with the size of farm. The proposals included three major policy reforms: compulsory set-aside with compensation; major reductions in price supports; and compensation to be limited to small farms. These proposals induced a vitriolic response from the House of Lords (1991), which argued that it would adversely affect the United Kingdom with its structure of large farms.

The proposals were modified in the summer of 1991 (13042(COM)(91)258), but they remained unpopular in the United Kindgom, and the House of Commons (1991) attacked them for increasing the budget by possibly 3.5 billion ECU, criticised the principle of modulation because it rewarded poor farm structure, and argued that compensation should be based on cross-compliance, for example, by insisting on a conservation gain in return.

Eventually, a much modified reform was agreed in May 1992 with three main objectives: 1) sufficient income support to allow farmers to remain on the land, as there was no other way to preserve the natural environment, traditional landscapes and a model of agriculture based on the family farm as favoured by society generally; 2) control of production to bring markets back into balance; and 3) guaranteeing competitiveness and efficiency required to maintain the Community's role in international agricultural trade. These objectives (Bowers, 1995) entailed a recognition that farmers fulfil a dual role of production and environmental management and the policy changes to bring them about contained three key features: 1) across-the-board compensation; 2) moving away from production support of farm*ing* to direct support to far*mers*; and 3) moving away from support paid by consumers in the form of high food prices, to support paid for by taxation.

Four key mechanisms were then set in motion: 1) compulsory set-aside for all those who wished to take advantage of CAP support and compensation payments – with the exact percentage to be set each year depending on the level of production – it began at 15 per cent; 2) cuts in support prices between 1993 and 1996 – for example, cereals by 29 per cent and beef by 15 per cent; 3) compensation to be paid in two forms: a) variable rates per hectare for set-aside depending on the region of the EU, and b) area payments for those crops still being grown on land not in set-aside to compensate for lower prices; and 4) an administration scheme (the Integrated Administration and Control System (IACS)) for farmers to register claims and for administrators to check for fraud by satellite photos. The reforms were to be instituted over the period between 1992 and 1996, when they were due for review. In the meantime the agrimonetary system was reformed in 1995 with the abolition of the inflationary 'switchover' mechanism. Although the replacement system allowed for a degree of currency fluctuation without a revaluation of 'green' rates the new system still provided considerable protection for farmers against currency induced falls in incomes in strong-currency Member States (Cm 2803, 1995).

In more detail, the initial minimum rate of compulsory set-aside was set at 15 per cent, although farmers can set aside more if they so wish, even the entire farm. In 1993 the initial set-aside payment was set at £218 per hectare in England. On the remaining arable land compensation payments (area aid) for reduced support prices began at the level of £121 per hectare in England in 1993, but a major devaluation of sterling in September 1993 meant that by 1994 that predicted set-aside payments had risen to £313 and Area aid payments to £247 for the growing year 1994–5. In order to qualify for the payments, however, farmers have to comply with a list of environmental conditions.

The reforms thus meant a major shift from market support to direct subsidies. Copus and Thomson (1993) calculated that total UK expenditure on market support would fall by £494 million to £292 million, while in contrast expenditure on direct subsidies would rise by £913 million to £1569, thus incurring a net increase of £419 million. In a wider critique Robinson and Ilbery (1993) argued that the measures remained overfocused on farmers, and that the CAP should evolve towards a wider rural development strategy with encouragement for sustainable environmentally-friendly agriculture and special help for poorer farmers and marginal areas.

The May 1992 reforms also included a series of agri-environment/forestry packages which were to be submitted by each country in a further attempt to take out more land and in this case to gain some environmental benefit, by planting trees or regenerating habitats. Thse measures are discussed more fully in Chapter 3.

Within the United Kingdom the first five years of the 1990s also saw some major reforms, notably the winding up of the near monopolistic producer's cooperative, the Milk Marketing Board, and its replacement by a number of competing wholesalers, notably a producer's cooperative, Milk Marque (Cm 2803, 1995). This process raised milk prices considerably and finally offset the deleterious effects of milk quotas. In the uplands farmers were dealt a double blow, as payments for upland livestock under the Hill Livestock and Compensatory Allowances schemes (HLCAs) were not only reduced, but also restricted to various maxima per farm and per area (Cm 2803, 1995). New tenancy arrangements were introduced in 1995 under the Agricultural Tenancies Act 1995 which created a new flexible tenancy, 'farm business tenancies', intended to break down the disincentive of guaranteed security of tenure through the generations which had been putting landlords off letting land, and to provide greater flexibility especially with regard to non-agricultural diversification.

Given this long litany of reforms, it is not surprising that confidence in farming reached all-time lows between the mid 1980s and mid 1990s, but fortuitously the 1993 devaluation of sterling and its continued fall between then and 1995 meant that UK farmers had by 1995 begun to feel that the worst of the transition was over and that a stable pattern was emerging. Whether this is the case or not will form one of the main threads of the rest of this chapter, but before this it would be wise to recapitulate our findings so far, in terms of the fairly discrete periods of agricultural policy shown in the box below, by examining Figure 2.2, and to ask if the pre-CAP system had any advantages which could be considered in any future policy reform.

1947–55	Postwar recovery. Quantity emphasis at almost any cost.
1955–64	Stability. Shift to quality emphasis and controlling inflation.
1964–72	Shedding labour for use in industry, thus emphasis on productivity.
1972–9	Transition to CAP. Continued expansion but with different mechanisms.
1979 to date	Restructuring via incremental reforms in response to growing costs and environmental concerns.

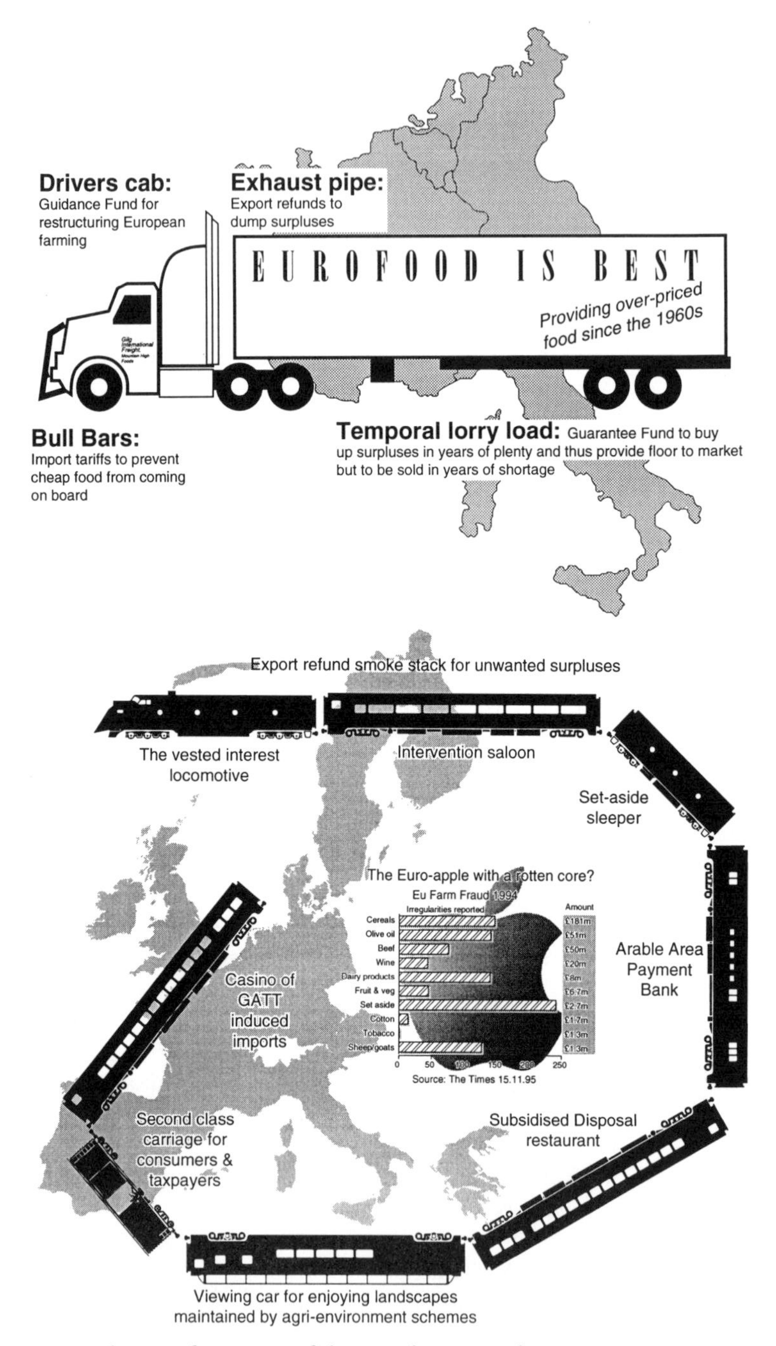

Figure 2.2 The transformation of the CAP lorry into the runaway Eurogravy train

Source: Author

If a return to the 1947 system were ever contemplated, its proponents could point out the following three advantages. First, surpluses, if they occurred, led to lower food prices, unlike intervention buying, which provided a rigid floor to the market; second, payment was limited to goods sold, unlike the open-ended nature of the CAP until the late 1980s when controls were introduced; and third, government liability was limited by standard quotas, unlike the CAP, which again was open-ended till the late 1980s. Tarrant (1992) has also shown that the cost of the UK system was roughly the same between 1950 and 1970, at around £300 million, but rose rapidly with the transition to the CAP between 1974 to 1986 to around £1,400 million, but then remained roughly the same to 1990.

THE SITUATION IN 1995

In 1995, according to its Annual Report (Cm 2803, 1995) MAFF was concerned with a diverse range of issues, such as the rural and marine environments, development of the rural economy, and policy on farm animal welfare and the food and drink manufacturing industry. However, its expenditure was weighted very heavily to supporting production and farm incomes as shown in Table 2.1. In addition to the expenditure shown in Table 2.1 MAFF's total expenditure amounted to some £3.2 billion in 1994–5, £0.8 billion on domestic programmes and £2.4 billion on EU programmes. In many instances MAFF's responsibilities are confined to England, while the other Agriculture Departments have equivalent responsi-

Table 2.1 Overall expenditure on agricultural planning between 1989–90 and 1994–5 and predicted for 1997–8

	1989–90 actual	*1993–4 actual*	*1994–5 estimate*	*1997–8 plans*
		£ million		
1 *Intervention Board and EC Expenditure*				
Market support	854	2077	1770	2103
1 *Total including items not shown*	*921*	*2153*	*1839*	*2169*
2 *Domestic Agriculture*				
To protect the public	73	106	117	118
To protect environment	55	75	106	110
To improve performance	75	99	118	105
R & D	60	112	120	123
Executive Agencies	80	4	–5	–6
2 *Total including items not shown*	*566*	*727*	*807*	*790*
Total of both 1 and 2	1487	2880	2646	2959

Source: (Cm 2803, 1995)

bilities in Scotland, Wales and Northern Ireland, although in some cases MAFF has UK-wide responsibilities. Thus, the division of spending across national territories, and the impact of EU repayments for certain programmes, mean that there can be very different sums of expenditure on agricultural planning, and so any data referring to agricultural planning expenditure should be treated with great caution and the small print should be examined carefully.

MAFF's work in 1995 was divided up by a financial planning system which categorised its expenditure by aim and sub-aim. The aim/sub-aim structure was then divided into 80 individual programmes, each of which was designed to achieve specific objectives subject to annual review and modification. In more detail the aims were: to protect the public (24 per cent of expenditure in 1992–3 and 15 per cent of staff in 1990–91); to protect and enhance the rural environment (21 per cent and 5 per cent); to improve the economic performance of agriculture (41 per cent and 48 per cent); to protect farm animals (2 per cent and 1 per cent); and to ensure the best use of internal resources in support of the Ministry's business (30 per cent of staff in 1990–91). The figures in brackets are derived from Cm 1903 and Cm 1503, since unfortunately Cm 2803 doesn't provide any breakdown by staff time. In addition 34 per cent of MAFF staff were employed in five executive agencies including ADAS, each of which produced its own annual report, but were expected to become self-financing by charging for their services by the late 1990s as shown in Table 2.1. Finally, the Intervention Board Executive Agency, while a separate government department in its own right with its own annual report, also came under the miscroscope of the 1995 report. The aims and the work of the separate agencies are now considered in turn, not by the public relations order ordained by MAFF, which coyly hides the vast majority of spending in the middle of the report, but by descending order of expenditure.

Improving the economic performance of agriculture. In 1995 this involved implementing MAFF's CAP obligations efficiently, and seeking a more rational CAP while avoiding discrimination against UK businesses. In more detail, expenditure under the CAP was roughly divided into two categories, market intervention and direct payments. Under the 1992 reforms direct payments had grown rapidly to account for 42 per cent of total CAP expenditure by 1994–5. Overall the Producer Subsidy Equivalent, a calculation of the value of transfers to EU farmers which included both intervention measures (measured by the difference between world and EU prices for agricultural goods) and direct payments to farmers, was projected to be 62,600 million ECU or 46 per cent by value of EC agricultural production, exactly the same percentage as 1990.

In more detail, direct payments to farmers were administered by the Integrated Administration and Control System (IACS), which was set up after the 1992 reforms as an anti-fraud measure. IACS required farmers to submit an annual application providing details of their farmed land on a field-by-field basis. In 1994 83,000 farmers submitted IACS applications for Arable Area Payments, the Beef Special Premium, Suckler Cow Premium Scheme, Sheep Annual Premium and Hill Livestock Compensatory Allowances. Typically payments to farmers were in the tens of thousands of pounds, but some big farmers gained cheques of several hundreds of thousand pounds, with rumours of even some large estates getting over a million pounds annually. Some cheeky journalists even obtained a large cheque before revealing their ruse. Thus the system was largely discredited in the eyes of the public for paying out obscene sums to farmers for doing nothing. In contrast, farmers objected to the intense form filling that IACS entailed until they remembered that they were being paid hundreds of times the rate for form filling paid to an average clerk.

Arable Area Payments, which began in 1993, offered in 1995 direct area payments to growers of cereals, oilseeds and protein crops. All but those claiming on a very small area had to set aside land each year to qualify for the payments. Payment rates differed according to the productivity of the region and the crop. In 1994 47,000 applications were made for 3.8 million hectares and 741,000 hectares were in set-aside. Options for the use of land in set-aside were widened in 1994 and the scheme was also extended to include a five-year scheme where farmers perceived an agronomic or environmental need. In 1994–5 the payments amounted to £1,074 million, all refunded by the EU, and were predicted to rise to £1,254 million in 1997–8.

Five-year voluntary set-aside and pilot beef and sheep extensification schemes date back to 1988 and 1990 respectively. In 1994–5 they cost MAFF £25 million, of which £10 million was refunded by the EU, but they were due to be phased out by 1996, as participants in set-aside either withdrew entirely or moved into the Arable Area Payments Scheme. In 1995 only 35,000 hectares out of the original 152,000 hectares were left in the scheme. The extensification scheme programmed to run from 1990–1 to 1995–6 was limited to only 76 participants in 1995.

Intervention support for cereals cost an estimated £104 million in 1994–5 and was expected to rise to £115 million in 1997–8. Intervention cereal stocks rose from 400,000 tonnes in January 1993 to 1,400,000 tonnes in October 1993 but had fallen back to below 800,000 tonnes in December 1994. Other arable intervention expenditure was estimated to be £19 million in 1994–5, and forecast to fall to £16 million in 1997–8. In both cases EU receipts cover all or more than the cost. Intervention in the sugar market, which is complicated by the Lomé Convention, cost £131 million in 1994–5, and was predicted to fall to £109 million by 1997–8.

Beef production was supported in 1995 by market support and headage payments. The cost of both schemes was estimated to be £344 million in 1994–5, and was expected to rise to £559 million in 1997–8. Direct payments were expected to rise rapidly as export subsidies and intervention buying ceased, since between January 1993 and December 1994 intervention beef stocks fell from 160,000 tonnes to only 20,000 tonnes. Sheep production was supported in 1995 almost entirely by the Sheep Annual Premium Scheme and marginally by market support at a total cost of £410 million, which was forecast to rise to £482 million in 1997–8. Support payments for both beef and sheep were made subject to certain limits in the 1990s, however, in order to prevent environmental damage by overstocking (Cm 3016, 1995). In contrast to the large sums spent on sheep and beef, support for pigmeat production only cost £2 million in 1994–5, and similarly support for eggs and poultry cost just under £3 million.

Market support for milk and milk products totalled £223 million in 1994–5, with projected expenditure set to fall to £200 million in 1997–8. Between 1993 and 1994 intervention stocks remained pretty consistent at around 11,000 tonnes in spite of continuing quotas and regular price cuts of between 1 and 2 per cent.

The 1995 Annual Report also considers a number of other measures under the aim of improving the economic performance of agriculture, under the sub-theme of 'creating the conditions in which efficient and sustainable agriculture can flourish' (62). Support for potato production under the 1947 Act was due to be phased out in 1997, and had been much reduced by 1995, with, for example, a substantial loosening of quota controls. A draft EU proposal to bring potatoes under CAP support was, however, on the table. Support for other horticultural production was still available in 1995 at a total cost of £7 million. A new programme aimed at encouraging 'alternative crops' for non-food purposes began in 1995 but its estimated budget was only £275,000 for 1995–6.

Improved marketing continued to be a major expenditure item totalling an estimated £84 million in 1994–5, rising to a projected £91 million in 1997–8. The objective was to foster efficiency and competitiveness by marketing products more effectively. Examples included: marketing grant schemes; export refunds; supporting the Food from Britain promotional organisation; supporting speciality food producers via the Processing and Marketing Grant Scheme; implementing UK food quality schemes and EU regulations protecting food names, like Stilton cheese (Cm 3016, 1995).

Support for training the labour force was transferred to ATB-Landbase Limited from the Agricultural Training Board in 1994, in the expectation that this would reduce costs from £4 million in 1994–5 to £2.6 million thereafter. The wider rural economy was also aided to an estimated £2 million in 1994–5, which was expected to rise to £21 million in 1997–8, mainly due to the planned expenditure of £64 million under the Objective 5b programme of the EU, discussed more fully in Chapter 5. Programmes were also in place to safeguard the future of farmland by, for example, influencing the town and country planning system of development control also discussed in Chapter 5. In 1993–4 MAFF dealt with some 3,500 planning applications.

The application of science which has so transformed agriculture continued to be a major plank in agricultural policy in 1995, but funds have been reduced, and the emphasis has shifted from production for its own sake to wider issues. Nonetheless £49 million of the total £128 million spent on research in 1992–3 went on improving the economic performance of agriculture, only £6 million less than the £55 million spent on research into protecting the public. Significantly, research into enhancing the rural environment and protecting farm animals only accounted for £20 and £4 million respectively. Production is also supported by the continuation of extensive controls over pests and diseases at an estimated cost of £19 million in 1994–5.

Protecting and enhancing the rural environment. Most of these schemes are considered in Chapter 3, but some of them which seem to be optimistically placed in this category by MAFF are considered here, since the environmental aims of the schemes seem to be more of an adjunct than their prime purpose, for example, Hill Livestock Compensatory Allowances (HLCAs). These are intended to ensure the continuation of extensive livestock farming in the Less Favoured Areas (LFAs), thereby helping to maintain a viable population and help conserve the countryside. The payments had been implicated in overstocking and thus environmental damage, and so in the 1990s headage limits were imposed. In 1994–5 it was estimated that the scheme would cost £111 million, falling slightly to £105 million in 1997–8.

Protecting the public. Programmes under this aim included food safety, and promoting action to alleviate flooding and coastal erosion mainly by providing grants for flood defence and coastal protection.

Agencies. The oldest agency, the Agricultural Development and Advisory Service (ADAS), has been transformed in the 1990s into a virtually self-funding agency. In 1994–5 its costs of £82 million were offset by receipts of £74 million, although only 63 per cent of its consultancy services were expected to be received from commercial charges. The basic work of ADAS, however, remained consultancy advice based on research and development.

The Intervention Board for Agricultural Produce, which was set up in the 1970s to administer the financial support schemes of the CAP, became the Intervention Board Executive Agency in 1990. In 1995 it had the following objectives: to implement the CAP; to ensure all concerned are aware of their rights and obligations; and to advise the government on existing schemes and proposals for new schemes. In support of these objectives in 1995 it made the physical, administrative, legal and accounting arrangements required to buy, store and sell various commodities, to provide aid for production, processing and storage, to administer quotas, to licence and aid imports and exports, and to enforce the regulations and maintain financial controls. In 1994–5 it was estimated that the Board would account for gross expenditure of £3,100 million. Its separate annual report provides detailed information on the nuts and bolts of CAP policy as well as the evolution of surplus stocks.

The situation in Scotland and Wales. These countries have their own agricultural departments and their work is set out in the annual expenditure white papers for each country. For example (Cm 2814, 1995) set the work of the Scottish Office Agriculture and Fisheries Department (SOAFD) into the Scottish context when it stated that its primary aim was 'to foster efficient and competitive Agriculture, Fisheries and Food industries, having regard to the needs of sparsely populated rural communities, of consumers, and of the environment and of animal health and welfare' (19). In more detail SOAFD estimated that it would spend £486 million in 1994–5, mostly on market support, £305 million, but also £69 million on structural/ environmental measures. These measures are more important in Scotland because of its remoteness, and high environmental quality. In Wales (Cm 2815, 1995), the aim of the programmes was to foster: 'efficient and competitive agricultural industries, while encouraging the protection of the environment and the health and welfare of farm animals' by commodity support measures (£159.6 million in 1993–4), structural measures (£34.6 million), environmental measures (£1.5 million) and other measures at a total cost of £209.1 million. By far the biggest item was the £138.6 million spent on the Sheep Annual Premium Scheme, reflecting the upland farming of Wales.

In conclusion, in spite of the hyperbole claiming that MAFF is a multivariate agency covering all aspects of food production and distribution, its budget remains enormously weighted towards supporting production through intervention, and supporting farmers via direct payments. This is not to say that large numbers of staff are not employed in wider concerns and that these wider concerns are not gradually gaining in importance, only to remind one that the financial activity of MAFF is still dominated by traditional, albeit modified, support programmes. Whether the approximate £2,500 million represented by this support could ever be diverted to the wider concerns on which only about £450 million is spent is of course the $64,000 dollar question, which is considered at the end of this chapter, and in the final chapter of this book. Before then attention needs to be turned to various alternative interpretations of the efficacy of postwar planning for agriculture as a productive enterprise.

D: ANALYSIS OF THE IMPACT OF PLANNING FOR AGRICULTURE AS A PRODUCTIVE INDUSTRY BETWEEN 1945 AND 1995

The purpose of this analysis section is to take a long-term view and not be seduced by the excitement of any recent events, especially bearing in mind the key theme of this book, the incremental nature of policy development. As Chapter 1 pointed out, there are a number of ways in which policies can be analysed in general, but for agriculture in particular, a number of routes are possible. First, the classical route followed by agricultural economists (Hill and Ray, 1987) in which complex econometric models of whole economies or individual sectors are constructed, thereby allowing the effect of different variables to be measured, for example, increasing support prices for wheat. A similar route is followed by empirically minded social scientists seeking to find trends and links but not to subject them to over-rigorous statistical procedures. Second, the route followed by political economists seeking to find structural links between different groups in society, notably those involved with capital accumulation. And third, those post-modern followers of Derrida who see no real pattern at all in spite of the fact that planning is a quintessential exercise in modernism. These three routes are now used in turn.

Traditional meta-narratives based on empirical observation

The most obvious and still probably the most valid route is to take the policy goals set out by governments at face value and to assess their impacts by a variety of quantitative techniques. This process can, however, suffer from a lack of focus and is akin to not seeing the wood for the trees. Thus it is much improved if some conceptual rigour is imposed at the outset. Accordingly, this section sets out the aims of policy and then attempts to group these into distinct categories before any analysis is made. It will be recalled that the aims of agricultural planning in the UK are based on the Agriculture Act 1947, and the aims of the CAP are set out on pages 40 and 41.

Hill and Ray (1987) have argued in practice that six policy topics have been pursued in the postwar period, namely: securing a reliable food supply; encouraging trade; ensuring low food prices for consumers; improving the marketing structure of agriculture; encouraging improvements in productivity and efficiency; and providing a fair standard of living for the agricultural community. A more conceptual approach to agricultural policy in general has been provided by Pierce (1992) which divides the effects into: resource effects; economic effects; and social effects, in addition to a number of impacts on the ground, for example, concentration of landholdings. If these categories are combined with the aims of the 1947 Act and the CAP then a more rigorous policy evaluation system can be set up as follows:

Policy impacts within agriculture
Production aim with sub-elements of quantity/quality and stability
Productivity aim with sub-elements of effect of inputs
Farm structure aim with sub-elements of farm size/tenure
Incomes aim with sub-elements of land values/capitalisation/return on capital/different types of farm
Farm community aims with sub-elements of farmers/farm workers
Policy impacts in the wider economy
Self-sufficiency aim with sub-elements of stable supply/food costs
Geographical aim with sub-elements of intended differentiation, for example, Less Favoured Areas
Reducing cost of food to consumers aim with sub-element of opportunity costs/alternative policies

These aims are now evaluated in turn.

Policy impacts within agriculture

Production aim with sub-elements of quantity/quality and stability. This aim has been met extremely well, as any reading of either MAFF's *Annual Review of Agriculture* or the annual report on *Agriculture in the UK* will show, so much so in fact that surpluses have become a problem, even within some sectors of UK agriculture. In more detail cereal production rose from 7.19 million tonnes in 1947 to 19.4 million tonnes in 1993 (a figure actually reduced by CAP reforms from the 22.4 million tonnes of 1985), and the number of chickens increased from 65 million in 1947 to 130 million in 1993. Quality has also been vastly improved in terms of the eradication of poor grade or diseased food, but the subjective elements of variety, taste and flavour have perhaps suffered. Stability has also been achieved and since the early 1960s markets have been remarkably stable, except for a few crises induced by external events or by freak growing conditions. However, there is much concern as to whether this production is sustainable given its dependence on oil, agrochemicals, and energy inputs which are often several factors more than the energy output created, notably for protein products based on animals.

Productivity aim with sub-elements of effect of inputs. Increased production has been achieved virtually entirely by productivity gains which show no signs of abating. Much of this has been achieved by better crop varieties and selective livestock breeding, but equally important has been the elimination of pests and diseases and the application of growth agents like fertilisers. In more detail, milk yields rose from 2183 litres per cow in 1947 to 5278 in 1993. Machinery has also changed agriculture out of all recognition, and between 1971 and 1991, for example, the labour requirement for producing winter wheat fell from 18.5 hours per hectare to 13.1 hours and for dairy cows from 54 hours per cow to 36 (Munton, 1992). Between 1973 and 1988 labour productivity rose by about 2.5 per cent a year, and land productivity by around 1.5 per cent to give overall gains of 3 per cent (Tarrant, 1992).

Farm structure aim with sub-elements of farm size/tenure. British farming was comparatively well structured before 1945, but the incentives and stability offered by policy have perhaps slowed down the trend to increasingly big farms. Many farmers if exposed to a free market would have been forced to leave farming without the safety net of regular milk cheques or the extra subsidies paid in marginal areas. In addition, there does appear to be a farm size above which economies of scale are offset by the sheer size of the farm, making day-to-day management difficult. In an evaluation of the £1.6 billion spent on structural measures between 1980 and 1988 the National Audit Office (1989) concluded that some projects would have proceeded without assistance, and that they had been unable to measure the impact of structural support on the efficiency of British agriculture. The trend from rented to owner-occupied tenure has continued, but various Tenancy Acts have tried to make renting more attractive to both landlord and tenant, without much success.

Incomes aim with sub-elements of land values/capitalisation/return on capital/different types of farm. Incomes have improved, but at very different rates over the period, with in general a decline since the early 1980s, but a marked recovery in the early 1990s. In terms of 1982 values, overall farm incomes fell from £1,700 million in 1982 to £800 million in 1988, but recovered to an estimated £1,600 by 1993. Land values rose dramatically between 1950 and 1973 from £500 per hectare to £2,000 (at 1975 values), but between 1973 and 1990 they fluctuated around £1,500, with a gradual fall during the 1980s (Munton, 1992). Many farmers thus became 'cash rich but income poor'. The return on capital has thus been poor in relation to both land and the expensive buildings and machinery that have been invested in the industry, which in 1981 amounted to a working capital of 1,850 ECU per hectare (Munton, 1992). Farmers are willing to accept such returns because it is the only way of life they know, most of them having been born into the job. Outside investors who were encouraged into the industry by the heady returns of the early days of the CAP in the 1970s (Munton, 1977) have departed, except where land is seen as a social asset. There have been enormous variations between types of farm and farm regions. Big arable farms in the east have done well, whereas livestock farms, notably in the north-western uplands, have become very dependent on support. Farm workers, although they have acquired considerable skills, remain one of the worst paid sectors of the workforce, although some workers whose incomes are supplemented by the dubious privileges of a rent-free tied cottage, and prodigious overtime, are able to reach a near-average standard of living.

Farm community aims with sub-elements of farmers/farm workers. The farm community as a whole has been reduced dramatically, from 6 per cent of the workforce in 1945 to only 2.5 per cent in 1989 (Tarrant, 1992). It is very difficult to estimate how the number of farmers has changed since the definition of a farmer by the agricultural census has been subject to a good deal of change. However, most experts estimate that the number of farmers has fallen only slowly to the current estimate of around 280,000, although the figures hide a good deal of change within farm households, with much more part-time farming or substitution of paid labour by farm family members. In many livestock farming areas, farm workers have almost ceased to exist, and even on arable farms, they are few and far between. Accordingly, it is farm workers who have suffered catastrophic falls in their numbers. In more detail, the number of 700,000 full-time workers in 1947 had shrunk dramatically to only 85,000 by 1993. In terms of employment and GDP (Cm 2803, 1995) agriculture in 1993 accounted for 2.2 per cent and 1.4 per cent respectively, but the wider food industry accounted for 11.7 per cent and 7.4 per cent respectively, made up as follows: food and drink manufacturing (2.0 and 2.9); food and drink distribution (5.7 and 2.6); and catering (4.0 and 1.9). In England it has also been estimated (Cm 3016, 1995) that farm diversification activities like recreation had created 80,000 jobs. These data thus confirm the concept of the food system, outlined in the introductory section A of this chapter which predicated the transfer of food production employment away from the farm.

Policy impacts in the wider economy

Self-sufficiency aim with sub-elements of stable supply/food costs. In spite of EU surpluses the UK is still only 57 per cent sufficient in all agricultural products, albeit this is an enormous improvement on the 30 per cent of the 1930s. In addition, 100 per cent self-sufficiency is impossible, because certain products have to be imported, for example coffee and bananas. Nonetheless, there is still room for improvements in temperate food self-sufficiency, which although it reached nearly 80 per cent in the mid 1980s, had fallen to just over 70 per cent in 1993. These recent falls in self-sufficiency have been caused by the need to cut surpluses and thus production in the wider European context, but they make little sense in many cases. For example, Britain is not self-sufficient in butter and cheese, but since 1984 milk production has been curbed with large numbers of workers being made redundant in creameries. This is hardly sensible in a wet mild country ideally suited to dairy farming. In terms of the balance of payments, food and drink have accounted for a large percentage of the visible trade gap, but this percentage has fallen largely thanks to a dramatic rise in food and drink exports. However, only cereals and alcoholic drink show a net surplus. Vegetables and fruit account for nearly half the crude food and drink trade gap, with meat accounting for one-fifth of the gap, which in 1993 totalled £5,890 million, out of a total trade gap of £16,468 million (Lund and Price, 1995). Food supplies have been stable since the mid 1960s, but although their cost has fallen, for example, the real cost of food fell by around 8 per cent between 1985 and 1993, food costs are still much higher than they would have been without the high support prices of the CAP, and import controls which raise much lower world market prices to CAP levels. In 1995 MAFF officials estimated the cost to be £4 per person a week, of which £3 was due to higher food prices (Anon, 1995), but the 1995 'Rural White Paper' put the overall cost at £4 billion or £80 per head, although this only included taxpayer costs, and not consumer costs (Cm 3016, 1995). The May 1992 reforms should in theory reduce prices nearer to world prices, and thus consumers in general should benefit, while taxpayers will lose out. This should be 'progressive' in that poorer people who pay few or no taxes will gain, in contrast to the formerly 'regressive' policies which kept food prices high and thus bore most heavily on poor people.

Geographical aim with sub-elements of intended differentiation, for example Less Favoured Areas. The overall policy aims do not have a specific geographical content, although these were added later, notably policies for the Less Favoured Areas (LFAs) since the mid 1970s, and the Objective 5b regions under the reformed Structural Funds policy from 1987. Nonetheless, various studies have noted changes in the geography of agriculture as a result of general and specific agricultural policies (Coppock, 1964 and 1971, and Bowler, 1976a and 1983a), notably a reduction in enterprise diversity on the farm and farm enterprise diversity within regions, and an acceleration of the division into an arable east and a livestock west, with mixed farming declining in the middle in favour of arable farming. This latter trend has been confirmed across the period between the 1950s and the 1980s by both Ilbery (1992c) and Walford (1995). Bowler (1976b, 1976c, 1979, 1981 and 1983b) has also noted significant variations in the uptake of various schemes, Ilbery (1992b) has noted considerable variation in the uptake of the Farm Diversification Grant and Farm Woodland Schemes, and Ilbery (1992a) and Briggs and Kerrell (1992) have also noted considerable variations in the uptake of set-aside.

Reducing cost of food to consumers aim with sub-element of opportunity costs/alternative policies. It has already been noted that food prices, although reduced, have not fallen as far as they could. Another viewpoint is whether the money that has been spent on farm support could have been spent better elsewhere. For example, as long ago as the mid 1950s, Wibberley (1959) calculated that the money could have been better spent on manufacturing industry, and a return made to importing food. Today, although expenditure looks high, in terms of overall government spending and the contribution of agriculture to the GDP of Britain, which fell from 6 per cent in 1948 to 2 per cent in 1988 (Tarrant, 1992), it does not look out of line. Nonetheless, the aim of succesive Conservative governments between 1979 and 1996 has been to cut agricultural spending whenever her European partners would allow, since there has been a long-standing criticism of agricultural policy, that it 'featherbedded farmers' in the 1950s and provided them with a 'gravy train' in the 1980s and 1990s. The fact that the 1992 reforms increased agricultural spending, and diverted support from farming to farmers shows how these assertions of 'favoured status for farmers' remain valid. Buckwell *et al.* (1982) have calculated that a return to the free market would cost producers 24 billion ECU, but that consumers and taxpayers would gain 37 billion ECU and 14 billion ECU respectively to give a net gain of 27 billion to the United Kingdom. Finally, international comparisons by Tarrant (1992) show that the cost of UK policies rose roughly in line with those in the United States, thus suggesting that wider factors were at play. One major opportunity cost is of course environmental damage to wildlife, landscapes and pollution. These are discussed elsewhere, notably in Chapter 3, but also in Chapter 6.

Some further considerations. In addition to the studies that have focused on the overt impact of policies, a number of other studies have examined the wider factors at play, because agricultural policy does not work in a vacuum and economic and technological change takes place at the same time, albeit often encouraged or discouraged by the policies. As section A of this chapter has already shown, agriculture has been subject to both a trading and technological revolution in the postwar period, and this has elicited, according to Bowler (1992a), a set of primary responses and a set of secondary consequences to *industrialisation*. In particular, the three related processes of *intensification, concentration*, and *specialisation*.

In more detail, the characteristics of *industrialisation* include the creation of scale economies at the farm level (larger farms), the increased reliance on purchased inputs (machinery, fertilisers, feed, agrichemicals), resource substitution (capital for land and labour), the implementation of organisational features associated with the concept of the 'firm', specialisation of the labour function within the farm business, and mechanisation of the production process. *Intensification* is the use of rising levels of purchased non-farm inputs in agriculture (capitilisation) and the resulting increases in output per hectare of farmland. *Concentration* describes the process whereby productive resources, and the output of particular products, have become confined to fewer but larger farm businesses, and to fewer regions and countries. Finally, *specialisation* is measured as the proportion of the total output of a farm, region or country accounted for by a particular product. This type of analysis naturally leads into the political economy approach which forms a key element of the next section.

Structural analyses focused on political economy analyses and regulation theory

These analyses can be divided into macro and micro scale. At the macro scale Le Heron (1993) has provided useful analyses of the last three rounds of GATT: the Kennedy round (1964–7); the Tokyo round (1973–9) and the Uruguay round (1986–2). Le Heron begins by noting that the GATT system rested on four features: liberalism, transparency, stability and non-discrimination. Liberalism favoured market allocation of resources and the relaxation of trade barriers. Transparency led to tariffs being the main sort of protection. Stability was promoted via constraints on protection measures, and non-discrimination derived from the principle of comparative advantage led to the concept of Most Favored Nation status, that is, those allowed to join the GATT club. The history of GATT, according to Le Heron, has been a diary of push and pull between purely national mandates and attempts to transcend such interest with arrangements of a more internationalist stamp. Most of the time domestic stabilisation has taken precedence over liberalisation, and MacLaren (1991) has demonstrated how a descriptive political economy approach can explain the reluctance of national governments to reduce the levels of protection and shun rational economic advice.

In the 1980s progress was made with the development of new measures for comparing the rate of agricultural subsidies between countries. These estimated and monitored the transfer of taxpayers' monies to farmers via agricultural assistance of all kinds (labelled producer subsidy equivalents) and the transfers from consumers to farmers through inflated food prices (termed consumer subsidy equivalents). These measures made the level of subsidies employed more transparent and increased already severe doubts about the motivations and degree of political commitment to the GATT ideals between participant countries. Le Heron thus argues that analysis should move away from nation–nation interactions to look instead at what the gains from freer trade might be, rather than describing a profile of winners and losers.

Assessing the overall impact of GATT, Le Heron argues that transnational corporations may now be able to adapt to any rules that governments can lay down, and that such firms are building up globally integrated production and distribution systems in advance of further GATT liberalisation. Therefore the slow pace of GATT has delayed but not prevented global agriculture, and politicians have had to accept economic forces rather than mould them. In the meantime regional trading blocs, like the EU and NAFTA (the North American Free Trade Area) (Troughton, 1991) represent an interim stage in globalisation.

Within the context of the EU Le Heron, from a political economy analysis of the CAP, identifies five important features. First, the CAP structure was a progeny of US development advice and embodied principles relating to pricing, preference and financing which were seen as essential ingredients of a strong postwar Europe. Second, the initial decision to set support prices at the high end has been crucial in charting the subsequent evolution of the CAP. Third, an inspection of CAP policy debates reveals a remarkable degree of contingency in decision making, and the continuation of considerable national autonomy. Fourth, nation states still have considerable discretion in implementation, and thus the continuation of national agricultural trajectories overlain by the CAP. Fifth, the commitment to achieving an 'internal market' from 1993 onwards encapsulates the very essence of creeping globalisation. Earlier in his book, however, Le Heron argues that the efficacy of the CAP is notoriously difficult to assess, because of ambiguities surrounding policy objectives and instruments, but does note that the marked turnaround from the world's top food importer to the world's second food exporter is one of the more significant developments.

At the UK level, Smith (1990) using a model of structural power and the concept that power is exercised by keeping some issues of the political agenda, and confining the scope of decision making to relatively safe issues, has developed a closed community explanation of the development of UK agricultural policy. This argues that agricultural policy has not been the result of pressure-group lobbying, but was evolved by a closed policy community which excluded certain groups and issues from the process. This enabled MAFF to develop the necessary capabilities for intervening in agriculture by incorporating farmers via the NFU lobby. This was not, however, a response by government to NFU lobbying, as has been traditionally argued, but a decision by government that it needed to expand agriculture, and thus needed a natural ally who could provide the illusion that there was a genuine policy debate. This increased its autonomy because it closed the policy off to other departments and interests, but also reduced its autonomy because it depended on farmers to support it. Assuming this support was forthcoming, the closed community reinforced itself through institutional and ideological means which made change and outside interference difficult. However, if a pressure group or new constraint result in the questioning of the ideology because its model answers no longer solve the problems, the instititional structures become vulnerable to a prolonged battle in which a period of pluralism is experienced before a new closed community develops.

Smith's main argument is that a closed policy community operated between 1945 and 1980, when a period during which the main tenets of agricultural policy went unquestioned even during the transition to the CAP in the 1970s when pluralism could have been expected to operate. Thus there was a change in the mechanisms, but the aims remained the same: high support to encourage high production. So while the EC operated a formal pluralist system of policy making, pluralism cannot explain agricultural policy making during this period, and though there were clear, well-articulated grievances, the policy structure and ideology prevented them being raised or seriously considered. In the 1980s, however, as the agenda of expansion drifted further from reality it was increasingly questioned and the consensus that there was only one possible policy disappeared. The NFU and MAFF drifted apart and a new period of pluralism began. However, it was not pressure groups that caused the policy to be questioned, for this was brought about by the structural constraints of increased costs, over-production, and international trade issues. Thus, taking Smith's ideas forward into the 1990s, it is possible to argue that pressure groups are now lobbying to fill the vacuum created between the demise of one closed community, and the rise of another, possibly the DOE and the green lobby, and that the MAFF may fail in its bid to become a countryside, rather than an agricultural, ministry.

At the micro scale, political economy research has been dominated by a team of workers who worked together for several years between the mid 1980s and early 1990s, notably under the umbrella of the Economic and Social Research Council's Countryside Change Initiative. Their two core contributions have been, first, the concept of 'subsumption' whereby farm businesses become inextricably dependent on outside capital in a process whereby capital does not eliminate the family farm, but its central role in the food production process (Goodman and Redclift, 1985), and, second, the changing role of agriculture in the economy from being a solely productive factor to being a provider of environmental goods. In other words the farm is transformed into countryside which can be consumed and thus be bought or sold as a commodity, and thus agriculture in the second concept is 'commoditised'. Early work by members of this team, Marsden *et al.* (1986a), used a longitudinal study of farmers between 1970 and 1985 to show that 20 per cent of farm businesses had been subsumed by outside capital. The level of subsumption was then found by Marsden *et al.* (1986b) to be dependent on the type of strategy the business adopted, with a fourfold division into agriculturalists (34 per cent); survivalists (32 per cent); accumulators (26 per cent); and hobby farmers (9 per cent).

Whatmore *et al.* (1987a) developed these findings to produce a fourfold typology, with at one extreme the 'closed enterprise', where the family farmer owns and manages the business and land, through to the other extreme, where 'subsumed enterprises' are effectively run by outside interests. This typology has been tested in the field by Whatmore *et al.* (1987b) and evidence for a transition from 'closed' to 'subsumed' enterprises was found, albeit not as fast, or as homogenous through time and space as it had been expected to be. Further work by the team has confirmed a continued but uneven resistance to subsumption. For example, Munton *et al.* (1992), in an analysis based on the level of family commitment to the farm and the type of area, found that most farms fell into the transitional category between independence and subsumption. From this evidence they argued that external capital and state farm policy need the family farm to continue, and that though the family farm will increasingly lose its independence of action, it will continue to be a cog in the food production machine.

The team (Marsden *et al.*, 1993) has also examined how the farming community as a general lobby has managed to maintain considerable influence by employing another core concept, 'representation'. At the beginning of the twentieth century landownership was 'represented' as a custodian of rural society and the nation's heritage. In the productivist period between 1940 and 1980, existing owners sought to argue that only those with a long and detailed knowledge of the land could ensure its efficient but caring management. As the productivist regime declined between 1980 and 1990 the farming lobby turned to the maintenance of their property rights, especially the potential threats to their freedom of action represented by extensions to environmental and planning legislation. In the 1990s, having found it difficult, but not without some financial reward, to sustain their claim to the moral high ground of environmental custodianship, it has been the creation of new and diversified markets that has beckoned.

The political economy school has thus made a great contribution by placing emphasis on the wider forces behind farm decision making, but by trying to see everything their empirical observations attempted to show an over-complexity which could not easily be captured (Munton *et al.*, 1992), and thus tended to reveal less than more limited studies. In their early work they also overstressed the power of capital accumulation and class divisions, which is not to say that outside capital and global influences are not having an ever-increasing impact on farming, because clearly they are, and recent work by some of the now dismembered team has focused instead on the less rigorous theoretical constructs of regulation theory (Flynn and Murdoch, 1995 and Marsden and Arce, 1995). However, within the great structural forces represented by global capital and government agricultural policies, individual farmers still exercise considerable free will in deciding which of the inducements on offer they will take. Accordingly, much work has been done on the reaction of individual farmers to changing circumstances.

Human agency, behaviouralist and post-modern perspectives

This approach was pioneered by Gasson (1973) and Ilbery (1977 and 1978) in a reaction to the over-deterministic models of agricultural economics, which predicated rational economic reactions to price signals from the market and the Government. Instead they argued that farmers placed great value on farming as a 'way of life', and that profit maximisation was not the only goal. From their empirical surveys of farmers they found that farmers ranked factors like 'independence' and 'doing the work you like' above 'making a reasonable living'.

The behaviouralist approach became less fashionable in the 1980s with the advance of the political economy approach, but in the 1990s there has been a resurgence of interest, as the political economy school retreated from their arrogant position, and new empirical research confirmed the idiosyncratic behaviour of farmers. In particular, Potter and Gasson (1988), Morris and Potter (1995) and Potter (1986) have found considerable behavioural independence in farmers as revealed more fully in Chapter 3. In addition, Gilg and Battershill (1996) have found that farmers in the south-west of England act in idiosyncratic ways that cannot be classified into any of the neat boxes of either classical economics or structural/class-based analyses. However, most of these studies have been concerned with small and traditional farmers in areas of high environmental value, and studies of larger farms in high farming areas, like East Anglia show a more calculating and economically rational approach to decision making.

Taking a wider perspective, the incremental and accretionary growth of the CAP and its complex mechanisms can be represented as a random sequence of events shaped by accidental events such as unpredictable weather, the collapse of communism in 1989, the advent of Mrs Thatcher and so on. If this line of argument is taken it is not too difficult to build a post-modern account of agricultural policy, in that no rational explanation linked to one set of causes can be found. In contrast, the more one studies the CAP the more bizarre, Byzantine, and irrational it appears to be, notably the way in which the reforms of 1992 did not produce the predicted catastrophic results for British farming, but instead some of the best profits for well over a decade.

Le Heron (1993) has indeed even argued from a different perspective that:

> The troubled history of the post war era of agricultural support policies is unequivocal evidence that the policy framework was a flawed framework, built upon incorrect premises and problem conception, with little regard to the degree to which policy actually straddled the policy problem.
>
> (193)

Some concluding thoughts. Le Heron's globalisation thesis does appear, however, to offfer some way forward, both for the evolution of agricultural policy and for our conception of that policy. In particular, he argues that much thinking about agriculture is still based on a myth that agriculture differs significantly from other activities, whereas in reality globalised agriculture is an intrinsic element of contemporary capitalism. Therefore globalised agriculture encapsulates three interwoven ideas: increasingly the historical forces organising agriculture are global forces; increasingly the main organisations bringing about significant changes are organised globally; and increasingly the historical form of agriculture in this era involves the organisation of integrated, often global, agro-production systems. The lack of any really coordinated response to these changes means that agriculture is being increasingly unplanned and moving towards a global free market, in spite of short-term trends which might imply the opposite, for example the extra layer of controls and subsidies imposed by the May 1992 reforms.

For the future Le Heron sees agriculture being shaped by the interaction of two main actors, governments and global corporations. This should not be addressed by the current wholesale retreat from public–private institutional relationships, but by innovative institutional forms.

E: PROSPECTS FOR THE FUTURE

The really big question for the remainder of the 1990s is whether the 1992 reforms represent a breathing space before farm support policies are dismantled or whether they represent another incremental ratchet to the farm racket. Within this agenda are the sub-questions of 'decoupling' farm income support from price regulation, so that direct income support to farmers can be provided and 'cross-compliance' under which environmental and social gains will be demanded in return for the continuation of farm income support (Russell and Fraser, 1995).

The first three years of the 1992 reforms suggest that the reforms had gone badly of course, notably in the United Kingdom by mid 1995. First, wheat prices did not fall to world-level prices as predicted in 1992 of around £85 per tonne, but instead world market prices rose to EU levels of around £110 per tonne, and in the autumn of 1995 to over £120 per tonne. Second, the continued depreciation of the pound and upward adjustments by the EU raised set-aside payments from an average £210 per hectare to £340 per hectare, and compensation payments intended to make up for lower support prices from an average £120 per hectare to £270 per hectare. Of course only arable farmers gained from these changes and set-aside was predicted to fall from 15 to 10 per cent in 1996, in response to a tighter world market, and should to some extent offset the changes. Nonetheless the 1992–5 period proved yet again that planning policies which appeared at the time to make sense, albeit only incremental sense, can easily be blown off course by unpredicted events. Overall, *Farmers Weekly* (1 September 1995) calculated the following changes in profit on three model farms between 1991 and 1995, which speak volumes about the unexpected effects of the 1992 reforms:

120-hectare clay farm 1991 profit £18,581 1995 profit £32,930
240-hectare loam farm 1991 profit £43,321 1995 profit £62,450
480-hectare brash farm 1991 profit £71,525 1995 profit £136,550

In spite of rising grain prices in 1995 in response to poor worldwide harvests and increased demand, most experts still agree with long-standing forecasts that food production in Europe will continue to increase ahead of demand, thus creating a surplus of land. Estimates for the year 2000 by Edwards (1986) have forecast a UK land surplus of up to 33.4 per cent if productivity grows by 2.5 per cent per annum. Such increases seem perfectly attainable, not just with existing science, but notably with the advent of biotechnology and transgenics which offers the possibility of transferring useful genes from one species to another, for instance from mice to chickens to make them resistant to salmonella. Another factor will be the growth of the EU, and in addition to the enlargement to 15 countries in 1995 when Austria, Finland and Sweden joined, the former Communist countries of eastern Europe are expected to join early in the twenty-first century. This will impose enormous new strains, since the modernisation of farming in eastern Europe and Russia, once the breadbasket of Europe, in response to the free market reforms which followed the collapse of communism in 1989, will potentially provide a source of plentiful food at prices way below western European levels. When these countries join the EU the CAP will come under a probably unbearable strain.

A number of possible reforms to agricultural policy have therefore been proposed and analysed by Gilg (1991a and 1992a). These can be divided into those within agriculture or outside agriculture. Within agriculture proposals have included: cutting support prices drastically and returning to the free market (North, 1989); encouraging extensive farming; diversifying within farming (Selman, 1988b); and diverting support to small farms on social grounds (Sinclair, 1985). Outside agriculture proposals have included: increased afforestation notably in the lowlands (Forestry Commission, 1978); more land given over to conservation, recreation and tourism, and urban developments thus transforming the countryside into a manicured suburban space. Most forecasts of the land surplus created by either a market-led or extensive-led policy are, however, in excess of the demand for land that could be created by alternative uses to productive agriculture, and North (1989) has forecast that out of the existing agricultural area of 17.65 million hectares, only around 12 million hectares will be needed for agriculture. The surplus of nearly 6 million hectares can be reduced by other land uses such as forestry and recreation, but North still predicts a potential surplus of around 3.5 million hectares.

The exact surplus will be determined by the individual decisions of farmers and land managers. Bowler (1992b) has argued that there are three pathways that farms can follow. First, they can continue in agricultural production either by continuing with existing systems or by redeploying resources into new agricultural products. Second, they can diversify the income base by redeploying resources into non-agricultural products, or by redeploying resources into off-farm activities. Third, they can marginalise farming, either by resorting to traditional farming with lower income/inputs or they can become hobby, part-time or semi-retired farmers. Contemporary research by Ilbery *et al.* (1995) suggests that just under one-third of farmers are adopting the second route, and that farms in marginal areas (Gilg and Battershill, 1996) are resorting to traditional farming encouraged by the environmental incentives outlined in Chapter 3. In good agricultural areas farmers will probably continue to farm as productively as they can. Three types of farming area and policy can thus be visualised. First, 'high' farming areas where support can be massively reduced since farms are genuinely profitable. Second, more physically marginal farming areas, but with some locational advantages, for example, near a big city or in a holiday area, where both on- and off-farm diversification is a possibility, but where support may still be needed. Third, marginal or remote areas where for social or environmental reasons considerable support is still needed, but where in return farmers will have to provide considerable environmental benefits. Whether these scenarios come to pass depends on the strength of sterling, inflation and interest rates, changes in CAP support generally and the set-aside rate specifically, and world market supply and demand.

The need for reforms that would bring this tripartite division into existence are, however, far from agreed upon. Officials in the EU, including the Agriculture Commissioner, Franz Fischler, for example, went on the record in 1995 to claim that the 1992 reforms provided the foundations for an efficient European agricultural industry on the evidence of much reduced intervention stocks and improved incomes since 1992. In contrast the UK Government, a long term critic of the CAP, has the following strategic objectives for the CAP : to reduce costs to taxpayers and consumers; to bring agriculture closer to the market; to avoid discrimination against UK agriculture and food sectors; to ensure that environmental concerns are integral to the CAP; and to reduce bureaucracy (Cm 2803, 1995a).

A case for radical reform of the CAP has been produced by a 'think tank' for the Ministry of Agriculture (1995a). The paper made four criticisms of the policy. First, it is a waste of resources; second, it saddles the food industry with a high cost base; third, it imposes bureaucratic controls over what farmers can produce; and fourth, environmental objectives are inadequately reflected. The paper thus advocated radical reforms based around a more open world trading environment, more competitiveness promoted by reductions in end-price and other production-related support as well as removing quotas and other controls on production such as set-aside, better targeted and more effective policies, and specific measures to address socio-structural difficulties, the environment and rural development. The consequences of removing all support would cost UK farmers £5,600 million, which would be offset by lower and cheaper inputs and access to higher world prices, leaving a net shortfall of some £3,000 million. This would cut farm income by about two-thirds, cut land prices by 40 per cent and reduce the labour force by up to 100,000. The 'Rural White Paper' (Cm 3016, 1995) was less explicit but argued that the case for CAP reform was 'unanswerable' (53) because: the CAP maintains a high-cost agriculture which is not competitive on world markets; disadvantages consumers and industry by forcing them to pay higher prices; is unduly expensive and wasteful; and has provided incentives for production, without due regard for environmental considerations. The 'white paper' thus argued for progressive reductions in production-related support, the eventual abolition of supply controls, and for a higher proportion of direct payments to farmers to be directed towards the encouragement of environmentally beneficial and sustainable farming. Elsewhere the US Farm Bill for 1995 included proposals for ending set-aside after 60 years, cutting out many subsidies and reducing farm aid by about 20 per cent.

In conclusion we can return to the three concepts at the beginning of this chapter: the world food system; reasons for state intervention into agriculture; and mechanisms for state intervention into agriculture. The evidence of this chapter is that the world food system will continue to grow in influence but will become bipolar. At one end of the spectrum accounting for maybe 80 per cent of production and 70 per cent of value, day-to-day foods will continue to be produced and distributed by multinational corporations via supermarkets. At the other end of the spectrum, niche products for weekends, quality restaurants and special events will continue to be produced by traditional regional systems based on the concept of the French *appellation* certificate. Being of higher value, these products could account for only 20 per cent of production, but 30 per cent of sales. It is forecast that these niche products will expand for a number of reasons. First, consumers are more aware than ever before of animal welfare, environmental issues and food quality rather than merely quantity and price. Second, modern communications can bypass the supermarket and wholesaler and contact the producer direct. For instance, a specialist home food storage system could monitor the rate at which foods were consumed and feed the information back to a cooordinating warehouse which would replenish stocks via home delivery on a regular basis. A computer prompt could then ask the consumer to order less mundane products for, say, Christmas up to two years ahead, so a consumer wanting a particular breed of turkey, fed in a certain way, and killed and dressed and stuffed to order, could do so as part of their regular computer housekeeping. In time such computer systems could obviate the need for both public and private planning of agricultural production by providing farmers with long-term order books.

State intervention systems will therefore deviate from food production concerns into environmental and social concerns and intervention will have to be justified on two counts. First, controls and subsidies to prevent pollution and misuse of resources, and second, positive encouragement to farm in environmentally-friendly and sustainable ways. This in turn means that the mechanisms for state intervention will have to be radically altered. To some extent the period between 1980 and 1995 has seen a marked shift in this direction already, and so the next chapter traces this transition in more detail. But before this, the chapter closes with the author's own evaluation in the accompanying checklist.

Policy evaluation checklist derived from Chapter 1.

1 *Were the problems understood and which value system dominated?* Initially the problem was one of food shortages and the value system was one of production, but since the 1970s the problems have been related to different perspectives leading to complex debates exacerbated by failure to rethink the issues.
2 *Which goals were to be achieved and were the policy objectives clearly spelt out?* The main goal was increased production and this was well set out by the 1947 Act and the CAP, but by 1995 the priority accorded to objectives in the MAFF annual report no longer matched actual expenditure.
3 *Were the policy responses accretionary, pragmatic and incremental?* The initial policy response was radical and robust, but in the 1980s and 1990s a series of often crisis-induced or politically driven incremental changes has transformed a sensible system into a collection of contradictory measures.
4 *Which powers in the Gilg/Selman spectrum were mainly used?* The main measures have involved financial incentives allied to exhortation, advice and demonstration, but since the mid 1980s regulatory controls and financial disincentives have been added to control production.
5 *Were planners given sufficient powers/resources and sufficient flexibility for implementation?* The 1947 Act provided potentially immense powers, and the CAP until the late 1980s had open-ended resources. The systems are however too inflexible, except at the margin, and between countries.
6 *Were performance indicators built into the policies?* Not originally, but in the 1990s the MAFF annual report has set out clear performance targets. Unfortunately, the CAP remains difficult to evaluate because its goals are still couched in very general terms, although the 1992 reforms did set some targets.
7 *Have the policies been underresearched?* In some ways there has been a plethora of research, but it has focused largely on macroeconomic issues and on the policy mechanisms. There has been little research on the effect on individual farmers and regions.

8 *Have the policies been effective overall?* The initial goal of production and self-sufficiency has been met at the European level, but the United Kingdom still remains a major food importer, in spite of having to cut back production progressively since the mid 1980s, so the policy has become sub-optimal.
9 *Have there been uneven policy impacts, by group or by area?* Yes, arable and large farmers have done best, while upland and livestock farmers have fared less well. Dairying has had mixed fortunes. Marginal and small farmers have lost out, but the major group to suffer are farm workers.
10 *Have the policies been cost-effective and are they environmentally sustainable?* In the 1960s the policies were extremely cost-effective, but in the 1980s and 1990s far too much money has been paid on curbing production, and current policies are neither financially nor environmentally sustainable.
11 *What have been the side-effects?* The main side-effect has been the degradation of the countryside (see Chapter 3). Other side-effects have been regressive effects on farm workers, and higher food prices for poorer urban consumers, although the 1992 reforms are less regressive.
12 *Were the policies modified by unforeseen events/policy changes elsewhere?* Initially, the policies were resistant to outside forces, but in the 1990s the collapse of Communism, and continued strains between currencies, threaten to overwhelm the CAP unless it is drastically rethought.
13 *Were the policies overtly sectoral and were they compatible with other policy areas?* Initially, agriculture took precedence over all other rural policies, in both Britain and the EEC. In the 1990s the policy is transforming into a general rural policy in a bid for survival and to outflank the competition.
14 *What have been the effects of 16 years of Tory rule?* Far less than if Britain had been outside the EU. The Tories have attempted to radically change the CAP many times but have been thwarted repeatedly, leading to calls to repatriate agricultural policy by the Euro-sceptics.
15 *Should the policies be modified?* No, they should be scrapped and a fresh start made, but because this is politically unacceptable, every effort should be made to drastically streamline the policy to provide a simple stabilising mechanism based on free access to much lower world market prices.
16 *What alternatives are there?* Unfortunately, the current Agricultural Commissioner believes that the 1992 reforms are working and that eastern Europe can be added without pain and so the realistic option is either a gradual return to sanity or a catastrophic budget crisis.

17 *Overall verdict.* Given the condition of Britain and Europe in 1945 the policies were initially brilliantly successful. Their goals having been largely achieved by the mid 1970s, no one knew how to turn the policies off, and since then the nightmare has become increasingly worse as contradictory and obscene policies based on paying farmers not to produce on the one hand, but still encouraging production on the other, have snowballed out of control. However, the policies account for only about 2 per cent of Government spending, which is a small price to pay for well-stocked supermarkets.

3

PLANNING FOR FARMERS TO MANAGE THE COUNTRYSIDE

A: INTRODUCTORY CONCEPTS

Until the 1950s and the advent of agri-chemical/ hi-tech farming it would have been axiomatic that agriculture was a land-managing activity, embodied in the centuries-old concept of *land stewardship*, or the aphorism

Live each day as if it were your last
Farm your land as if you were going to live forever

Since the 1950s, however, short-term considerations driven by the imperatives of outside capital, government policies, and a general cost-price squeeze on farming balance sheets have been instrumental in farmers 'mining' the land for production, rather than husbanding its resources. This meant that the fortuitous coincidence by which agriculture had coexisted with environmental quality, and had probably increased it by producing high-diversity plagio-climaxes, as shown in Figure 3.1, rapidly became a thing of the past as agresssive farming practices caused a rapid drop in biodiversity.

Thus the first concept in this chapter is that farming, if practised in certain ways, can be beneficial to the environment, and that policies need to be redirected so that the high-diversity plagio-climaxes shown in Figure 3.1 can be recreated. The potential plagio-climaxes, however, vary between the three core landscape regions which characterise the British countryside and that have been derived from their agricultural history as, for example, portrayed by Hoskins (1955) and Rackham (1986). First, the arable areas where the virtual removal of all tree cover by the medieval open-field system was partially ameliorated by the eighteenth-century enclosure movement which created the so-called traditional landscape of hedges and fields. Second, in the wetter west or marginal fringe areas, more tree cover was retained and pastoral /mixed farming landscape was maintained within a matrix of semi-natural woods and hedges. Third, in the uplands or in very dry or very wet lowlands open landscapes of low vegetation cover, such as heathlands or grasslands, were created by the removal of woodland

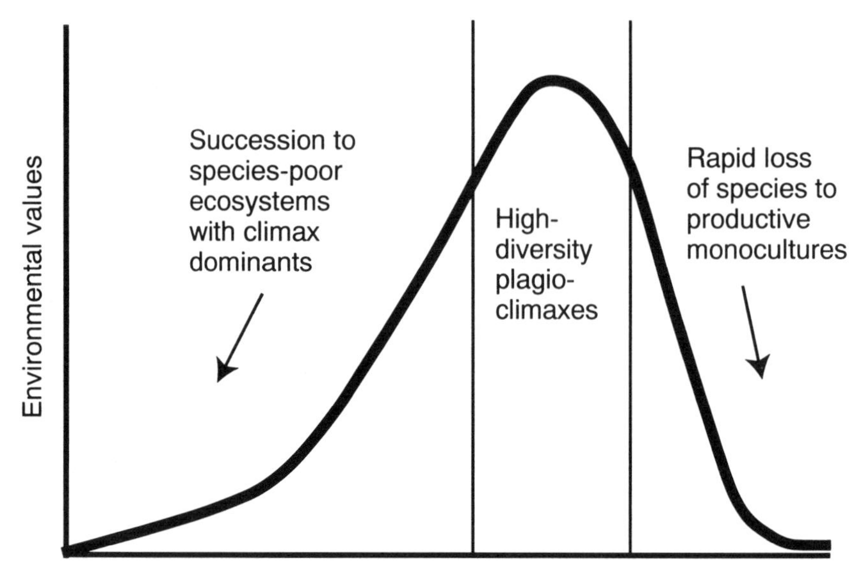

Figure 3.1 The effect of agriculture in establishing and maintaining diverse ecosystems

Source: Gilg, 1991a, after Green, 1989

by a mixture of felling, fire and grazing. There is thus virtually no natural vegetation in the British countryside.

The second concept relates to perception. In just one generation, many farmers have ceased to think of themselves as stewards, and have begun to see themselves as business managers, maximising profits, and using sophisticated computer programs to manage their land and livestock by ruthless scientific and accountancy formulae. Daisy the cow is now production unit 354 with a daily monitorable profit margin. However, these same approaches can have environmental sustainability and landscape and habitat management built into them. The real conceptual problem is that farmers have always seen themselves through the lenses of other farmers, and a good farmer is perceived to be one who runs a good business and derives a good income from farming, by which is meant, producing goods and selling them at a healthy profit. Farmers have constantly denied the massive contribution made to their incomes by the state, and by the same token they have decried any role that sees them as 'park keepers'. The key conceptual difficulty is thus one of persuading farmers to derive peer-group support and self-esteem from trading off production for environmental goods.

The third concept is that managing the land is very different in the lowlands compared to the uplands. In the lowlands, agriculture can be potentially profitable without subsidy and the land is intensively farmed so that conservation gains can only come from less intensive methods or from land removed from agricultural production. Farming in the uplands cannot, however, survive without subsidy and since upland farmers are therefore heavily subsidised by public monies, many people have argued that these subsidies should be paid not to support farming by itself, but should be paid instead to manage the uplands for public amenity. This, however, raises problems, for example: should we subsidise landscape or nature, and in either case should we try to retain the man-arrested wet desert found in most of the uplands or seek to return to the native woodland that existed before agriculture, or even allow recreational uses such as golf courses, bearing in mind that such a manicured landscape is universally more popular than bleak uplands? In other words, since the uplands will continue to be heavily subsidised, should the general public decide on their future, or the current elite who manage them either as landowners or as members of National Park Authorities?

This chapter will thus focus on these three key concepts: the potential coexistence of farming with environmental quality; farmer perceptions; and what type of landscape/nature do we want and who should decide in both the lowlands and the uplands. Overlying all of these is a fourth concept, which is the ideological one of 'voluntary' action, under which farmers are not to be forced into doing things that are good for the environment, but will be encouraged first, and then if this fails will be encouraged a little more by financial inducements

B: EVOLUTIONARY EVENTS

The key process in the evolution of policies designed to influence the way farmers manage their land and thus the countryside has been one of incremental policy making in reaction to changes in agriculture induced by the policies discussed in Chapter 2. It is thus necessary to outline changes in agriculture before considering the policy responses that have been made. This section is divided into three sub-sections: agricultural changes in the lowlands and the 'voluntary response', 1970–84; agricultural changes in the uplands and the 'financial inducement' response embodied in the Wildlife and Countryside Act 1981; and policy responses between the mid 1980s and mid 1990s.

Agricultural changes in the lowlands and the 'voluntary response', 1970–84

Although there had been some concern expressed in the 1960s by academics and pressure groups about the increasingly destructive practices of postwar farming, it did not really become a political issue till the 1970s, when the newly created Countryside Commission published the results of a study on New Agricultural Landscapes (NAL) carried out for them by the consultants Westmacott and Worthington (1974). The brief for the study was typical of the then contemporary attitude that

> 'agricultural change is inevitable. Therefore how can agricultural improvement be carried out efficiently in such a way to create new landscapes no less interesting than those destroyed in the process?'

The brief thus conceded the principle of change and any power over controlling change, for example, by development control, and instead only offered the planning option of 'guidance'.

The approach of the NAL was to to study, first, farmers' motives and attitudes to change which acknowledged the need to work with farmers rather than against them; second, the different roles of each landscape feature; and third, the anatomical structure of the landscape. Seven 'typical' areas were selected for detailed study. Within each area four methodologies were employed: questionnaire interviews to ascertain the changes made and the motives behind them; a landscape survey; a wildlife habitat survey; and a land-use analysis.

The following major trends were noted. First, arable land had gained at the expense of grassland, creating a less varied and less wildlife-friendly landscape. Second, change had often occurred at a change of tenure. Since the twentieth century has witnessed a massive change of tenure from renting to owner-occupiership, this has been particularly important. For example, rented land contains restrictions in the lease and removes the incentive to improve someone else's land. In contrast owned land can be changed at the owner's discretion, and there is every incentive to improve the land and thus its value. In addition a tenure change brings a new ambitious view of what the land could yield. Third, the move to arable farming and the advent of larger machinery had either made hedges redundant or forced their removal to make space for the larger machines. In many areas the removal of hedges and trees had been almost total, leaving hedges and trees only alongside roads, ownership boundaries, parish boundaries, and around farmsteads. Fourth, efficient but indiscriminate hedge and ditch maintenance machinery had turned once valuable habitats into shadows of their former wealth, and a general attitude that tidy farms were good farms had cleared up and eliminated formerly valuable features like ponds, spinneys and copses. The net result as portrayed by Gilg (1985a) was that some lowland landscapes changed more radically in the 50 years between 1930 and 1980 than they had since the enclosure movements of the eighteenth century.

Having looked at past trends, the NAL study went on to examine a number of possible futures which predicted a gradual loss of hedgerow and park trees as they died out and were not replaced; a reduction in the variability and diversity of hedges as field shapes and sizes were further rationalised; and the loss in time of all internal farm hedges, with the result that the landscape would become more open and relief orientated. Alarmed by this scenario but constrained by the belief that farming change was inevitable, the Countryside Commission came up with a limited response based on three related principles. First, the policy had to be voluntary, thus, second, it had to appeal to farmers, and so, third, it had to be directed at non-productive land. Some hypothetical examples were produced which showed that re-creating habitats on as little as 2 per cent of a farm, but in long linear strips, especially along natural features like waterways, could have a landscape and wildlife benefit out of all proportion to the small amount of land used.

This policy of re-creating habitat was then implemented by two methods, both based on advice and persuasion. Published advice given by the Countryside Commission (1977) exhorted farmers to stop the unnecessary clearance of features of landscape value and to plant trees and shrubs in greater numbers. However, farmers are known to react more positively to on the ground experience and word of mouth as demonstrated by Hagerstrand's (1967) well known 'Innovation Diffusion Theory' and gentle persuasion via soap operas. Indeed, BBC Radio's *The Archers* began in the 1950s as a thinly veiled means for the MAFF to promote modern farming methods, and careful listeners can still spot the intruded 'public information story lines' except that they now cover wider general countryside issues. Accordingly the Countryside Commission (1984a) also set up a Demonstration Farms Project in which 10 farms were chosen to show the extent to which landscape conservation could be combined with commercial farming and the extent to which multiple land use management could be feasible on commercial farms, and to demonstrate to working farmers how they could achieve landscape conservation on their farms. Another approach to spreading the vision of habitat recreation was provided by the setting up of Farming and Wildlife Advisory Groups (FWAGs) in most counties during the 1970s (Cox *et al.*, 1990) which provided advice to farmers on how to reconcile conservation and modern farming.

The NAL approach was evaluated in a series of publications by the Countryside Commission (1984b) in the mid 1980s, by means of a re-survey of the original areas. The key findings of both the 1974 and 1984 publications have been analysed by Gilg (1985a). In essence the second survey found: a slowing down of field size increase but a switch to such increases in the west and north; a slowing down in the rate of hedge removal, albeit some areas had little left to lose; the loss of hedgerow trees was also down and, encouragingly, more saplings were found for the future. Although the results showed a slowing down in the rate of change and the continuation of quite large areas of habitat, the results were disturbing enough, especially when allied to continuing pressures for change, for the Countryside Commission to shift the emphasis of policy from the planting of often insensitive species to more emphasis on conserving the remaining, basically irreplaceable heritage, via the concept of farm conservation schemes which set out features on the farm to be conserved or created and the management work that would be needed to achieve it: for example, managing streams and their immediate habitat to create more habitat, while leaving the vast majority of the farm under commercial modern management. The approach was thus low key and piecemeal.

Further studies, this time with a national coverage and based on remote sensing surveys, were carried out in the mid 1980s for the Countryside Commission and the Institute of Terrestrial Ecology (Barr *et al.*, 1986) and these confirmed the need for a more robust approach, since they showed that a quarter of the semi-natural landscape had disappeared between the 1940s and the 1980s, and that hedgerow loss was accelerating from 2,000 miles a year to 4,000 miles per year. Looking to the future the Nature Conservancy Council (1977) had already argued that the landscapes being created under the NAL policy did not contain the basic requirements which are essential for many species, and that if the policies were continued 80 per cent of birds and 95 per cent of bird species would disappear. They thus concluded that a positive rural land-use strategy was becoming increasingly necessary, albeit one based on the hierarchical system of nature reserves discussed in Chapter 6.

Seven years later the Nature Conservancy Council (1984) further investigated the link between the nature reserve approach and the conservation of wildlife in the wider countryside by arguing that there had been a number of successes. For example, nature reserves and other protected sites had by and large protected their habitats, but there were too few of them. However, there had been more failures, notably: the integrated concept of conservation had not been implemented; there had been a severe loss outside the protected sites, for example, herb-rich meadow was down by 95 per cent, limestone grassland by 80 per cent; and 10 out of 55 butterfly species were threatened with extinction. Thus site protection was deemed to be satisfactory but the protection of the wider habitat in the farmed countryside was argued to be hopelessly inadequate. Therefore the Council proposed that strict protection in the protected sites should continue and be extended from 7 to 10 per cent and that in the remaining 93 per cent (90 per cent in due course) they should persuade the Government to amend rural policies away from damaging practices towards socio-economic activities supporting nature conservation.

A similar approach was also advocated in the same year by the Countryside Commission (1984b) who, while applauding and recommending a continuation of the 'voluntary approach' as exemplified by demonstration farms, the work of FWAG, and the safeguarding of key sites under the management agreement approach, also argued that the voluntary approach was at stake and that if it failed a greater reliance would have to be placed on legislative action.

Thus between 1970 and 1984 lowland farmers became aware that conservation had a place in farm management, but apart from exhortation and advice, they had not been subjected to any specific policies. This was to change dramatically from the mid 1980s, as a result of a series of events triggered by landscape changes in the uplands.

Agricultural changes in the uplands and the 'financial inducement' response embodied in the Wildlife and Countryside Act 1981

In the uplands two major landscape changes in the postwar period had been set in motion by productivist planning policies, reclamation of moorland as farmland and coniferous afforestation. Both processes were widely criticised by certain groups as not only changing the landscape and ecology but also restricting access to formerly open land. Crucially, both processes were dependent on grant aid from the Government.

The issues raised reached boiling point in the mid 1970s when moorland reclamation in Exmoor caused the meetings of the National Park Committee to break down in acrimony. In order to resolve the issue the Government called in the establishment figure of Lord Porchester (now Lord Carnavon but then racing manager to the Queen and a former chairman of Hampshire County Council). He collected data, met individuals and published a report (Environment, 1977a) which found two contrasting views: first, the conservationist view appalled at the conflict of MAFF grants undermining the aims of the 1949 National Parks Act and Section 11 of the 1968 Countryside Act (both discussed in Chapter 6); and second, the farmers' view that Exmoor could only survive by change and as a living working community. He found that various voluntary agreements to conserve the moor had had some effect but that the future of the moor was under threat because some 5,000 hectares of the 24,000 hectares of moorland in 1947 had been lost by 1977 (4,000 hectares to reclamation, 1,000 hectares to afforestation), and that between 32 and 64 per cent was under threat of further reclamation/afforestation.

Lord Porchester recommended a survey of the moor to produce two maps. A Map 1 which would show all existing moorland and heathland and where any proposed changes would be subject to consultation, and a Map 2 which would show the 'core' moorland that should be preserved for all time. In Map 2 areas MAFF would withhold grant aid, thus removing most of the incentive to reclaim, and would pay a once and for all compensation in the form of Moorland Conservation Orders at a total cost of £750,000. The Countryside Bill 1978–9 followed this socialist and sensible long-term solution based on US free-market precedents, and put forward by an impeccable Tory in full. But when the Labour Government (1974–9) fell over the Scottish devolution issue in March 1979 the incoming Tory Government radically altered the financial aspects of the Bill in the Wildlife and Countryside Act 1981.

The 1981 Act contained provisions for Map 1 and Map 2 areas to be defined, but in all National Parks and for all areas. This was similar to Porchester, but in complete contrast to Porchester, the mechanism to safeguard moorland was quite different. In essence, under Section 43 of the Act if a farmer applied for a grant to reclaim moorland from MAFF, and if MAFF withheld the grant after advice to do so by the National Park or any other objecting authority, the objecting authority had to offer a management agreement restricting the use of the land and offering compensation based on either a lump sum calculated on the capital loss involved or an annual index-linked payment calculated on lost profits, in both cases including the grant aid that would have been paid (Brotherton, 1988; Curtis, 1984; Kinniburgh and Marshall, 1988; and Leonard, 1982). During the debates on the Act this principle of 'financial inducement' was extended to all cases where farm grants were refused on conservation grounds in not only the uplands but also the lowlands (Cox and Lowe, 1983).

This principle lies at the root of almost all the policies outlined in the rest of this chapter and is fundamental to an understanding of them. In essence the principle states that farmers still had the right to use their land as they saw fit, and that if they were to be denied this right, they must be compensated. But as Chapter 5 clearly demonstrates, land-use planning can only work once the right to use land has been bought by the state as it was by the 1947 Town and Country Planning Act. This enormous concession to farmers can be likened to students denied university entrance who would under this principle be granted compensation not only for the financial reward they may have gained from a university degree, but also the grant aid they would have received while students.

The need for some sort of response to reclamation was, however, further highlighted by research for the Countryside Commission (1983) which showed steady rates of moorland reclamation and afforestation across all the uplands, although Parry (1982), taking a longer term perspective, has shown that the uplands have always been a marginal zone of land use, dependant on medium-term climate change and food prices. In the 1980s, however, the key variable controlling the way in which farmers managed land in both the uplands and lowlands was the level of farm support prices and agricultural grant aid, and as Chapter 2 has already shown the 1980s were a period when these began to be seriously questioned on several grounds as the next sub-section discusses.

Policy responses between the mid 1980s and mid 1990s

During the 1980s according to Tarrant (1992) a convergence took place between three strands of the policy-making process. First, the costs of existing farm support programmes became financially unsustainable. Second, the persistence of food surpluses drew a recognition that less farmland was needed in production. Third, concern with the conservation of the environment turned to the damaging impacts of modern agriculture. Policy makers began to perceive a linkage between these three strands: surplus farmland could be turned to conservation uses, with farm subsidies redirected from price supports to encouraging 'environmentally friendly' farming (Gilg, 1991b).

In more detail, it was estimated in 1993 that a family of four paid over £1,000 per year to support farming, that each farmer was subsidised to the tune of more than £10,000 per farmer, each hectare was subsidised by £630, and that spending on farm support was way out of proportion compared with spending on other rural policies (Gilg, 1993). The environment had also been damaged not just by the removal and changes in habitat outlined earlier in this chapter, but also because of increasing use of chemicals designed to control pests and diseases. For example, between 1974 and 1988 the amount used rose from 4.08 kg per hectare to 6.76 kg per hectare (Pitman, 1992). In addition fertiliser usage had risen from 33 kg per hectare in 1950 to 182 kg per hectare in 1990 for winter wheat and from 25 to 92 kg per hectare for spring barley. Total use of nitrogen had risen from 229,000 tonnes in 1950 to 1,469,000 tonnes in 1989/90, leading to serious problems with nitrates in drinking water and other pollution of aquatic environments which doubled in the 1980s (Ward *et al.*, 1995).

The policy response to these changes was patchy and incremental on the whole and largely based on one-off policy responses to individual issues or crises, rather than by a reform of primary legislation. Nonetheless, the Agriculture Act 1986 did mark a significant shift when Section 17 redefined the duties of MAFF by imposing on agricultural ministers the duty to:

> have regard to and endeavour to achieve a reasonable balance between the following considerations: a) the promotion and maintenance of a stable and efficient agricultural industry; b) the economic and social interests of rural areas; c) the conservation and enhancement of the natural beauty and amenity of the countryside (including its flora and fauna and geological and physiographical features) and of any features of archaeological interest there; and d) the promotion of the enjoyment of the countryside by the public.

The Act also included other land-management provisions, for example, Environmentally Sensitive Areas (ESAs), which are considered later in the chapter, and thus marked a recognition that agricultural policies had to be balanced with social, conservation and recreational policies (Gilg, 1987). Further moves to a more diversified agriculture were provided in the Farm Land and Rural Development Act 1988 which introduced Farm Diversification Grant Schemes, and the Agricultural Act 1993 which allowed the Government to phase out marketing boards. Between the mid 1980s and 1990s there was therefore a policy retreat away from agricultural fundamentalism/productivism and a move towards a new philosophy based on a still productive countryside, but one in which specific policies would try either to check over-aggressive farming or to encourage more environmentally friendly farming practices. The development of these specific policies are now considered.

Farm grants. The first move away from purely productive and basically habitat-destroying grants came in 1984 when the then Agricultural and Horticultural Development Scheme and the Agricultural and Horticultural Grant Scheme were modified by not only reducing the level of aid available, but also by taking out damaging reclamation activities. In 1985 both schemes were replaced with the Agricultural Improvement Regulations and the Agricultural Improvement Scheme, which had an emphasis on diversification but for the first time allowed expenditure on conservation if it was a secondary item of expenditure related to diversification. In 1988 these schemes were replaced by the Farm and Conservation Grant Scheme which actually introduced grants to regenerate woodland, replant hedges and orchards, and generally attempted to restore by grant aid much of the habitats lost by grant-aided destruction throughout the previous three decades. In addition the scheme provided grants for pollution prevention, notably from slurry lagoons associated with intensive dairy and pig production units. These grants and most other grants under the scheme were however withdrawn at the end of 1994, and thus drew the real sting out of the 1981 Act. In addition, the obligation to offer a management agreement if a capital grant was refused on environmental grounds was withdrawn.

Management agreements. These date from the 1968 Countryside Act (see Chapter 6) and allow any local authority/National Park to conclude an agreement with a farmer to manage land according to the agreement, in return for some financial reward. Lack of public cash and interest in the policy among farmers have meant that few agreements have been made (Kinniburgh and Marshall, 1988) except for those special agreements made under the subsequent provisions of the Wildlife and Countryside Act 1981.

Environmentally Sensitive Areas (ESAs). These areas were first provided for in EC Regulation 797/85 on 'Improving the Efficiency of Agricultural Structures'. Over 60 candidate areas were proposed in 1985 (Gilg, 1986b) and between 1986 and 1987 11 ESAs were designated. In ESAs farmers can volunteer to enter 10-year management agreements which set out various environmentally friendly practices, normally based on lower stocking rates, lower levels of chemical use, and delayed farm operations in order to allow birds to nest or wild flowers to set seed. The degree of conservation commitment is usually divided into tiers. In addition the schemes vary widely in detail, notably between the lowland schemes in England, and the upland schemes in Scotland (Robinson, 1994). In return farmers receive payments ranging from around £8 per hectare to over £400 per hectare depending on the tier of environmentally friendly activity agreed to. The ESA scheme was inspired by plans to drain part of the Norfolk Broads which resulted in a sucessful scheme, the Broads Grazing Marshes Conservation Scheme which prevented reclamation by offering farmers sufficient inducements to stop reclamation (O'Riordan, 1985b) and celebrated its 10th anniversary in 1995. In essence therefore the ESAs are a variation of the 'voluntary financial inducement approach' pioneeered by the Countryside Act 1968 and the Wildlife and Countryside Act 1981. By 1994 38 ESAs had been designated, covering over 2.2 million hectares or 15 per cent of the United Kingdom. The estimated average additional net income from being in an ESA scheme is not high, for example, £880 in the Llyn Peninsula and £3,025 in the Cambrian Mountains (Crabtree and Chalmers, 1994). In order to qualify for designation an area must pass four tests. First, it must be of national environmental significance. Second, its conservation must depend on adopting, maintaining or extending particular farming practices. Third, farming practices in the area must have changed, or must be likely to do so in ways that pose a threat to the environment. And fourth, each area must represent a discrete and coherent unit of environmental interest. Originally, the ESAs were very unpopular with farmers, but as the squeeze on agriculture began, and as the generosity of the payments on offer began to be appreciated, the take up has been very high, and the ESAs have proved popular with farmers and conservationists alike. Nonetheless, Colman (1989) has argued that public purchase even over only 20 years could be a cheaper option, and that in the long term ESAs are a very expensive commitment offering conservation only while the farmer volunteers stay with the scheme.

Woodland schemes. These grant schemes traditionally focused on afforestation programmes in the uplands, but in the 1980s they were refocused and extended to encourage new planting on lowland farms, and the regeneration of farm woodlands which by and large had been very neglected. In particular the Farm Woodland Premium Scheme was introduced in 1992 to provide even more incentive to plant new, notably broadleaved, woodland on farms. These schemes, which include a combination of planting, establishment and maintenance payments, are discussed in more detail in Chapter 4.

Nitrate Sensitive Areas (NSAs). Modern agriculture has become very dependent on nitrogen fertiliser and is also a major producer of nitrogen via animal effluents and crop residues (Cook, 1993). Much of this has found its way into watercourses and caused eutrophication and the slow death of wildlife, notably in the sluggish rivers of the Midlands and eastern England (Pitman, 1992). In addition, high levels of nitrates in drinking water have been implicated with stomach cancers and problems with child health, notably babies (House of Lords, 1989). In 1989 the Water Act set up pilot NSAs and in 1991 the EU agreed to cut nitrates in water to a maximum of 50 mg per litre, and to impose experimental/pilot NSAs where these limits were exceeded (Seymour and Cox, 1992). In these areas farmers must reduce their use of nitrogen or cut nitrogen-producing activities, in return for compensation payments (Foster and Ilbery, 1992). In 1995 the 10 pilot NSAs from 1990 were joined with the 22 new NSAs under the agri-environment scheme (see below) to form a unified scheme of 32 NSAs. In 1994 72 compulsory Nitrate Vulnerable Zones covering over 650,000 hectares, were proposed under which maximum nitrate levels would be set, without compensation, thus suggesting a radical shift to a new post-compensation pitch, and in 1995 after consultation the Government began detailed planning for their designation. However, the level of control would be less than in the NSAs.

Countryside Commission and similar schemes. In the 1970s the Countryside Commission pioneered small payment schemes administered by applicant-friendly field officers in the Upland Management Experiments and the urban fringe 'Groundwork' schemes. These involved farmers applying for small grants to repair countryside features. This concept was extended in 1991 by the introduction of Countryside Stewardship agreements under which farmers can apply for money to recreate or improve habitats in certain target areas and also improve access for the public (Bishop and Phillips, 1993a). At the same time a Countryside Premium Scheme was set up to provide environmental benefits on set-aside land (Bishop and Phillips, 1993b). Later in the 1990s a Hedgerow Incentive (later Restoration) Scheme was also set up. Payments under the schemes range from £15 per hectare to over £200 per hectare. Similar incentive schemes have also been made available from the mid 1980s (Bishop and Phillips, 1993b) by, for example, English Nature, but the limited budgets of these schemes means that they are limited in their impact. They are nonetheless important because they demonstrate what can be done, can be used to pioneer new ideas, and collectively cover a wide variety of schemes (Countryside Commission, 1995b).

Schemes arising from or modified by the May 1992 Reforms to the CAP. The May 1992 reforms included a measure (Regulation 2078/92) whereby national Governments would submit to the EC their proposals for an 'agri-environmental' package. The UK Government submitted a £30 million package in 1993, which was gradually introduced between 1993 and 1995. The consultation paper produced by the Ministry of Agriculture contained an interesting analysis of the Government's philosophy on countryside management when it stated:

> farmers meet some three-quarters of the nation's need for temperate food. This is a major achievement reflecting great technical advances and investment over the last 30 years. But farmers are not only food producers. They are also responsible for the care of the great majority of the countryside. They therefore need to reconcile the demand for efficiently produced food with the demand for the countryside to be protected and cared for. It is the prime responsibility of farmers and other land managers to care for their land and Government policy assists them to reconcile agricultural and environmental objectives through a combination of guidance, protection measures and financial incentives.
>
> (Agriculture, 1993, 3)

The objectives of the package were thus to: protect and enhance wildlife, habitats and natural resources; conserve and enhance the most attractive landscapes; and promote new opportunities for enjoyment of the countryside by the public.

The package included seven proposed measures. First, six more ESAs. Second, new payments to improve opportunities for access to ESAs. Third, a new Moorland Scheme aimed at protecting and improving the condition of heather and other shrubby moorland by offering payments of £15 per ewe to farmers who reduced stocking levels. Fourth, a new Habitat Scheme to create or improve a range of wildlife over a 20-year period, with grants of £360 per hectare per annum for 20 years. Fifth, 30 new NSAs with payments ranging from £55 to £590 per hectare. Sixth, a new Organic Aid Scheme to encourage farmers to convert to organic methods of production by offering grants over a five year conversion period starting at up to £70 and falling to £25 per hectare. Seventh, a new Countryside Access Scheme to increase opportunities for public access on non-rotational set-aside land.

Hedgerows. There were several attempts to bring hedgerows under the development control powers discussed in Chapter 5, during the 1980s and 1990s, and these finally achieved success in the Environment Act of 1995, which provided ministers with powers to protect important hedgerows in England and Wales. These were expected to be brought into operation in 1996 (Cm 3016, 1995).

Set-aside. The initial set-aside scheme which ran from 1987 to 1992 automatically involved some conservation gain, since land was farmed less agressively. In order to build on this the Countryside Commission developed a Countryside Premium Scheme, which provided grant-aid for

Table 3.1 The range and scope of key incentive schemes

Source: Swales, 1994

Specific schemes	*1*	*2*	*3*	*4*	*5*	*6*	*7*	*8*	*9*	*10*	*11*	*12*	*13*	*14*	*15*
Objectives															
Maintain nature conservation	•	•	•				?		•	•	•	•	•		•
Enhance nature conservation		•	•			•	?		•	•	•	?	•	•	•
Restore nature conservation			•		•		?		•	•				•	?
Maintain income	•	•	•												
Produce food							•								
Produce timber												•			
Reduce production			•		•	•	•								
Archaeology	•	•	•						•						
Access				•				•	•	•					
Landscape	•	•	•			•			•	•		•	•	•	
Target:															
Habitat	•	•	•		•	•	•		•	•	•	•	•	•	•
Species											•	•			•
Whole Farm	S	S	S							•					
Part Farm	•	•	•	•	•	•	•	•	•				•	•	•
Administrator:															
Agriculture department	•	•	•	•	•	•	•	•					•		
SNH/CCW/EN										•	•				
Countryside Commission									•						
ADAS/SAC	•	•	•												
Forestry Authority												•			
Non-government organisation															•
Payments:															
Annual	•	•	•	•	•	•	•	•	•	•	•	•	•		•
Capital	•	•	•	•					•	•	•	•	•		
Positive	•	•	•	•	•	•		•	•	•	•	•		•	•
Length: (Years)	10	10	10	10	20	5	5	5	10	10	3–21			10	10

Key

1 ESA Tier 1
2 ESA Tier 2
3 ESA Tier 3
4 Access in ESAs
5 Habitat Scheme
6 Moorland Scheme
7 Organic Scheme
8 Countryside Access Scheme
9 Countryside Stewardship
10 Tir Cymen
11 Wildlife Enhancement Scheme
12 Woodland Grant Scheme
13 Farm and Conservation Grant Scheme
14 Hedgerow Incentive Scheme
15 Corncrake Initiative

ADAS: Agricultural Development and Advisory Service CCW: Countryside Council for Wales EN: English Nature S: Scotland SAC: Scottish Agricultural Colleges SNH: Scottish Natural Heritage

managing set-aside land overtly for conservation. The short-term nature of set-aside, however, undermined these schemes, until the mid 1990s when the Commission introduced 20-year set-aside, and also allowed land being transformed into woodland, or coming under the NSA or Habitat Schemes, to count as part of the set-aside requirements of the 1992 reforms (Cm 3016, 1995). Thus, set-aside provides a potentially rich source of new habitats in the arable areas where such habitat is most badly needed.

It is very difficult to summarise the large number of schemes that have been developed or modified between the mid 1980s and 1990s but Swales (1994) has produced a useful summary of the position in the mid 1990s as shown in Table 3.1.

C: THE SITUATION IN 1995 (MAFF SCHEMES ONLY)

In common with Chapter 2, this section is derived from the 1995 report of the Ministry of Agriculture (Cm 2803, 1995) and thus the expenditure totals normally refer to England only. This section uses the same order as the report in order to convey the priority that MAFF places on each programme. All the programmes are contained in Chapter 3 of the report, which is entitled 'To protect and enhance the rural and marine environment'.

Environmental Protection (nutrients). The objective of this programme was to minimise the potentially polluting loss of soil nutrients from agriculture. This was being pursued through the voluntary/pilot NSA scheme, the phased implementation of the EC Nitrate Directive (91/676), and by a substantial research programme. In 1994–5 £1.5 million was paid to farmers in the pilot NSAs who had registered 87 per cent of the eligible 9,300 hectares. Payments were expected to increase as a further 22 NSAs covering 25,000 hectares under the EC Agri-Environment Regulation of 1994 became operational. Fifty-four per cent of eligible land was entered in the first year. In addition agreements under the pilot NSAs could be extended for a further five years as they expired in 1995 or 1996. Overall expenditure was expected to rise from £5.3 million in 1994–5 to £12.6 million in 1997–8.

Environmental Protection (General). This programme was designed to reduce farm waste by helping farmers to prepare farm waste management plans. Expenditure was expected to rise slowly from £1.61 million in 1994–5 to £1.67 in 1997–8.

Farm grants. The Farm and Conservation Grant Scheme was much reduced in 1993 and 1994 by withdrawing horticultural, land improvement and farm pollution grants from the scheme, so that in 1995 only conservation investments were eligible for grant aid. Accordingly expenditure was expected to fall from £44 million in 1994–5 to £17 million in 1997–8.

Environmentally Sensitive Areas. By the end of 1994 nearly 6,500 farmers had signed or applied to join ESAs in England. Each ESA was reviewed for its effectiveness every five years, with a biannual review of the payment levels. It was expected that this rolling review programme, together with more farmers joining, and more farmers moving to a higher tier of environmental activity, would lift expenditure from £49 million in 1994–5 to £77 million in 1994–5.

Conservation, Habitat Scheme and Access. The objectives of this programme were to encourage farming practices which conserve and enhance the natural beauty and amenity of the countryside, its wildlife and features of historical interest, and to promote enjoyment of the countryside by the public. This was achieved by an estimated number of 4,500 farm visits by ADAS during which free conservation advice was dispensed. In addition Farming and Wildlife Advisory Group (FWAG) advice was aided by MAFF. Since 1994 the Habitat Scheme offered long-term management agreements targeted at land in the initial five-year set-aside programme. Around 4,800 hectares were accepted into this scheme in 1994–5. Total expenditure in these programmes was expected to rise from £1.8 million in 1994–5 to £7.4 million in 1997–8 partly due to the addition of the Countryside Stewardship Scheme which was due to be transferred from the Countryside Commission to MAFF in 1996 under the provisions of the Environment Act 1995.

Farm woodlands. This programme's objective was to encourage the planting of woodland by farmers who would remain in farming, thereby enhancing the farmed landscape and environment, and providing a productive alternative land use to agriculture. It offered 10- or 15-year payments to offset farm income foregone, and between 1992 and 1994 the Farm Woodland Premium Scheme, which replaced the pilot Farm Woodland Scheme in 1992, attracted over 3,000 succesful applications to plant nearly 20,000 hectares (75 per cent broadleaf) in the United Kingdom. The programme operated under EC Regulation 2080/92, which requires all Member States to operate schemes to encourage afforestation of farmland, was expected to cost £8.0 million in 1994–5 and £14.4 million in 1997–8.

Countryside management. This programme aimed to control mammals and birds, to control heather and grass burning by publicity, closed seasons, and codes of practice, and to control straw burning by a virtual ban, implemented in 1994.

Countryside Access Scheme. The objective of this programme was to encourage farmers who intended to enter suitable land into the guaranteed set-aside option of the Arable Payments Scheme (outlined in Chapter 2) to make this land available to the public for walking and other forms of quiet recreation for a period of five years. Expenditure was expected to rise from £0.1 million in 1994–5 to £6.7 million in 1997–8.

Moorland Scheme. This scheme, which was due to begin in 1995, was intended to encourage the conservation and enhancement of the quality and extent of semi-natural heather and shrubby vegetation of moorland to the benefit of its wildlife and landscape by paying farmers to remove an expected 128,000 ewes from moorland by 1977 in return for a payment of

£25 per ewe. Expenditure was forecast to rise from £0.1 million in 1994–5 to £10.7 million in 1997–8.

Organic farming and agricultural inputs. This programme included establishing and maintaining a framework to enable organic farming to respond to consumer demand, and setting and implementing appropriate standards for organic foods and their production. A scheme to aid the conversion to Organic Farming began in 1994. Overall expenditure was expected to rise from £1.9 million in 1994–5 to £2.7 million by 1997–8 with about half being spent on organic farming.

During 1995 the Ministry of Agriculture (1995a) also issued a consultation document on the future of environmental land-management schemes. This proposed a twin-track approach. First, ESAs should continue to be the main mechanism for protecting large areas, and second, that Countryside Stewardship (due to become a MAFF scheme in 1996) should be targeted at valued environments in the wider countryside. In due course the Farm and Conservation Grant Scheme should be merged with Countryside Stewardship, but other mechanisms like the Woodland and Nitrate schemes should remain separate since they were too specialised. The document also proposed a 'one-stop advice shop' where farmers could obtain information on what would still be a bewilderingly wide range of schemes as shown in Table 3.2. Most of these themes were taken up by the 'Rural White Paper' (Cm 3016, 1995), which targeted the environmental land-management schemes on valuable countryside in decline or under threat, areas with significant potential to provide new environmental benefits, or areas in need of positive management. Countryside Stewardship would become the flagship scheme outside ESAs, and would grow, first by merging it with conservation grants under the Farm and Conservation Grants Scheme, and second, by possibly merging with the Habitat, Countryside Access and Morland Schemes when they completed their pilot phases in 1998–9. The white paper also welcomed the LEAF (Linking Environment and Farming) initiative and its use of environmental audits as a valuable checklist for farmers reveiwing their operations. Finally, the Environment Act 1995 imposed a statutory duty on the Countryside Commission, English Nature and English Heritage to advise MAFF on all its environmental grant schemes.

D: ANALYSIS

Traditional meta-narratives based on empirical observation

Assessment of policy changes. Between 1980 and 1995 agricultural policy apparently moved away from policies which encouraged maximum production with little regard to the environmental consequences to policies which controlled production within environmental parameters. These

Table 3.2 National incentive schemes which encourage environmentally beneficial management of land in England

	£ million		
	Forecast payments in 1994/5	*Forecast payments in 1995/6*	*Forecast payments in 1996/7*
Ministry of Agriculture			
Countryside Access Scheme	–	4.5	4.5
Environmentally Sensitive Areas	20.1	43.3	43.1
Farm & Conservation Grant Scheme (conservation grants)	4.0	4.0	4.0
Farm Woodland Premium Scheme	1.4	3.8	5.1
Habitat Scheme	–	3.0	3.0
Nitrate Sensitive Areas	1.5	8.4	9.7
Moorland Scheme	–	1.6	3.2
Organic Aid Scheme	0.1	1.5	1.5
Countryside Commission			
Countryside Stewardship	10.5	11.4	11.7
Other grants	13.4	10.7	10.0
English Heritage			
Management agreements	0.1	0.2	0.2
English Nature			
SSSI Management	7.5	7.7	7.8
Wildlife Enhancement Scheme	1.1	1.2	1.4
Reserves Enhancement Scheme	0.5	0.7	0.8
Conservation Grants Scheme	1.8	1.8	1.8
Forestry Commission			
Woodland Grant Scheme	13.1	16.1	18.5
Total	**75.1**	**119.9**	**126.3**

Source: Agriculture, 1995b

policies evolved along two twin tracks according to one MAFF official (Waters, 1994). First, the introduction of schemes which offered financial incentives for positive environmental management, and second, enhanced penalties for negative actions, like farm pollution from slurry or nitrates. It is the purpose of this section to analyse the degree to which this was a real change in policy and the degree to which farmer attitudes and land-use change on the ground have been affected by these policy changes. We can begin the analysis by comparing two policy statements in the period: first, section 17 of the 1986 Agriculture Act; and second, the policy statements made in MAFF's 1995 Annual Report, as outlined in the accompanying box diagram.

Section 17 of the Agriculture Act 1986 imposed a duty on agricultural ministers and their officials to try to achieve a reasonable balance between: a) the promotion and maintenance of a stable and efficient agricultural industry; b) the economic and social interests of rural areas; c) the conservation and enhancement of the natural beauty and amenity of the countryside; and d) the promotion of the enjoyment of the countryside by the public.
MAFF's Aims and Sub-Aims, from 1995 Annual Report (Cm 2803, 1995): To protect the public; to protect and enhance the rural environment; to improve the economic performance of the agriculture, fishing and food industries; to protect farm animals; and to ensure the best use of internal resources in support of the Ministry's business.

These statements appear to portray a move to a balanced policy approach, but when actual and planned expenditure in the 1990s is actually analysed from MAFF's 1995 Annual Report (Cm 2803, 1995), a very different picture appears. First, market support was expected to rise from £921 million in 1989–90, to £1,839 million in 1994–5, and to £2,169 million in 1997–8. Second, expenditure on protecting and enhancing the rural and marine environment was expected to rise from £55 million in 1989–90 to £106 million in 1994–5, and to £110 million by 1997–8. Thus, not only was actual and expected expenditure on environmental measures tiny in comparison with expenditure on production support and direct subsidies to farmers, but its growth rate was far less than that for market support. In more detail, expenditure on market support was due to rise by a factor of 2.35, while expenditure on the environment was due to rise by only a factor of 2.0. It should be remembered, however, that market support is now a bit of a misnomer since, as Chapter 2 has shown, there has been a massive shift in the way that farming is supported, from support from production to support for farmers. However, in spite of much rhetoric to the contrary there has not been a shift away from supporting farming to supporting the environment. Environmentalists can only look with envy on the £2,000 million-plus spent on farming and farmers compared with the £100 million-plus spent on the environment.

Empirical observation of changes on the ground and scheme effectiveness. The agricultural census is still concerned with production totals rather than environmental use, and so there is no comprehensive record of landscape and wildlife change. Nonetheless, work carried out by the Institute of Terrestrial Ecology (1993) for the DOE as part of a continuing *Countryside Survey* has used satellite imagery and a field survey of 508 one km squares to reveal a number of alarming trends which showed that landscape and habitat loss was still continuing. For example, the survey revealed a reduction of 23 per cent in hedgerow length and 30 per cent fewer species on arable land between 1984 and 1990. Continuing work, for example a 1993 DOE hedgerow survey, has shown that losses are still occurring, but at a reduced rate.

In the uplands a Countryside Commission (1991a) study of land use in the National Parks using aerial photos from the 1970s and late 1980s revealed that coniferous high forest had increased by 178 sq km, and cultivated land and improved pasture by 74 and 66 sq kms respectively, mainly at the expense of rough pasture and grass moor (down by 133 and 49 sq km respectively). However, the study concluded that the landscape of the National Parks had changed little in the two decades of the 1970s and 1980s.

In summary, surveys of land-management practices on farms have come to the broad conclusions shown in the accompanying summary box.

1 A massive growth in the use of fertilisers, pesticides and other inputs has led to direct and indirect attacks on wildlife. For example, direct attacks include the elimination of any competing species in arable areas, while indirect attacks include the pollution of water courses by dairy slurry as a result of intensified dairy production. The consequences have been widespread throughout the ecosystem, with some species suffering catastrophic losses or even extinction.
2 There has been a widespread removal of habitat and thus loss of species. In some cases entire destruction of entire landscapes. A more insidious trend has been neglect, which leads to gradual death by decay, notably in hedgerows and farm woodlands.
3 Conservation is often only tolerated where another priority like fieldsports or commercial recreation prevails. This does preserve some specialised habitats but some competing species may be eliminated. There may also be moral objections to fieldsports and problems with general recreational access.

Turning to an anlysis of the schemes, Swales (1994) has provided a matrix of 15 schemes as already shown in Table 3.1. Swales argues that incentive schemes alone cannot guarantee the future protection of the countryside, and that though the incremental development of incentive schemes and the concept of cross compliance has helped, what is now needed is a new framework which recognises the value of producing environmental goods as well as food. A fundamental shift in thinking is now required according to Swales, making environmental considerations integral rather than peripheral to land-use policy objectives.

A classical and exhaustive economic analysis has been provided by Colman (1994), who evaluated the various schemes according to their capacity to protect and enhance; their timeliness; their targetability; their monitorability; their cost-efficiency; their political acceptability and transparency; and their promotion of conservation-mindedness. He concentrated on ESAs and found that they were basically paying farmers to do what they would do in any case. This finding extended to other schemes where farmers who were already interested in conservation were the ones taking up the schemes most often. In terms of environmental change, Colman found that the rate of damage had slowed, but that positive increases in environmental quality were few. He therefore concluded that schemes should be targeted on only those farmers who might contemplate undesirable environmental damage, and that there were in principle cheaper ways of achieving the policy targets.

A further analysis, this time along the lines of the Gilg/Selman spectrum outlined in Chapter 1, has been developed by Bishop and Phillips (1993a). From this perspective they pick out six policy instruments: regulation; advice and information; grants; compensation; support for environmentally friendly farming; and cross compliance. All of these have some shortcomings and have thus opened the door to a seventh mechanism, the market-led approach. Bishop and Phillips (1993b) have concluded that this market-led approach is distinct from other policy tools for reconciling conservation and agricultural interests. First, there is no obligation on either side to participate. Second, payment is for the product, not the process. Third, the thrust is moving away from protection of existing sites, to re-creation of certain landscape or habitat types. And fourth, they are being used to promote multi-purpose land management.

In more detail, Garrod *et al.* (1994) and Garrod and Willis (1995) have carried out contingent valuations of two Environmentally Sensitive Areas and found them both to be providing benefits well above the costs incurred. However, aggregate user benefits for the South Downs ESA were nearly five times as great as for the Somerset Levels ESA. This was partly due to the fact that just over 3 million households were estimated to visit the South Downs, compared to the 1.7 million estimated visitors to the Somerset Levels. In addition, people were prepared to pay about £10 a year more for the South Downs, and the South Downs cost the exchequer a little under £1 million a year, compared to the £1.86 million for the Somerset Levels. In both cases, though, the ESAs are providing good value for money.

In another study of ESAs, Froud (1994) found that average incomes per hectare entered had increased on average by £35 in the South Downs to £106 in the Breckland ESA, but at a net cost to the exchequer, since savings in support costs had not outweighed the cost of the schemes, even before administration had been taken into account. The impact of the schemes on farm practices have been limited, however, except where arable has been converted to grass. The impact of ESAs on the environment is much harder to gauge, and Froud argues that contemporary assessment based on net budgetary costs and farmer acceptability are inadequate, and that economic methods, such as those outlined in the previous paragraph, need to be treated with extreme caution. The problems involved in the future will be threefold. First, disentangling effects of the schemes from underlying trends; second, identifying the environmental changes which result from the schemes; and third, placing a value on these changes.

Structural analyses focused on Political Economy and Regulation Theory

Private property rights are a core concept in many political analyses (Marsden *et al.*, 1993) and so the erosion or purchase (Crabtree and Chalmers, 1994) of these by the restrictive practices imposed by, for example, ESA agreements is of particular interest. In more detail, Crabtree and Chalmers argue that 'standard payments' used in ESAs, Countryside Stewardship and similar schemes have a number of desirable economic characteristics, in that they are voluntary, transparent, and equitable. However, in many situations land purchase would be a more cost-efficient option for environmental protection than the transfer of a perpetual stream of standard payments, and certainly better than capital grants which, they argue, have a limited role in conservation.

Moreover Paice (1994) has reported work by financial consultants which shows that many environmental schemes have made substantial differences to land values. In more detail, three schemes have increased land values: the Countryside Access Scheme by £125 per hectare; the Arable Area Payment Scheme by up to £750 per hectare; and milk quota by £5,000 per hectare in addition to the value of the quota, which is itself a tradeable asset worth 55 pence a litre in November 1995. In contrast, three schemes have caused a loss in value: ESAs by up to £1,000 per hectare; NSAs between £250 and £500 per hectare; and Nitrate Vulnerable Zones by between £375 and £750 per hectare. Finally, Countryside Stewardship has caused a loss or gain of between £375 per hectare or £75 per hectare.

However, all the schemes in total cover at best only 20 per cent of all farmland (with 15 per cent accounted for by ESAs) and it is easy to concur with Le Heron's (1993) analysis that the EC has essentially opted to link environmental policy to the 'margins' of agriculture, without seriously addressing the fuller connections between the intensification of agriculture and environmental change. All this means that while an overall commitment to environmental objectives is present, it still stands as a secondary-level political choice, circumventing and perhaps constraining, but hardly overturning, the techno-economic trajectory of mainstream EC agriculture.

Le Heron's political economy conclusion to the reforms of the early 1990s thus concurs with Tarrant's (1992) classical conclusion that attempts to merge agricultural and environmental policies during the 1980s were disappointing. In particular the agricultural interest has proved resistant to the switching of funds, and so the adoption of environmental measures has remained voluntary, and therefore most environmental measures have been bolted on to existing farm policies which have remained essentially unreformed (Robinson, 1991). However, Russell (1994) argues that agri-environmental policy has now evolved in response to concerns that the complex relationship between agriculture and the rural environment has changed, from being characterised by a complementary symbiosis to displaying features of a competitive trade-off. In particular, a twin-track approach (Waters, 1994) is evolving, based on, first, positive incentive payments, and, second, a growing use of the sort of regulatory and penalty-based options characteristic of environmental policy in the wider economy and based on the 'polluter pays' principle, outlined in Chapter 6. In addition, the principle that no compensation should be paid when developments contrary to the public interest are prevented, and which underpins the town and country planning system discussed in Chapter 5, is gradually entering the equation in the form of controls over hedgerows and the proposed financial arrangements for Nitrate Vulnerable Zones, which do not include a compensation element.

Human agency, behaviouralist and post-modern perspectives

A continual theme throughout this chapter has been that farmers will remain the key decision takers, albeit influenced by government signals and outside capitals (Pile, 1992). Accordingly, studies of how farmers might and have reacted are an essential form of analysis. Many studies (for example, Potter and Gasson, 1988; and Potter and Lobley, 1992) have emphasised the idiosyncratic nature of farmers and their attachment to farming as a way of life, and decision-making processes based on 'satisficing' rather than profit maximisation.

For nearly 10 years research has also emphasised that farmers are remarkably resistant to change and have only espoused schemes when they fortuitously fit in with what they were doing or planned to do anyway. This 'selectivity' effect has been a constant feature of much behavioural research which does, however, take into account Potter's (1986) division of farmer decision making into the three types of managerial, investment, and strategic, and the three investment styles of incrementalists, programmers, and mixers. Potter has also concluded that:

> it is more important to relate the pattern of landscape change in a locality to the process of farm business growth and development rather than to typical farm or farmer characteristics such as farm size . . . the process of countryside change has an inbuilt momentum and will continue to be driven forward by factors which are embedded within the present structure of the industry and the value systems of individual farmers. These will be slower to change than the policy setting which nourished their development. Countryside change may prove to be less manageable than many expect.
>
> (194)

This was confirmed by work by Potter and Gasson (1988) which showed that farmers, when asked to set the price at which they would consider diverting land from production, set this price well above politically acceptable levels, notably for farm woodland. The lack of interest in non-productive uses was also noted by Eldon (1988), who used Potter's (1986) categories to show a generally low level of interest, among farmers in seeking ADAS advice about farm conservation practices. In addition Adams *et al.* (1994) have found that those few farmers who still had islands of less intensively farmed habitat, in the much damaged east of England, were reluctant to look after these islands without financial incentives. Nonetheless, Crabtree and Appleton (1992), although they found that farmers expected incentive schemes to add to their net incomes, also found that half the entrants to the Farm Woodland Scheme had suffered a financial loss, but that this was compensated by the perceived welfare gained from amenity and conservation, and potential land value increases.

This low level of interest could be attributed to the period when the research was done, just as the environmental policies were beginning to be introduced and before the CAP reforms. However, research published in the mid 1990s confirms that farmers still remain resistant to these schemes unless their farms already contain environmental interest, and/or they were thinking of improving or creating habitats. For example, Gilg and Battershill (1996), from a survey of 122 farmers in 14 environmental schemes, found that most conservation schemes had only had a marginal impact on farming practices and environmental outcomes compared with what the participants would have been doing in any case; that environmentally friendly farmers were distinguished by behavioural or attitudinal traits rooted in their personal histories and circumstances; and that they were either practically committed to traditional farming or had strong pro-conservation attitudes. Similar findings have been reported by Tarrant and Cobb (1991) and by several case studies of ESAs (Whitby, 1994).

These results are disturbing since they imply that unless farmers are in the 20 per cent of the countryside where traditional farming has persisted (mainly the marginal fringe) or have a personal commitment to conservation, then existing schemes will have little effect on their behaviour. More encouragingly, Ward and Lowe (1994) have demonstrated that fewer farms are being managed for succession, and that this has encouraged a more outward-looking perspective and a greater receptiveness to environmental concerns.

This process is, however, a slow one and in a superb exposition of recent trends and analyses Morris and Potter (1995) argue that in essence current schemes are only temporary bribes, are shallow in operation and will be transitory in their effect. In addition they are the result of political expediency and they will not influence long-term attitudes. They base these conclusions largely on a survey of 101 farmers which revealed that 45 per cent were non-adopters. This group could then be subdivided into resistant non-adopters (63 per cent) and conditional non-adopters (37 per cent). Fifty-five per cent of the sample were, however, adopters subdivided into 51 per cent passive adopters, and 49 per cent active adopters. For the 51 per cent of passive adopters there were, however, few additionality effects, and no difference to attitudes after having participated. Morris and Potter thus conclude that if schemes are to develop further farmers will need to be pushed along the participation spectrum, and they propose some ways in which this might be accomplished, for example by identifying the 'resistance' factors among farmers.

In the same vein Ward *et al.* (1995), from a survey of farmers' attitudes to farm pollution in east Devon, encouragingly have noted that the increased numbers of non-farming middle-class people living in rural areas could lead to farmers adapting their values to the changing values of society around them and thus become more conservation conscious.

From a post-modern perspective there is also the worrying concern that a minority of farmers have often acted perversely to policy signals and that they can have a disproportional impact on the landscape. An example is provided by one Kent farmer, Hughie Batchelor, who in the 1980s continued to cut down trees on his large, very visible landholdings along the North Downs even after being sent to prison. Returning to one of the core concepts at the outset of this chapter, we still have little idea just what landscapes and wildlife habitats either farmers or the public prefer and it could be that the eclectic irrational mix of policies that grew up incrementally from 1981 onwards may just provide the best recipe for continuing the rich mosaic that has long been the hallmark of the British landscape.

Notwithstanding the possibility that a policy based on serendipity may just work, this section should conclude with some words of caution from one of the key actors in the NAL approach and its successors, Adrian Phillips (1993), a former Director of the Countryside Commission, who has concluded that the Countryside Commission was unable:

to do much more than point at the changes brought about by modern farming, for example in its New Agricultural Landscapes report of 1974 . . . had to rely on underfunded programmes of tree planting and cosmetic conservation – modest gestures before a tide of destruction which it could not stem.

(66)

E: COMMENTARY ON THE FUTURE

It is tempting to see the 1990s as a crucial turning-point in farm policy in which the crisis in policies based on production can only be solved by devoting the same formidable resources to repairing the damage done in the postwar period. Gilg (1991) has provided a theoretical set of possibilities as shown in Figure 3.2. In this figure, point 1 represents extra food output but a loss of environmental quality (the postwar scenario). Point 2 represents an environmentally neutral strategy in which food output growth is not achieved at the expense of the environment (by adopting a more precise form of scientific farming). Point 3 represents environmental gains and also

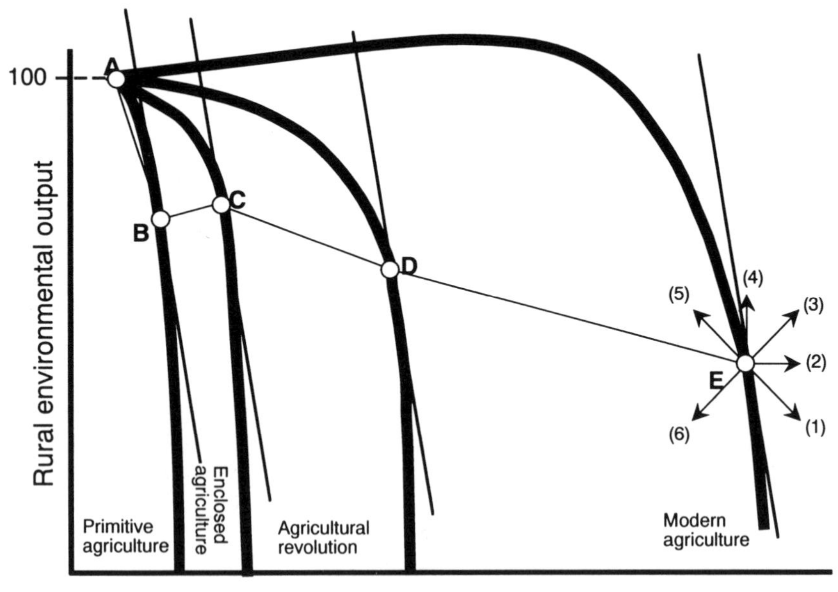

Figure 3.2 The evolution of the agriculture–environment production possibility frontier

Source: Gilg, 1991a

food production growth (the view of the scientific optimist via biotechnology). Point 4 represents environmental gains with no loss of food production (the NAL solution of the 1970s and 1980s). Point 5 represents environmental gains with a loss of food production (the ESA and organic food scenario) and point 6 represents the worst possible scenario, a loss of environmental output and also food output (the doomwatch scenario in which the Earth bites back at its misuse).

Whichever scenario is chosen the Organisation for Economic Cooperation and Development (1989) has provided a model of the type of approaches and mechanisms that may be needed to integrate agricultural and environmental policies successfully. In particular they see set-aside as offering an interim solution during which the most suitable land can be targeted for conversion to woodland or other semi-natural habitats. This is also the implicit assumption of current EU policies at both the EU-wide and Member State Level, as Wilson (1994) has shown in his study of German agri-environment schemes. Webster and Felton (1993) have argued for a regional dimension which would target certain habitats and farm systems, Potter *et al.* (1988) have demonstrated which land if targeted could provide the best long-term environmental gain in set-aside, and Potter *et al.* (1993) have provided an assessment of how effective agri-environment schemes have been in targeting the best areas. Craighill and Goldsmith (1994) have also stressed the need for set-aside land to be managed for at least 20 years in order to achieve habitat improvement, and for longer if natural regeneration is to take place, and have set out various options including non-food crops, biofuel production, habitat improvement, and natural regeneration. In order to achieve this they advocate a move away from farmers' property rights, towards that of landscape property rights, namely transferable conservation obligations (using US concepts). These conservation obligations would be rewarded by compensation payments which would place land management on a long-term community based footing. Such obligations would be a logical progression of transferring the emphasis of policy from one of reducing surpluses to one of enhancing nature conservation under transferable conservation obligations. However, Whitby (1994) has argued that land ownership, preferably by amenity groups, would be a better long-term solution in sensitive areas, like ESAs.

Such solutions do not offer much hope for the majority of farmland, since ESA schemes only cover 15 per cent of farmland and set-aside only between 10 and 15 per cent of land. This leaves about two-thirds of land subject to continued hi-tech farming with only the NAL/Countryside Stewardship solution of 'odd-corner' planting, which has been criticised because non-native species have been planted, and because the habitats that have been created have been too small to create meaningful environments. For example, in the first four years of Countryside Stewardship less than 90,000 hectares had been signed up. However, Oglethorpe (1995), in a survey of farmers' behaviour, confirmed previous work which showed that they had built up a high aversion to risk in the 'feather-bedded' climate of the postwar period. Ironically, the 1992 reforms have made farming even less risky by providing the payments already shown in Table 3.2. Accordingly, Oglethorpe recommends that further policy reforms should create a more variable, less regulated, price-support regime and thus a far more risky economic climate. He argues that most farmers would react by moving to less risky, less intensive production based on prewar mixed farming so that bets could be hedged between enterprises, and thus risks reduced in the riskier climate. Oglethorpe claims that this would kill the three problems of current policies with one stone, namely, reduce expenditure; cut surpluses, and prompt a return to less intensive, less specialised production, and thus re-create traditional high-diversity plagio-climaxes.

Bowers (1995) is, however, less optimistic about the transition to sustainable agriculture, and argues that the process would not be costless, and will not be brought about by the May 1992 reforms since these are principally supply-control measures, with environmental 'add-ons'. In addition, the cost of the CAP is unsustainable in the long-term, as are cross-compliance schemes. Bowers concludes that the concept of sustainable agriculture opens a wide agenda, and although the long-term trend to lower prices will open the door to conservation bodies to buy farmland, it still leaves open the very big question of what sort of environment we might want to create.

Furthermore, Moxey et.al. (1995) have argued that even if we could agree on the desired ecological state we want, estimating the incentives to achieve it via altered farm-management practices would not be straightforward. First, ecological responses are spatially variable so that universal formula do not apply; second, spatial and temporal variations in agricultural productivity generate uneven responses by farmers; and third, the new ecology in the incentive area may have consequences in other areas.

In conclusion, it is most likely that muddling through incrementalism will continue and that the countryside will become more rigidly divided into three habitats. First, in most of the lowlands, hi-tech farming, punctuated by NAL-style planting and recreational sites will continue. A gradual reduction in farm subsidies will reduce the potential to influence landscapes in these areas. Second, areas of high-quality environments where existing schemes will evolve to produce a hybrid type of landscape in which farming is practised mainly as a form of landscape management. These will be highly subsidised but farmers may be able to earn premium prices from high-quality value-added food products produced by traditional and/or cruelty-free methods. Third, in the less favoured areas, mainly in the uplands, severely degraded and farm-arrested habitats will continue to be heavily subsidised for social reasons, and for the political reason that the electorate doesn't like change, and thus any rational policy to re-afforest these areas with semi-natural broadleaved woodland would be politically impossible, especially given the twentieth century experience of insensitive reafforestation of some of the uplands with non-native, poorly landscaped conifers which is considered next in Chapter 4, after the accompanying checklist, which provides a summary of this chapter.

Policy evaluation checklist derived from Chapter 1.

1 *Were the problems understood and which value system dominated?* Conservationists took a long time to demonstrate there was a problem, and the farming community even longer to accept that it was the cause. The value system remains that farming is the best model for managing the countryside.
2 *Which goals were to be achieved and were the policy objectives clearly spelt out?* The main goal was one of halting destruction, allied to some reparation in selective areas of a farm.The goals were thus limited and policy objectives have confused curbing production and encouraging conservation.
3 *Were the policy responses accretionary, pragmatic and incremental?* Extremely so, and a main criticism from the farming community is that there are a bewildering variety of schemes to choose from, and too many offices and forms to visit. MAFF (1995) has thus proposed a 'one stop shop' solution.
4 *Which powers in the Gilg/Selman spectrum were mainly used?* The policies have evolved in a classical development from exhortation in the 1970s, to financial incentives preventing bad things in the 1980s, to financial incentives for doing good things and regulatory controls in the 1990s.
5 *Were planners given sufficient powers/resources and sufficient flexibility for implementation?* Far too few powers have been given and the policies remain largely voluntary. In addition, resources are tiny compared to the main policy budget. However, most of the schemes allow discretion at the farm level.
6 *Were performance indicators built into the policies?* Not at first, but later, schemes like ESAs have been subjected to detailed targets, as have the two Nitrate schemes. The indicators are, however, restricted to financial costs and environmental indicators remain vague.
7 *Have the policies been underresearched?* No, this is one area where a good deal of research has taken place, but on a few key schemes and areas. In addition research has focused on financial and behavioural factors, and less on the environmental impacts.
8 *Have the policies been effective overall?* Given their small budgets,

the schemes have been effective where they have operated, and are slowly changing farmers' attitudes. However, they have paid farmers to do what they would have done anyway, and overall the effect has so far been cosmetic.

9 *Have there been uneven policy impacts, by group or by area?* The policies, especially the area-specific ones like ESAs, have been targeted at certain areas. Outside these areas, many farmers remain resistant to the schemes, and only conservation-minded farmers are involved confirming point 8.

10 *Have the policies been cost-effective and are they environmentally sustainable?* Yes, and ESAs in particular have been shown to represent value for money. However, other approaches, for example, land purchase, could have been more cost-effective. They are sustainable but only if the money keeps coming in.

11 *What have been the side-effects?* The environmental side-effects are not yet apparent, but the schemes have helped change attitudes, notably ESAs which have been praised by both the conservation and the farming community. There is, however, a danger of complacency and forgetting wider deterioration.

12 *Were the policies modified by unforeseen events/policy changes elsewhere?* Yes, the 1981 Act was radically changed after the election precipitated by Scottish devolution, and the 1992 reforms to the CAP led to another series of incremental changes, rather than a rethink from first principles.

13 *Were the policies overtly sectoral and were they compatible with other policy areas?* Yes, as Table 3.2 and other chapters show, almost every rural agency has an environmental scheme(s) and this is a major problem for potential users. However, the schemes do not often conflict with other policies.

14 *What have been the effects of 16 years of Tory rule?* The main effect was felt in the 1981 Act and since then in the voluntary-financial compensation approach. If the Tories had had a free hand in drastically reducing the CAP, one of the main driving forces for deterioration would have been halted.

15 *Should the policies be modified?* Yes, since in the words of Morris and Potter (1995) the policies are temporary bribes, which are shallow in their effect, and the result of political expediency. Most importantly, they breach the principle of paying compensation for preventing actions leading to private profit against the public interest which is the cornerstone of public planning.

16 *What alternatives are there?* There are a host, as shown by Gilg (1991a), but the main priorities are to replace the CAP with a general environmental policy which pays for conservation results as measured by, for example, the number of birds on a farm, thus

giving farmers a conservation crop to produce and a new form of motivation. This will need to be backed up by regulatory controls.

17 *Overall verdict.* The policies have helped to slow down environmental deterioration, and even reversed it in some areas. But the wider environment continues to deteriorate and current policies can only be seen as temporary and partial measures awaiting the fundamental reform of the CAP.

4

PLANNING FOR FORESTRY AND WOODLAND

A: INTRODUCTORY CONCEPTS AND QUESTIONS

The first concept is that Britain needs more trees. There are three enormously strong arguments why forestry and woodland should be expanded in the United Kingdom. First, the United Kingdom is one of Europe's least wooded countries; second, ironically it has one of Europe's best tree-growing environments; and third, woodland is Britain's natural vegetation.

In more detail, Great Britain, with about 10 per cent of its surface under trees, compares badly with a European average of 25 per cent and with countries like Germany with 30 per cent, and even Mediterranean countries like Greece and Spain with 20 and 31 per cent respectively (House, 1993). In contrast, only Ireland and the Netherlands with 5 and 7 per cent, come out with lower figures than Britain (Clout, 1984). Economically, this means that the United Kingdom has to import some 90 per cent of her timber needs, and accounts for 14 per cent of world timber trade, second only to Japan.

Being wet and mild throughout most of the year, almost every species found in the world will not only grow but flourish in Britain. Ironically, Britain is not only poorly treed, but has very few native species, commonly believed to be only the alder, ash, beech, birch, elm, lime, oak, Scots pine and willow (Eyre, 1963). This poor species diversity is largely due to Britain becoming a series of islands soon after the end of the last ice age around 10,000 to 20,000 years ago, and meant that only five distinct climax vegetation formations ever became established, before human interference took over and began virtually to eliminate all but 5 per cent of tree cover by 1900. According to Eyre (1963) two of these vegetation formations were dominant: the deciduous summer formation dominated by oak, with beech or ash as alternative dominant species; and the northern pine forest. Only in the very wettest lowlands or uplands would tree cover not be the norm without human interference. Thus, in theory, the planting of trees kills three birds with one stone: first, it increases and diversifies Britain's meagre tree resources; second, it produces maximum biomass and enhances soil development; and third, it returns the landscape to its former glory.

There does therefore seem to be a strong case for a re-afforestation of Britain on ecological grounds. There are also a host of other reasons for not only planting, but also regenerating, managing and conserving woods. The Countryside Commission (1993a) has summarised these as follows:

- To safeguard and enhance landscape
- To protect and enrich wildlife habitats
- To provide recreation and access opportunities for the public
- To diversify rural employment opportunities
- To provide new sources of income from the land
- To yield timber or biomass to satisfy a range of markets including energy
- To reduce the speed of global warming
- To provide an educational resource
- To create an appreciating asset for future generations

The Forestry Commission in its 1989 Management Plan (National Audit Office, 1993) listed nine similar benefits of forests and woodlands, but from a different perspective, as shown below:

1 Wood processing into sawn timber, wood-based panels, paper and other products.
2 A reduction in reliance on other countries for wood and wood products.
3 Employment and economic activity, which supports rural communities and diversifies their economic structure.
4 Opportunities for a very wide range of sporting and recreational activities for the owner and the public.
5 An alternative use for farmland which is no longer required for agricultural production.
6 Habitats for plants and animals.
7 Variety and beauty in the landscape.
8 Soil conservation and stabilisation in areas of coastal dunes.
9 A sense of continuity with the past, including the historical interest of ancient woodlands.

These two lists introduce two further concepts in forestry, multiple use and the long-term nature of tree planting. These have found practical expression for all this century in both Scandinavia and the Alps, where the powerful concepts of farm-forestry, and a mixed-species, age-diverse forest is managed for long-term sustainability, with only the annual increment of timber being felled each year. In addition since 1902 Switzerland has had a Forest Law which stipulates that if any forest is of necessity felled permanently a similar area must be afforested elsewhere (Gilg, 1985). These concepts are further set out in Figure 4.1. This shows that tree planting can be seen simply as an economic investment for its primary production, but that there are also issues of employment creation, and also future benefits that might accrue, for example, in recreation. In addition,

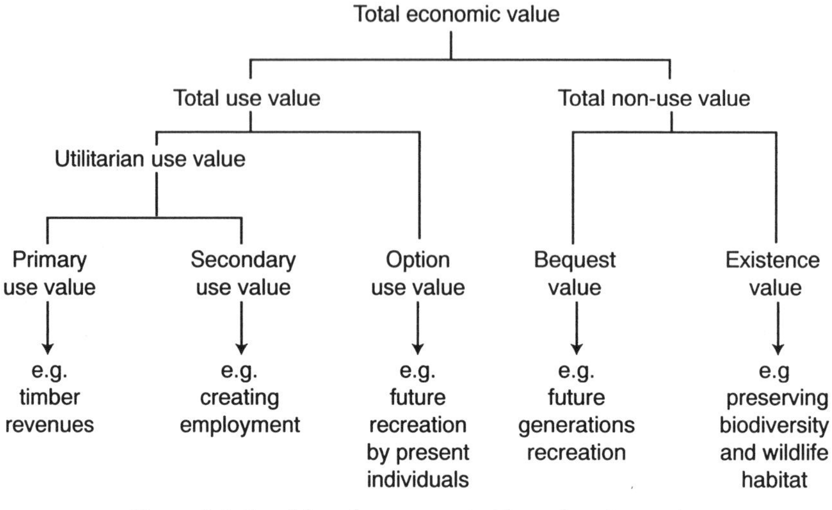

Figure 4.1 Possible values generated by a forestry project

Source: Bateman, 1995

there may be non-use values, notably in the future benefits being stored for future generations. This is a difficult argument since future generations may or may not hold our values when current planting matures in anything from 60 to 600 years.

The apparently clear-cut arguments for forestry set out at the beginning of this chapter therefore contain some very difficult conceptual questions about the detail of that planting. Seven in particular stand out:

1 Should forestry expansion be carried out by the public sector, forestry firms, or farmers?
2 Should forestry be expanded in the uplands or lowlands?
3 Should forestry expansion be carried out in large blocks, or in a mix of woods, shelter belts and copses?
4 Should forestry expansion be achieved by shorter-term, commercially more attractive, but to some eyes less attractive, and probably non-native coniferous forests, or by longer-term, less profitable, but probably more ecologically sound and attractive native deciduous woodland?
5 Can forestry ever be profitable and how on earth can profits/losses be estimated over the decades between planting and harvesting, or should forestry be seen as an investment in the future with many potential benefits?
6 How does one estimate for the losses involved when planting takes place, for example, land uses and landscapes lost, wildlife losses offset by some gains, and one rural economy replaced with another?

7 Should forestry be seen as a purely profit-making business with restricted access except on payment or should it be seen as a social and environmental good with virtually free access for public recreation?

All these questions were traditionally seen in either/or terms, but since the early 1980s forestry has come out of the closet of narrowly conceived upland coniferous afforestation to announce itself as the flagship of environmentally friendly multi-purpose land use management for the twenty-first century. The rest of this chapter seeks to discuss how this came about and whether indeed forestry provides a role model for future land-management policies, in the context of the three key concepts shown in the accompanying summary box.

Three key concepts in forestry thought

1 Britains needs more trees on ecological and economic grounds.
2 Forestry should be multi-purpose.
3 Forestry needs to take a long-term view and thus should contain the methodologies and techniques needed to provide a role model for 'sustainable' development of the environment.

B: EVOLUTIONARY EVENTS IN THE FORMULATION OF FORESTRY POLICY

Introduction

Forestry is unlike any other land use considered in this book, in that the state agency set up to influence it, the Forestry Commission, owns and manages significant amounts of land, so much so in fact that by the 1980s the Commission had become Britain's largest single landowner (Norton-Taylor, 1982). Accordingly, this section is divided up into two major sub-sections: first, the evolution of forestry policy and the role of the Forestry Commission as an executive agency; and second, forestry policy and the private sector. In addition, there is a concluding sub-section on other policy influences on forestry.

Evolution of overall policy and the role of the Forestry Commission as an executive agency

During the First World War the Government realised that Britain, then with only about 5 per cent of its land under, mainly poor-quality, timber, was very vulnerable to any interuption in supplies of imported timber.

Accordingly, following the recommendations of the Acland Committee the Government passed the 1919 Forestry Act. This set up the Forestry Commission with two primary objectives:

1 strategic (to reduce dependence on imported timber); and
2 economic (to reduce the cost of softwood for manufacturing) and three broad powers:

a) to plant new forests;
b) to encourage private forestry by grant aid and advice; and
c) to help maintain an efficient timber trade.

Thus, unlike the model for agricultural planning used in the aftermath of the Second World War, the 1919 Act set up a bilateral planning system involving direct land management in both the public and private sectors.

The interwar period saw the ground rules of state planting set in motion but none of the plantations planted then were of any use by the time the Second World War reinforced the need for forestry expansion. This was set out in a White Paper (Cmnd 6447, 1943) which argued for a doubling of the forest area by 1993 to 2,023,000 hectares, from 5 per cent in 1900 to 10 per cent in 1990 at a rate of 20,000 hectares per annum. This expansion was to be achieved by state planting and a series of carrot-and-stick measures in the private sector which were introduced in a series of Acts. First, the Forestry Act 1947 introduced a system of Dedication Agreements which offered grant aid in return for a commitment to long-term forestry. Second, the 1947 Town and Country Planning Act introduced the concept of Tree Preservation Orders (TPOs) controlling the felling of trees on mainly amenity grounds, while third, the 1951 Forestry Act introduced felling licences controlling the felling of commercial woodland, mainly by the power to insist on a replanting condition if felling was allowed. The further evolution of these powers is considered in more detail in the next sub-section, and TPOs in Chapter 5.

The 1967 Forestry Act reaffirmed the general roles of the Forestry Commission but widened them to include: the provision of recreational access and the enhancement of natural amenity. In addition the Act reinforced the division of the Commission into an executive body (Forestry Enterprise) and an advisory agency (Forestry Authority). The Forestry Enterprise was expected to: a) further develop the forest estate; b) protect and enhance the environment; c) provide recreation; d) stimulate the local economy in areas of depopulation; and e) further integrate forestry with the environment. The Forestry Authority was expected to: a) advance the science of forestry which was then still very basic; b) undertake research; c) develop the nation's timber industry; and d) administer private woodland and ensure land-use integration. Finally, the Act pegged the gross planting rate to 18,000–22,000 hectares per annum.

The first serious questioning of the policy of expansion came in the early

1970s with the publication of a report by the Treasury (1972) which used cost–benefit analysis to argue that new planting could no longer be justified on: a) the economic rate of return which was only about half the minimum test rate of the Government for public investments; b) import saving, since investments in other goods for export produced higher trade deficit savings; and c) the costs of job creation which was very high per job created. The findings of the report were accepted by the Government in 1974, but rather surprisingly at the same time the Government also accepted the proposal of an earlier document (Agriculture, 1972) that the existing planting rate of up to 22,000 hectares a year should be maintained for rural employment and recreational provision. In the private sector a new Dedication Agreement scheme was set up.

In 1980, in another review of forestry policy against the background of rising world demand for the diminishing world timber resource, the Government argued that a continuing expansion of forestry was in the national interest, both to reduce dependence on imported timber in the long term and to provide continued employment. The 1981 Forestry Act, however, modified existing policies. First, the Act saw an attempt to privatise the Commission by a thousand cuts. A target of 4 per cent of the Commission's assets of land and plantations were to be sold off during 1981–4 thus raising £40 million. In practice sales were lower and in 1984 the disposal policy was changed from reducing the Commission's call on public funds to one of rationalising the Commission's estate by allowing the sale of remote and/or small inefficient plots. In 1989 the Government set a further target of 100,000 hectares to be disposed of between 1989 and the end of the century. However, in 1990 this was made more difficult when it was stipulated that all new sales whenever feasible had to be made subject to access agreements with local authorities, and further safeguards to ensure that access would continue were implemented in 1994 (Cm 2814, 1995, 168). Between 1981 and 1989, 72,000 hectares of plantation or plantable land and 68,000 hectares of other land were sold.

Also as a result of the 1980 review, the Dedication Agreement schemes were closed to new entrants in 1981, but as the next section shows they were replaced with a variety of schemes over the next decade. In addition, felling licence procedures were simplified in the spirit of Mrs Thatcher's early 1980s drive against bureaucracy.

In the mid 1980s the Wildlife and Countryside (Amendment) Act 1985 amended the 1967 Act so that the Commission was now required to try to achieve a reasonable balance between the interests of forestry, the conservation and enhancement of the countryside, and the conservation of wildlife. In 1987 the Alternative Land Use and Rural Economy (ALURE) package (Cloke and McLaughlin, 1989) set a target of an extra 100,000 hectares of private commercial afforestation between 1987 and 1990 to soak up surplus farmland, and an overall medium-term target of 33,000 hectares per year for all tree planting. Thus by the late 1980s the objectives of the Commission (Cm 56, 1987) were to promote the interests of forestry, the establishment and maintenance of adequate reserves of growing trees, the production and supply of timber, and the development of the recreational and conservation potential of its forests.

In 1988 a new dimension was added with a European Commission proposal for a medium- and long-term strategy for forestry in Europe (10753 8415/88 COM(88)255) with eight objectives including: encouraging the development of rural life; ensuring a secure supply of renewable raw materials; contributing to environmental improvement; and extending the use of forestry as a natural setting for recreation. Expenditure was programmed to cover both the afforestation of farmland to reduce farm surpluses, and to improve the management of existing woods which were often neglected. The policy was given a further boost in 1992 when farm forestry was included as a key feature in the agri-environment package which accompanied the reforms to the CAP. The 1980s thus witnessed the first serious attempt to move away from only afforesting the uplands, which had till then been considered the only suitable areas.

In the 1990s there has been no primary legislation but there have been a series of policy and administrative changes partly in response to a series of critical reports (discussed below in the analysis section D) about the cost-effectiveness and harmful environmental impact of forestry policies. In particular, a report by the House of Commons Select Committee on Agriculture (1990) was instrumental in setting out a consolidated government policy (Forestry Commission, 1991). The two main aims of policy thus became: a) the sustainable management of existing woods and forests; and b) the steady expansion of tree cover to increase the many, diverse benefits that forests provide. The indicated level of 33,000 hectares per annum set in 1987 was to continue to be achieved largely by private owners under incentive grant schemes. At the same time the Commission was further divided into a Forestry Authority and a Forest Enterprise.

In 1994 it was decided, after considerable debate, centred around a 'Forestry Review', not to privatise the Commission but to retain the Forestry Authority while transforming the Forest Enterprise into a Next Steps 'trading body' run by a chief executive (Cm 2644, 1944). The target rate of planting to be achieved almost entirely by the private sector was confirmed at 33,000 hectares per year. As a wider review of forestry policy (as part of the 1992 Rio agreement) the Government further modified forestry policy (Cm 2429, 1994) to be largely as in 1991, namely a) the sustainable management of existing woods and forests, and b) the steady expansion of tree cover to increase the many diverse benefits that forestry provides, but added the role of acting as a sink for carbon to counter the greenhouse effect. In more detail, eight roles were set out: protecting our forest resources; enhancing the economic value of our forest resources; conserving and enhancing biodiversity; conserving and enhancing the physical environment; developing the opportunities for recreational enjoyment; conserving and enhancing our landscape and cultural heritage; promoting appropriate management; and promoting public understanding and participation.

The net result since 1919 has therefore been a developing role from a single-purpose agency to a multi-purpose rural land management organisation, albeit weakened by a reduction of its own planting programme in the 1980s and the virtual division of the Commission into two separate bodies in the 1990s.

Forestry policy and the private sector

Private woodland is valuable because it is found mainly in the lowlands, often in small plots scattered through otherwise spoilt English landscapes of the type discussed in Chapter 3, and, in England at least (Norton-Taylor, 1982), is most likely to be broadleaved and thus probably a semi-natural native woodland. In addition private woodland has frequently been neglected and thus provides wildlife havens even in the form of transitional scrub, which represents great potential if it were to be properly managed.

There are three main issues that have affected forestry policy in the private sector: a) incentive schemes, b) taxation and, c) planning controls. These are now considered in turn.

Incentive schemes. These date back to the introduction of Dedication Agreements in 1951 and between then and their phasing out from 1981 onwards, three schemes known as Basis I, II and III were developed incrementally. Under the Basis II scheme set up in 1974, for example, owners, in return for accepting a contractual obligation to manage their woodlands in accordance with sound forestry practice and good land use, including effective integration with agriculture and paying attention to such environmental benefits as the provision of recreation and public access, could receive grants of £44 per hectare for softwood and £122 for hardwood planting. In 1977 the Basis III scheme increased the rates to £100 and £225 respectively.

The Dedication Agreement schemes were replaced in 1981 with a Forestry Grant Scheme. Rates of grant varied from £230 per hectare for planting conifers on 10 or more hectares to £850 per hectare for planting broadleaves on less than 1 hectare. In 1985 the rates of grant were increased to £240 and £890 respectively, and a new scheme, the Broadleaved Woodland Grant Scheme, was instituted to maintain and create broadleaves, with grants for planting and natural regeneration varying from £1,200 per hectare for small woodlands below 1 hectare to £600 per hectare for woodlands above 10 hectares. The new scheme was backed up by a stricter felling licence policy and annual payments.

In 1988 both schemes were replaced with the Woodland Grant Scheme with grants ranging from £615 per hectare for planting conifers on 10 or more hectares to £1,575 per hectare for planting broadleaves on less than 1 hectare. Although these were very big increases compared to the rates shown in the previous paragraph, they were meant to compensate for the loss of tax concessions taken away in the 1988 budget, which are discussed more fully in the Taxation sub-section which follows. Also in 1988, a related scheme, the Farm Woodland Scheme was set up for three years with the aim of diverting around 36,000 hectares of farmland out of agricultural and into timber production. The Scheme supplemented grant aid available under the Woodland Grant Scheme, by paying an annual sum intended to bridge the gap between planting and the first income from thinnings. For example, 40 years for oak, but 20 years for conifers, and only 10 years for coppice. After three years the Scheme had however only attracted applications to plant 16,000 hectares.

In 1990 a scheme intended to improve existing woodland rather than aid new planting was set up under the Woodland Management Grants Scheme, marking a new era in grant schemes. In 1994 the schemes were modified and expenditure was increased by a planned £4 million (Cm 2645, 1994) in ways that further emphasised the principles of cross compliance with extra-forestry goals and payment by results as shown in the accompanying summary box which sets out the main changes made in 1994.

Summary of main changes made to forestry grant schemes in 1994

1 Conifer planting grant under the Mark III Woodland Grant Scheme increased from £615 to £700 per hectare.

2 Broadleaved planting grant rate £1,350 per hectare below 10 hectares (formerly £1,175) and £1,050 per hectare in areas above 10 hectares (formerly £975).

Both the conifer and broadleaved grants paid in instalments, 70 per cent after planting, 30 per cent after five years.

3 Restocking grants under the Mark III Woodland Grant Scheme were reduced by 50 per cent to £325 and £525 per hectare for conifers and broadleaves respectively to be paid 70 per cent on completion of planting, 20 per cent after five years, and 10 per cent after 10 years. However, the following supplementary payments were made available:

a A better land supplement for conifer and broadleaved planting where arable or improved grassland was taken out of production up from £400 to £600 per hectare.

b A community woodland supplement for woodlands within five

miles of a town of £950 per hectare was payable if public access and suitable parking facilities were provided.

c Livestock exclusion supplement for exclusion of livestock from old-established or native woodlands threatened by grazing of £80 per hectare per year for 10 years.

d Aged woodland supplement of £35 per hectare for those aged woodlands requiring extra management input.

4 Natural regeneration, a discretionary grant of 50 per cent of the approved cost. After regeneration further grants of £325 and £535 per hectares for conifers and broadleaves respectively.

5 £1 million a year for pilot schemes in priority target areas.

6 A new discretionary Woodland Improvement Grant to improve environment and visitor facilities.

7 New grants for short-rotation coppice at £400 per hectare on set-aside land and £600 per hectare elsewhere to run for five years initially. (In 1995 set-aside rules were relaxed to allow woodland to count in the set-aside area, thus boosting this and other forestry grants.)

8 Pilot scheme under which payments for larger Woodland Grant Scheme applications will be determined by negotiation rather than by reference to specific rates of grant.

9 Annual payments under the Farm Woodland Premium Scheme aimed to provide annual payments to compensate for profits foregone while trees grow in woods planted with grant aid under the Woodland Grant Scheme were increased.

10 Grant aid under the Farm and Conservation Scheme (discussed in Chapter 3) continued with for example, grant aid being paid at a percentage rate of the costs incurred as follows:

Shelter belt planting: outside LFAs 15 per cent; inside LFAs 50 per cent;
Shelter belt planting with more than 50 per cent broadleaved species: outside LFAs 40 per cent, inside LFAs 50 per cent;
Enclosing grazed woodland: outside LFAs 40 per cent, inside LFAs 50 per cent.

Collectively all the schemes outlined above contained a set of similar principles: first, they were paid in instalments, generally over 10 years; second, they were only paid if environmentally acceptable and some sort of access was normally provided; third, most of them contained supplements or cut-offs related to the type of woodland or farmer involved, for example, the percentage of trees under broadleaves or special help for young farmers; fourth, the grants were tapered to reflect economies of scale with small schemes being more generously aided; and fifth, the grants could be subject to taxation and be unavailable if other public funds were being used. These schemes also represented an attempt to switch afforestation from the uplands to the lowlands so as to soak up surplus farmland and to restore woodland to lowland England where it had been either cut down or neglected.

In addition to the main Forestry Comission and MAFF schemes outlined above, there were around 60 other grant schemes offering forestry as an alternative land use (Watkins, 1992). In particular three specific forest projects can be picked out: the New National Forest, Community Forests and the Central Scotland Woodland Project, all of which are attempting to achieve the benefits of multi-purpose woodland which Bishop (1992) has listed as being: timber production; a place for nature; benefits for private landowners; a better place to live and work; new environments for leisure; and educational value.

In more detail, the New National Forest, a 44 square-mile area between Derby and Birmingham first designated in 1992, plans to create a new forested area by merging two ancient forests – Needwood to the west and Charnwood to the east – to form a mosaic of farms, towns and villages. A National Forest Development Team which was set up in 1991 produced a draft strategy and a business plan by the end of 1992 (Countryside Commission, 1993c) which was finally approved in November 1994 (Countryside Commission, 1994a). These envisaged creating the forest mainly by initiating and encouraging predominantly broadleaved tree planting backed by selective site acquisition. It soon became apparent that landowners were not only suspicious of the project, but did not perceive the level of grants on offer to be sufficiently tempting (Bell, 1995). Accordingly the team has been seeking other sources of finance, notably a major National Forest project under the Millenium Fund created by the inception of the National Lottery in 1994.

Community Forests were launched in 1989 as a joint Countryside Commission and Forestry Commission venture (Countryside Commission, 1989a). By 1991 twelve areas around most of the major cities of England had been set up. Within these areas the plan is to create a multi-purpose forest designed, developed and managed to provide for the community's leisure needs. Each forest will cover approximately 10,000–15,000 hectares, within which 30–60 per cent of the land will be planted, predominantly though not exclusively with broadleaved trees. The vision is not of continuous forest but a network of community woodlands and other landscape features, such as farmland, heathland, meadows and water features. As in the National Forest everything will depend on the private sector, notably farmers, coming forward with afforestation schemes linked to farm diversification into leisure activities by using grant aid, notably the Community Woodland Supplement, for creating new woodlands for public access and the Management Grant for improving recreation facilities in existing woodlands.

The Central Scotland Woodlands project is a Scottish Office scheme launched in 1989 under the control of Central Scotland Woodlands Ltd, who have been given an annual budget of £2.5 million till around 2010. Central to the concept is economic regeneration by environmental improvement. Three main types of woodland are envisaged: Community Woodlands close to urban areas; Amenity Woodlands to improve the wider landscape; and Productive Woodlands located mainly on poorer land and providing the coniferous timber that industry needs (Scottish Office, 1989).

In conclusion, the accompanying summary box shows the main grant schemes on offer since 1945.

Summary of main grant schemes for forestry since 1945

Basis I to III Dedication Agreements 1951–81 replaced by:

Forestry Grant Scheme 1981–8
Broadleaved Woodland Grant Scheme 1985–8 widened grant aid

Both schemes replaced by:
Woodland Grant Scheme 1988 to date
Farm Woodland Scheme (FWS) 1988–91 provided supplementary grants
Farm Woodland Premium Scheme 1991 to date extended FWS with supplementary grants

Woodland Management Grants Scheme 1990 to date

National Forest 1991 to date
Community Forests 1989 to date
Central Scotland Woodlands 1989 to date

Taxation. The impact a general piece of public policy can have on other policy areas is well illustrated by the case of taxation. The main impact has been felt in those taxes that are levied on transfers from one generation to the next, specifically at death, and those that impose taxes on annual income.

In more detail, before 1975 forestry was favourably treated in that death duties were not payable on either forested land or the current value of the timber in the forest until the time of felling, thus encouraging long-term stewardship of hardwoods. In 1975, however, a general tax, Capital Transfer Tax, was introduced by the Labour Party for both lifetime and death transfers of wealth. In forestry this was levied on the value of the underlying land only, or at the time of felling, thus perpetuating some of the concession's incentive. However, enough of the concession was perceived to have been removed to lead to a drop of two-thirds in the rate of private planting during the next few years.

In 1986, partly as a result from lobbying by the powerful forestry land-owning group in the House of Lords and partly as a result of Mrs Thatcher's ideology of wealth transfer between generations, Capital Transfer Tax was replaced by Inheritance Tax, which worked in a similar way to the pre-1975 Death Duties. In particular, tax was only payable on assets of above £150,000, and if these assets were given away before death, tax was payable on a sliding scale as follows: if death within 0–3 years full rate; 4–5 years 60 per cent; 5–6 years 40 per cent; and 6–7 years 20 per cent. Therefore, apart from fear of replicating the King Lear effect, the new tax reintroduced a strong stewardship incentive to plant or keep trees as a means of transferring wealth from one generation to another, notably in landed estates.

In contrast to the landed estates, the *nouveaux riches* of the 1980s spotted that forestry provided a tax-avoidance scheme on annual income in that expenditure on afforestation, normally a massive loss in the first few years, could be offset under Schedule B against income. This allowed cash-rich individuals, notably entertainers, to offset very variable incomes against forestry tax losses. This was a double whammy to the average tax payer, who saw such individuals not only avoiding substantial tax bills but also gaining substantial grant aid. Thus an investment in forestry which cost £116, 175 could attract tax relief and grants worth £74,421, and thus actually cost only £41,754 for a 60 per cent taxpayer (*Sunday Times* 20 September 1987).

Environmentalists also claimed that the best land for this tax-avoidance scam was unfortunately in areas like the Flow Country in northern Scotland, which had high value as remote wilderness areas. The end result was a massive public outcry and a significant turning-point in the 'greening' of the Tory Party. Consequently Schedule B tax relief on commercial woodland was abolished in the 1988 budget, thus ending the tax incentive for cash-rich individuals to plant up land. At the same time a reduction in the top rate of tax from 60 to 40 per cent reduced the attraction of forestry as a long-term tax avoidance measure. In order to keep genuine foresters interested in forestry the various grant aid schemes already outlined above were increased pro rata to offset the taxation losses made by the 1988 budget. Crabtree and Macmillan (1989) forecast that the changes would lead to four possible effects: first, high tax investors would leave forestry; second, poor land in the uplands would not be planted; third, better land in the lowlands would be planted if farm prices fell; and fourth, broadleaves would gain at the expense of conifers.

Other forms of guidance. As the analysis section D below will show, afforestation has been the biggest single change in land use in the twentieth century and yet most of the decisions made about it have been made behind closed doors by non-elected officials. This is in complete contrast to the other major change of the century, the development of the built environment of the countryside which since 1947 has been subject to detailed public scrutiny under the Town and Country Acts, as shown in Chapter 5. Not surprisingly, many people have called for the introduction of a similar form of control over afforestation.

Nonetheless, the secretive system of deciding felling licences and approving plantation grants carries on, albeit modified over the years, as shown by Mather and Thomson (1995). Until 1987 the system was at its most secretive. Before that date applications for felling licences or grants were made to Regional Advisory Committees appointed by the Forestry Commission from lists of the 'great and the good' supplied by relevant organisations. Thus the system did represent sectional interests, but in a non-elected way. Another saving grace was that contentious or disputed applications were referred upwards for resolution and thus tended to become subject to open debate.

In 1987 two improvements were made. First, the Forestry Commission decided to advertise disputed applications for comment by the public and pressure groups. Second, the EC Directive on Environmental Assessment (85/337) became law in the United Kingdom. Under it major afforestation proposals can be subjected to an environmental assessment of their impact. In contrast, in 1989 the threshold at which MAFF had to be consulted over the afforestation of farmland was raised to 40 hectares, thus reducing outside influence on forestry decision making.

In spite of the lack of planning controls over forestry, non-statutory guidance has been made available. For example, in 1988 it was announced that in England afforestation, and thus grant aid, would not be acceptable in upland areas above 800 feet, except where it would be environmentally acceptable, as in the industrial Pennines. Locally, local authorities have been encouraged to follow the example set by the National Parks since the 1960s by dividing land up into three areas: a) areas acceptable for afforestation; b) areas where, though there is a presumption against new planting proposals, these might be acceptable; and c) areas where there are strong presumptions against afforestation. In the late 1980s the Scottish Office urged Scottish local authorities to use a similar system in 'Indicative Forestry Strategies' (Mather, 1991) by dividing areas into: a) preferred areas for new planting; b) potential areas where it may be acceptable subject to the resolution of particular issues; and c) sensitive areas where the pressure for new planting should be eased by the identification of preferred and potential areas.

In 1992 the Department of the Environment (1992a) followed suit in a

DOE circular which asked local authorities to prepare strategies which might divide land into preferred and inappropriate areas. In practice the strategy for the New National Forest has divided land up into: a) preferred areas; b) unsuitable land; c) existing woodland; and d) sensitive areas where new planting may be limited but not excluded by river floodplains, archaeological sites, historic parklands, or areas of ecological value and geologic interest (Countryside Commission, 1993d). In 1993 the Forestry Authority for England (1993) joined the bandwagon by providing guidance on 'Landscape Assessment for Indicative Forestry Strategies'.

In preparing their strategies local authorities were reminded by the 1992 DOE circular that most planting would be funded by grant aid and pointed out that the Forestry Commission, when processing grants, sends applicants details of the environmental standards required and notably Forestry Commission publications on good landscape design, nature conservation and water and recreational issues. Indeed since the 1960s the Forestry Commission has published increasingly sophisticated advice on how to produce high-standard multi-purpose woods in both public and private sectors (Forestry Authority for England, 1993). In addition in 1994 the Commission published guides for the management of the eight types of semi-natural (ancient) woodland which covers some 300,000 hectares or 13 per cent of the total woodland area (Cm 2814, 1995).

Nonetheless many calls for planning controls have been made over the years and Gilg (1991a) found in a trawl of over 1,500 documents relating to planning policies that it was the single most frequently called-for innovation. The case for planning controls quite simply is based on the absurdity of exempting a possible 1,000 hectare-plus forestry proposal from planning permission but not a small extension to a suburban house (Tompkins, 1989).

Other policy influences on forestry

Although the Forestry Commission and MAFF have been the main influences on forestry, a number of other organisations have had an impact. For example, local planning authorities, although they have never had the power of development control over forestry or forestry operations like forest road building, do however have power over forestry buildings, notably downstream processing facilities. Local planning authorities also have the power to issue Tree Preservation Orders preventing the cutting, lopping or wilful destruction of trees, groups of trees or woodland without permission. In addition, local planning authorities can also create a cultural climate in which forestry decisions are made via the attitude they strike to forestry in their plans. Advice on how to include forestry in local land-use planning has been provided in a number of DOE circulars, notably Circular 36/78 *Trees and Forestry* and 29/92 *Indicative Forestry Strategies*, outlined above.

Over the years a cultural climate has also been created by statements regarding forestry from the Countryside Commissions, the Nature Conservancy Council and its successor bodies. In addition these bodies have funds to grant aid to small woodland projects, and in some cases powers to designate woodland as a special site, for example, Sites of Special Scientific Interest, a Nature Reserve or Country Park. Another influential opinion former has been the House of Lords and Wilson (1987) has argued that the forestry lame duck was able to break the rigid rules of Thatcherite cuts in public subsidy, largely because of support for forestry in the House of Lords derived from the over-representation of large landowners in that chamber.

Nonetheless, actions when allied to words can make a powerful combination and forestry has also had a powerful lobby in the form of commercial forestry and timber companies or trade associations, for example, the Economic Forestry Group and Timber Growers UK. As a counterbalance the private charity the Woodland Trust, which was founded in the 1970s, has grown explosively. The Trust has acquired woodlands or planted woods using membership fees and donations in small pockets all round the country, and has prompted a number of villages to set up similar small schemes. In addition volunteer labour tends to find working in woods very attractive, and there is now a considerable grass-roots movement dedicated to restoring and managing a network of local woods and coppices.

In conclusion, the accompanying summary box shows the main evolutionary changes to forestry.

Summary of themes emerging

1 A gradual but accelerating shift from public to private forestry. As a consequence restocking has taken over from new planting as the main planting activity of the Commission.
2 A gradual shift from uplands afforestation on open land (conifers) to lowland woodlands on farmland (deciduous), notably in England.
3 A gradual shift from commercial to amenity woodland.
4 A continued reluctance to allow development controls over forestry.
5 Unfortunate side-effects caused by other policy changes, notably taxation.
6 Error of splitting Forestry Commission into two in the 1990s just when it was emerging as a genuine multi-purpose advisory and executive rural planning organisation.

C: THE WORK OF THE FORESTRY COMMISSION IN THE MID-1990S

This section is derived from the Forestry Commissions expenditure plans for 1995–6 to 1997–8 (Cm 2814, 1995) and from the Forestry Commission's (1994) Annual Report for 1993–4. Readers can use subsequent publications to update this section.

According to Cm 2814 (1995) the aims of the Government's policy in 1995 were: the sustainable management of Britain's existing woods and forests; and a steady expansion of tree cover to increase the many diverse benefits that forests provide. This multi-purpose forestry was intended to increase the domestic supply of good-quality timber and to maintain and enhance the contribution of woodlands and forests to the environment. The Forestry Commission was the government department with the lead responsibility for forestry. It reports to forestry ministers, namely the Secretary of State for Scotland (who takes the lead role), the Minister of Agriculture, Fisheries and Food, and the Secretary of State for Wales. The Commission was responsible for advising ministers on forestry policy and its implementation and had a Chairman and Board of Commissioners with statutorily prescribed duties and powers. Under the 1967 Forestry Act these were: a) promoting the interests of forestry; b) the development of afforestation including the establishment and maintenance of adequate reserves of growing timber; c) the production and supply of timber and other forest products; and d) endeavouring to achieve a reasonable balance between the needs of productive forestry and the environment.

These duties were delegated to two groups within the Commission, the Department of Forestry and Forest Enterprise. The Department of Forestry carried out the Commission's policy and regulatory roles in two groups. The Policy and Resources Group, with 11 per cent of the non-industrial staff, was responsible for support and advice to ministers, policy development and central services. The Forestry Authority, with 30 per cent of the staff, was responsible for implementing policy through regulation, advice, grant aid, research and the setting of standards. Forest Enterprise, with 59 per cent of the staff, was responsible for managing the Commission's estate. Forestry Commission expenditure was expected to rise from £166 million in 1989–90, to £219 million in 1994–5 and £239 million in 1997–8. With corresponding income of only £105, £124 and £165 million, the Commission had or expected to be grant aided to the tune of £61 million in 1989–90, £95 million in 1994–5, and £74 million in 1997–8. This expenditure should be seen against the background of capital assets which amounted to £1,530 million in 1994, of which £1,401 million was the forest estate.

Overall plans for forestry included replanting some 48,200 hectares in the three years between 1995–6 and 1987–8, and also new planting plans for 66,940 hectares (65,250 in the private sector and 1,690 by Forest Enterprise) in the same period. Under the plan to dispose of 100,000 hectares of Forest Enterprise land in the 1990s, 45,000 hectares were expected to be sold, raising some £60 million, between 1995–6 and 1997–8. Wood production is expected to increase as existing forests mature, and production, which only totalled 4 million cubic metres in 1976, had increased to 7 million cubic metres in 1993 and is expected to double by 2015. In terms of long-term planting the 1995 'Rural White Paper' (Cm 3106, 1995) set a target of doubling the woodland area in England over the next 50 years, because this would improve the landscape, create new jobs, enrich wildlife habitats, open up new opportunities for recreation, and help improve air quality.

The Forestry Authority. The Authority was responsible for: operating grant schemes for tree planting and woodland management; administering statutory controls in respect of the felling of trees, plant health and forest reproductive materials; providing information and advice; promoting sound forestry practice in relation to environmental, technical and safety standards; and forestry research.

In more detail, the Authority planned to increase grant-aided planting from 17,000 hectares in 1994–5 to an average of 21,750 hectares in each of the three years between 1995–6 to 1997–8, in spite of the failure to reach the planting rate set in 1989–90. It hoped that this would include an increase in conifer and mixed woodlands 'down the hill', while maintaining the increase in broadleaved planting from only 595 hectares in 1984–5 to 10,360 hectares in 1994–5. In addition 4,990 hectares of new native pinewoods had been grant aided between 1991 and 1994. In 1993–4 approval was given to 14,192 hectares under MAFF's Farm Woodland Premium Scheme, and for the future, provision has been made to grant-aid 2,250 hectares of short-rotation coppice as an energy crop by 1997–8. Similarly, Community Woodland Supplement was used to tot up payments in 1,740 hectares of woodland grant-aided under the Woodland Grant Scheme in 1993–4 and provision had been made for an extra 3,000 hectares between 1995–6 and 1997–8. Management grants under the Woodland Grant Scheme, which were first made available in 1992, had been approved for 169,000 hectares by 1994, and this was forecast to rise to 320,000 hectares by 1997.

The Forestry Authority also expected a considerable rise in expenditure, from £14 million in 1989–90 to £43 million in 1997–8, partly caused by increases in the rates of payment for new planting and replanting from £629 per hectare in 1989–90 to £1,136 in 1997–8. This is partly a consequence of support being switched in 1988 from fiscal incentives to grant aid, but also reflects the increased proportion of broadleaves planted, the higher proportion of planting on better land and an increase in small-scale planting, all of which attract higher grants. From 1994–5 the expected reduction in new planting grants takes account of the expected increase in the proportion of conifer planting, albeit offset by the increase in conifer grants stemming from the 'Forestry Review'. Reimbursements of certain grant-aid expenditure from the EC have risen from 25 per cent under the pre-1992 Forestry Action Programme to 50 per cent under the 1992 CAP reform package, and to 75 per cent in Objective 1 areas. Finally, total net expenditure on research amounted to £9 million in 1994–5.

The Forest Enterprise. This became a Next Steps Agency in 1995 and was given a target of becoming self-financing during 1995–6, but remained part of the Forestry Commission. In 1995 it had an estate of 1,100,000 hectares of which 874,000 hectares were forest and woodland, about 40 per cent of the total forest area. Much of this forest is now approaching maturity and thus provides an opportunity to develop a more diverse forest structure, giving a greater variety of colour and form, and introducing more open space. Forest Enterprise is taking this forward through the preparation of Forest Design Plans. However, there are few plans to increase the estate by new planting, and new planting was programmed to fall to an insignificant 500 hectares by 1997–8, most of which will be in the National Forest, Community Forests and the Central Scotland Woodland initiative. This will be more than offset by disposals which rose from 5,559 hectares in 1989–90 to 14,950 in 1993–4 and a planned 15,000 hectares in 1997–8.

Replanting will however remain steady, at around 8,500 hectares a year, as the forest estate moves towards the European concept of only harvesting the annual increment of timber each year. However, harvesting will continue to increase for several years as the benefits of planting in the 1930s onwards begin to come through. It seems ironic that as soon as the forest estate begins to show returns that such a great state asset should be sold off for individuals to enjoy.

Nevertheless, in the remaining areas the Forest Enterprise in 1995 still welcomed the public on foot to all its woodlands, subject only to considerations of safety, conservation and legal constraints. An estimated 50 million day visits are made a year. In addition, gross expenditure on recreation, conservation and heritage in 1993–4 amounted to £16.1 million, offset by charges which were made where practical and which amounted to £3.3 million. The Forest Enterprise also runs tourist enterprises and in 1994–5 it was estimated that these earned a return of 10.5 per cent on capital employed with occupancy rates as high as 85 per cent for cottages and houses and 63 per cent for forest cabins, but as low as 33 per cent for campsites.

In summary, the 1993–4 Annual Report (Forestry Commission, 1994) recorded the highest planting levels since 1988–9. Broadleaved planting had grown to a massive 61 per cent of the total area grant-aided. In addition approvals for private future planting had risen to 28,406 hectares compared to only 13,789 in 1991–2. Of the 24,352 hectares of grant-aided planting in 1993–4, 15,897 were for new planting, and 8,455 for restocking. The net result was that 821,838 hectares were in grant schemes in 1994, with 553,538 in the Woodland Grant Schemes, 198,931 in Basis I and II Dedication Schemes, 47,461 in Basis III Dedication Schemes, and 21,908 in Woodland Grant Schemes associated with the Farm Woodland Premium Scheme. The Forest Enterprise controlled 1,099,530 hectares, of which 873,804 were deemed to be forest land, and 225,726 other land. In 1993–4 a further 14,950 hectares were disposed of, gaining receipts of £17.4 million. The Forest Enterprise planted 9,289 hectares, but most of this, 7,904 hectares, was restocking. Finally, the Report records the provision of over 1,500 recreational facilities, including 741 forest walks and nature trails, 629 picnic places, 174 forest cabins and holiday houses and 143 cycle trails.

D: ANALYSIS

Traditional meta-narratives based on empirical observation

The simplest and yet most important analysis is to chart the rate of planting and clearance to assess the impact on the land use of Great Britain. This can be obtained from a number of sources, notably the annual reports of the Forestry Commission, infrequent censuses by the Commission in 1947, 1965 and 1980, and from the annual agricultural census carried out by MAFF. Figure 4.2 shows a compilation of how these data have been used in different ways. In more detail, Gilg (1978) (Figure 4.2a) has shown how planting grew steadily till 1950, but really took off in the 1950s and 1960s, and how coniferous planting was virtually totally dominant throughout the period. In addition planting was dominated by the Forestry Commission.

Sinclair (1992) (Figures 4.2b–c) has argued that the official data contain a number of inconsistencies and definitional changes over the years, but that a reasonable estimate can nonetheless be produced by a process of collation and adjustment. He shows for England only how all afforestation rose rapidly from 1945 to 1950, then remained at a very high level till the 1970s, when a combination of the 1972 cost-benefit study and the introduction of Capital Transfer Tax in 1975 badly hit planting till confidence returned with the taxation and grant changes of the late 1980s. In more detail, Sinclair shows how bare land afforestation was dominated by the Commission in the 1950s, but that since then its planting has been in terminal decline. In contrast private planting has experienced two booms, between the mid 1950s and mid 1970s, and in the late 1980s, when record rates of private planting were experienced.

Sinclair's data refer to England only. However, Tompkins (1993) has provided an analysis divided up between Scotland, England and Wales. This shows that planting of new conifers fell dramatically after the changes in fiscal and grant policy in 1988, but that planting of broadleaves grew rapidly, albeit to reach only a third of the conifer rate. The overall rate of

Figure 4.2 Various portrayals of the rate of afforestation and clearance in the twentieth century

Source: Gilg, 1978; Sinclair, 1992; Tompkins, 1993; and Forestry Commission, 1994

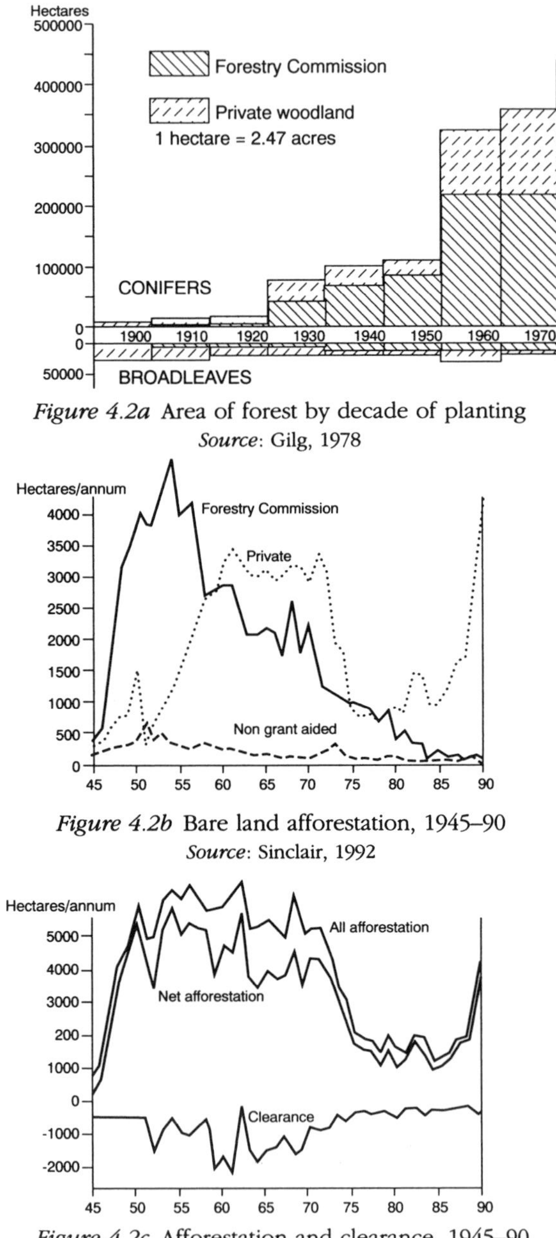

Figure 4.2a Area of forest by decade of planting

Source: Gilg, 1978

Figure 4.2b Bare land afforestation, 1945–90

Source: Sinclair, 1992

Figure 4.2c Afforestation and clearance, 1945–90

Source: Sinclair, 1992

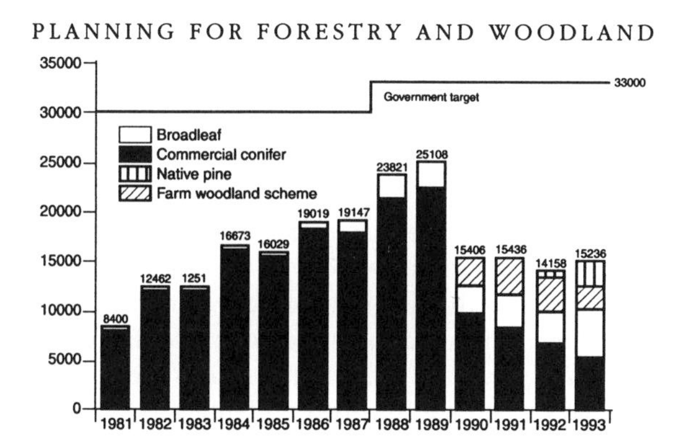

Figure 4.2d Hectares of new private planting in the United Kingdom
Source: Tompkins, 1993

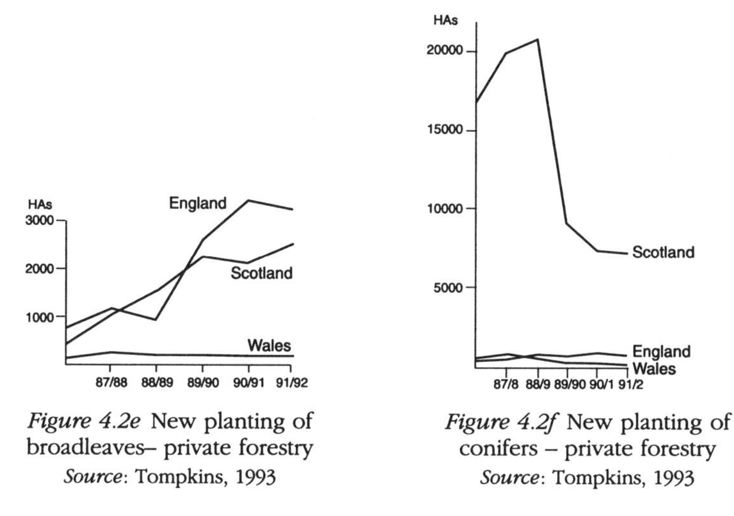

Figure 4.2e New planting of broadleaves– private forestry
Source: Tompkins, 1993

Figure 4.2f New planting of conifers – private forestry
Source: Tompkins, 1993

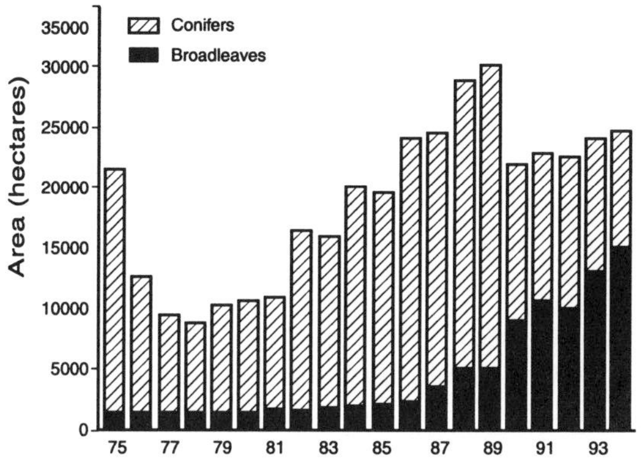

Figure 4.2g Area of grant-aided new planting, restocking and natural regeneration
Source: Forestry Commission, 1994

planting thus fell after 1989 and its composition changed markedly to become diverse instead of dominated by commercial conifer planting, as shown in Figure 4.2d. At no time, however, has the Government's target of 33,000 hectares a year been seriously threatened. Figures 4.2e and 4.2f show that Scotland continues to dominate planting, except in the case of broadleaves, where England has become the more important country. Finally, Figure 4.2g shows the rise in broadleaves in the 1990s. These data demonstrate two things. First, that planting rates have been very sensitive to financial signals, and second, that current rates do not appear to be providing a sufficient incentive for private planting.

Most planting, in spite of recent changes, has occurred in Scotland and Mather and Thomson (1995) have shown that the forest area in Scotland increased by 40 per cent between 1975 and 1990. However, they found that the effect on farming had been small, except in those areas where the forest cover had increased to 30 per cent or more. The effects of forestry have been offset by using existing farmland better, and because much afforestation took place on effectively unused land. The relationship between afforestation and farming was thus found not to be linear, but Mather and Thomson also concluded that the past could not be used to predict the future since policies had changed and the pattern of land use was also different.

The net result as far as the composition of Forestry Commission forests is concerned is the distribution shown in Table 4.1. This shows that most of the forests are less than 30 years old and thus that the production profile will rise from less than 2 million cubic metres in 1970 to over 6 million cubic metres in 2010 as these forests reach their maximum discounted revenue at around 50 years old (National Audit Office, 1993). In addition, Table 4.1 shows that the forests are dominated by alien coniferous species, with only Scots pine being native.

This sub-section has thus shown that the first three of the main changes outlined in the summary box on page 121, at the end of the section on evolutionary events, are correct, namely: a shift from the public to the private sector: a shift from the uplands to the lowlands; and a shift from commercial (coniferous) forestry to amenity (broadleaved) woodland. Empirical observations can throw little light on the remaining three changes, and so attention is now directed towards different analytical perspectives.

Structural analyses focused on political economy and regulation theory

Economic analyses. These have concentrated on the profitability of forestry. This is almost impossible to assess because of the long time gap between planting and harvesting, which can vary from at least 40 years to several hundred years. It is thus very difficult to compare costs across decades,

Table 4.1 The age structure and species mix of Forestry Commission forests

Age structure *Year planted*	*Area (hectares)*	*Species mix* *Species*	*Area (hectares)*
1981–92	139,110	Sitka Spruce	402,510
1971–80	170,023	Lodgepole Pine	97,232
1961–70	178,715	Scots Pine	77,479
1951–60	181,682	Japanese Larch	44,610
1941–50	61,223	Norway Spruce	43,174
1931–40	33,611	Corsican Pine	34,698
1921–30	16,244	Douglas Fir	22,044
pre 1920	8,320	Other conifers	33,349
		Broadleaves	33,842
Other areas	100,912	Other species	100,912
Total	889,850	Total	889,850

Source: National Audit Office, 1993

especially bearing in mind the high rates of inflation experienced several times this century. Nonetheless, some attempts have been made. For example, the National Audit Office (1986) have concluded that: 'There does not seem to be any financial case either in the private or public sector for supporting forestry on a) rural employment grounds; b) balance of payments grounds; c) strategic considerations; and d) non-monetary benefits.' In their commentary on this conclusion the Committee of Public Accounts of the House of Commons (1987) was very critical of the lack of criteria with which to measure forestry performance and concluded that the 'economic case for forestry is difficult to prove/disprove because of its long term nature'. Similar observations were made by the House of Commons Environment Committee (1993) and by the National Audit Office (1993), which argued that the problem was being exacerbated because trees were being felled too soon or too late.

Analyses by Macmillan (1993) of plantings in Scotland have found that 38 per cent had failed to achieve a 6 per cent rate of return and could be regarded as uneconomic, and that, depending on the measure used, between 12 and 48 per cent of existing forests could be deemed to be uneconomic. Turning to the narrower issue of whether forestry destroys or creates jobs when it replaces hill farming, Johnson and Price (1987) from a study in Wales found that in the short term forestry employment did not compensate for lost farm work. However, they also used an economic model to forecast that after 40 years (the period for a full rotation to harvest) forestry could provide more jobs if there were no large productivity increases. In contrast in Scotland, Mather and Murray (1987) found no significant difference between the employment created by hill farming or forestry but accepted that forestry employment might rise in the long term.

Political economy analyses. A number of authors have alluded to the political power of the forestry lobby, notably Wilson (1987), and Brotherton (1987), and Brotherton and Devall (1988) have conducted a series of studies on the bargaining processes employed during afforestation consultations which has demonstrated their political adroitness in reconciling the issues and keeping the threat of planning controls at bay. In addition, the political lobby at Westminster, particularly in the House of Lords, has been able to use to its advantage, potentially damaging legislation, notably the Wildlife and Countryside Act 1981, by inserting very favourable compensation payments when forestry grant aid is refused, as for example demonstrated in the case of Sites of Special Scientific Interest discussed in Chapter 6.

Crabtree and Appleton (1992), in an analysis of the extra costs imposed by certain forestry policies on both the Government and farmers, has shown that the Farm Woodland Scheme could have cost the average farmer £24 per hectare in lost income compared to the situation if farming had continued and could have cost the Government £58 per hectare if the scheme had not been set up. Crabtree and Chalmers (1994) have also found that farmers entering the scheme could on average suffer net income losses of £1,114 per year in England, and £208 per year in Scotland.

In a wider analysis Mather (1994) has argued that the Forestry Commission has failed to adapt to the changing world, suffers from having three different ministerial masters, has been reactive rather than proactive, and has no real conception of what forestry should be doing for the country. In particular he concurs with the House of Commons Environment Committee's (1993) complaint about the lack of a national forest strategy, and criticises the Government's defence that 'constructive tension' between different agencies involved in agriculture, forestry, conservation and recreation could ever be assumed or expected to achieve a satisfactory balance of interests. For example, 'constructive tension' failed to achieve a balance in the dispute over afforestation in the Flow Country of northern Scotland discussed earlier in this chapter. Mather concludes that the reforms of the 1990s will not achieve the radical change in institutional structures that are needed. In particular he doubts if single-sector, Britain-wide institutions are appropriate for the 1990s, especially in light of the continued reluctance of either the Government or the Forestry Commission to produce a national land-use or forestry strategy, in spite of repeated calls for such a strategy and the possible integration of not just forestry policy, but all rural policy into a combined ministry of environment, agriculture and rural affairs as advocated by the Environment Committee of the House of Commons (1993) and its chairman (Jones, 1994).

Human agency, behaviouralist and post-modern perspectives

Several studies so far in this chapter have referred to the questionable economics of forestry, not only for the Government, but also for those planting trees. This is particularly so for farmers in the lowlands, who have been specifically targeted since the mid 1980s, since they do not have the economies of scale of large upland farmers or the fiscal advantages of large forestry or financial firms like insurance companies looking for long-term returns. Little attention has, however, been paid to how farmers perceive forestry. The few studies that have been carried out have shown a strong dislike of the concept. For example, Potter and Gasson (1988), in their study of the incentives that farmers would need to be paid to transfer land out of agriculture, found not only that forestry needed the highest compensation payment, but that many farmers would not consider afforestation at any price. Similarly, Bishop (1992) has found that though poor economic viability was the main deterrent to forestry, a significant number of farmers also believed that farmers were not foresters, and did not wish to make any land available because of the restrictions this would impose on their profitability. Finally, Watkins (1994) found that farmers cited the following reasons in descending order of percentages: inadequate financial incentives (87); long pay-back horizon (47); loss of flexibility (40); public access concerns (33); and uncertainty (20). Although the main reason deterring farmers from planting trees is the poor return, the loss of flexibility, especially through the imposition of felling controls, has also been found by Lloyd *et al.* (1995) to be an important factor. Therefore Lloyd *et al.* suggest that abolishing felling controls on new woodlands may increase planting rates by reducing one of the arguments used by farmers against planting. Nonetheless, behaviourally it would seem that farmers are willing to embrace NAL-style forestry only and do not perceive themselves as farmer-foresters in the European mould.

Other forms of analysis

Because forestry is such a dramatic land use change it has attracted a good deal of general analysis from certain groups. For example, many recreational and environmental groups, notably the Ramblers Association (1980) and Tompkins (1986) writing on behalf of the Ramblers, have criticised twentieth-century forestry for: causing catastrophic damage to wildlife; creating monocultures; fundamentally altering rainfall and run-off patterns and exacerbating the effects of acid rain by raising acid levels in soils; creating ugly monolithic and geometric landscapes; restricting recreational access by creating dense woodland and removing open moorland; and by obliterating archaeological sites. In their defence the forestry industry has gradually become more sensitive to all these issues and has attempted to plant forests in more sensitive ways, but at the end of the day, afforestation will always represent a change from one habitat to another and will thus cause a good deal of heartache.

In spite of the criticism of the Ramblers Association the forest estate is not only being seen to be, but is increasingly being used as a major recreational resource. For example, Willis and Garrod (1992) using contingent valuation methods have calculated that the recreational value of the Forestry Commission estate could be £8, 10 or 50 million depending on the assumptions used. If the latter figure is correct, then the annual subsidy given to forestry of between £60 and £95 million is nearly justified on recreational grounds. Indeed, forestry can only probably ever be justified on wider holistic grounds, and so this chapter concludes with a brief discussion about how forestry policy might evolve after a brief evaluation of the second three themes outlined in the summary box on page 121.

First, the powerful forestry lobby has prevented development control over forestry by its influence at Westminster, notably in the House of Lords. Second, the unfortunate effects of other policies continue to be important, notably taxation and the general belief amomg farmers that forestry is not for them while farming is still supported so well. And third, the division of the Forestry Commission has exacerbated the sectoral tensions in rural planning as a whole, just when the Commission was emerging as a potential model for a multi-purpose countryside planning agency.

E: SOME THOUGHTS ABOUT THE FUTURE

Mather (1991b) has argued that a 'forest transition' has been occurring in many developed countries, in which shrinking forests are replaced by expanding forests. Mather argues that this follows classic models of resource management which follow a sequence of depletion followed by conservation and then restoration, although Grainger (1995) has argued for a delay to be inserted between when deforestation ends and reafforestation begins. Most European countries have now completed the model and have reached a steady state, averaging 25 per cent of land area. If Britain is to follow the model, its forest area, having doubled once this century from 5 to 10 per cent, will have to double again in the next, and this may well be forced on Britain by international forestry conditions (Dudley, 1994).

Whatever happens future forestry projects will be subjected to the holistic appraisal depicted in Figure 4.3. A key feature in such an appraisal will be the levels of supply and demand for timber. In the United Kingdom, forecasts of this equation show a demand of 8.5 million cubic metres in 2010 and a supply of 6.3 million tonnes, thus leaving a gap of 25 per cent to be filled by imports (National Audit Office, 1993). This equation, however, ignores the demand for timber for cellulose and paper products and Britain will continue to depend for over 80 per cent of its timber supplies from abroad.

Accordingly a number of forecasters (Gilg, 1991a) have called for an increase in afforestation rates. For example, Fargrieve (1979) on the grounds of favourable environmental conditions, import saving, job creation and

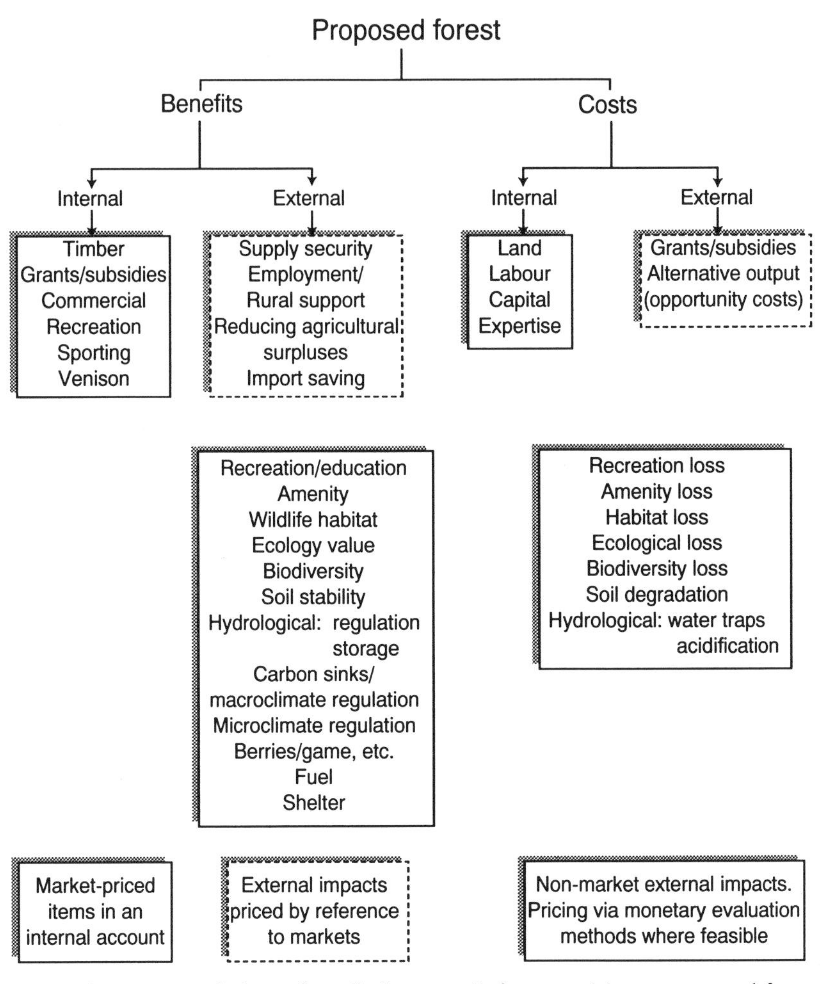

Figure 4.3 An extended cost-benefit framework for appraising a proposed forest

Source: Bateman, 1992

amenity has argued for a growth of 1.8 million hectares from 2.1 to 3.9 million hectares. Norton-Taylor (1982) has produced a map of areas suitable for afforestation which would result in all of upland Britain being covered in trees. Other forecasters have considered a range of options, for example Miller (1981). In more detail, the Centre for Agricultural Strategy (1980) has produced four options for 2025 ranging from an extra 1.96 million hectares to an extra of only 0.58 million hectares leading to self sufficiency ranging from 50 to 19 per cent. The Forestry Commission (1978) has produced three options ranging from no more planting to an extra 1.8 million hectares between 1980 and 2025. These forecasts have been criticised by Stewart (1985), Miller (1981) and Price (1987) and it would seem that no consensus can ever be achieved.

Another key factor in any appraisal will be how to build inflation rates into the equation via 'discount rates' and the rate of return that may be deemed to be acceptable. Kula (1988) has argued that there are three dominant factors in an economic analysis of forestry projects: the choice of discount rate; the choice of project-evaluation method; and the future price of timber. Under traditional methods of discounting, forestry comes out badly, and so Kula has attempted to build in future benefits instead of taking current values to show that forestry may make a profit. Price (1989) claims that the best method involves shadow pricing under cost-benefit analysis, but in the end argues that such methods are little better than instinct since they reach beyond the objectives of forestry and into the far future and thus the purpose of existence itself.

In conclusion, the range of issues raised by Figure 4.3 means that forestry proposals must inevitably be judged on gut feeling, albeit based on the best possible evidence. In the sense that forestry is the natural climax vegetation of the British Isles, and indeed Europe, it would seem to be a 'good thing' to reafforest with native trees as fast as possible. It is therefore possible to agree wholeheartedly with the Countryside Commission's (1993a) conclusion:

> In 1987 we concluded our policy statement by saying that the nation faced a new and exciting opportunity to achieve multi-purpose forestry. Some of the obstacles present then have since been remoced, and new tools exist to counter old problems. There are more reasons than ever for managing existing woods and planting new woods for the future; the plans and designs exist for some of the most adventurous projects of forest creation ever seen in Britain. We must all now seize the opportunity to put life back into our trees and woods, and to put trees and woods back into our lives.
>
> (27)

Policy evaluation checklist derived from Chapter 1.

1 *Were the problems understood and which value system dominated?* Yes, Britain is very short of trees, thus posing severe economic and ecological problems. The value system was however one of maximum economic production promulgated by large landowners.

2 *Which goals were to be achieved and were the policy objectives clearly spelt out?* The overriding goal was to increase production,

but as the cost-effectiveness of this was questioned from the 1950s more and more goals have been added and so the once-clear policy objectives have become confused.

3 *Were the policy responses accretionary, pragmatic and incremental?* Not until recently, although the remit of the Forestry Commission has been continually widened. Grant aid in the private sector has, however, been characterised by a mix of schemes with constantly changing parameters since the 1980s.
4 *Which powers in the Gilg/Selman spectrum were mainly used?* Forestry has been the one area of countryside planning to use public land ownership as a major tool, although this was abandoned during the 1980s. Grant aid has been the other main weapon, with regulation being conspicuously absent.
5 *Were planners given sufficient powers/resources and sufficient flexibility for implementation?* In terms of public forestry yes. In the private sector grant aid has either been inadeqate in the farm woods sector or overgenerous for high tax payers. The lack of development control is a major weakness.
6 *Were performance indicators built into the policies?* Target rates of planting were set early on, for example in 1943, and target rates of return have been set in the postwar era. However, the failure to reach these targets has not stopped the policy, only led to different justifications being sought.
7 *Have the policies been underresearched?* Yes and no. The economics of forestry and its social and environmental impact have been researched in detail. However, the way in which the policy has managed to survive so many attacks has been underesearched, as have the attitudes of farmers.
8 *Have the policies been effective overall?* They have achieved the limited aim of reafforestation remarkably well, but at a continuing high cost to the taxpayer, and Britain is still massively dependent on timber imports. There is still much to learn about how to reafforest a country.
9 *Have there been uneven policy impacts, by group or by area?* Yes, Scotland has borne the brunt of the policy, and the lowlands have hardly felt any impact until the 1990s. Large landowners and high taxpayers have been the main beneficiaries, with few gains for workers or small farmers.
10 *Have the policies been cost-effective and are they environmentally sustainable?* No, the rate of return has been low; however such long-term economics are prone to huge errors. Afforestation in single age/species blocks has led to many environmental problems and fears of unsustainability.
11 *What have been the side-effects?* The soil and water chemistry has

been radically changed with changes in aquatic life. Habitats have also changed and many species have been replaced with different ones. The landscape has changed and many ramblers dislike the change from open to confined space.

12 *Were the policies modified by unforeseen events/policy changes elsewhere?* Yes, notably the end of conventional warfare which reduced the strategic argument, and also the changing impact of taxation policy affecting forestry as a tax dodge. Nonetheless, the policy has a strong internal momentum.

13 *Were the policies overtly sectoral and were they compatible with other policy areas?* Initially, the policies were very sectoral and increasingly conflicted with other areas as the forested area and the trees grew. In the 1980s genuine attempts to be multi-purpose were made, but the 1990s has been more sectoral.

14 *What have been the effects of 16 years of Tory rule?* The decision to start selling off the forest estate, followed by the fiasco of tax breaks in the Flow Country, and then by the decision to make the Forest Enterprise a Next Steps Agency, have undermined the overview integrating role of the Commission.

15 *Should the policies be modified?* Yes, but only in certain respects since reafforestation is good.

16 *What alternatives are there?* The first alternative is to reinvent the Commission as a major purchaser of land, but with an environmental goal of recreating Britain's native vegetation in all the uplands and in certain lowlands. The Commission would thus become an environmental agency combining ecological land management with social aims. The second alternative would be to continue the unsatisfactory mix of commercial and amenity planting, but to make it more rationally integrated.

17 *Overall verdict.* At the simple level of planting, the policy has been a success and the forest area has doubled. However, the policy has been costly, has not been cost-effective, has not reduced the trade deficit, and has had bad side-effects on the landscape, wildlife, and soil and water chemistry.

5

PLANNING FOR THE BUILT ENVIRONMENT (TOWN AND COUNTRY PLANNING)

KEY CONCEPTS AND ISSUES

Concepts

The key concept in planning for the built environment is that planners try to guide and shape development by positive guidance backed up by negative control. Planners thus use two of the mechanisms set out in the Gilg/Selman spectrum in Chapter 1. First, the positive tool of exhortation and advice as represented by land-use plans and policy statements which set out where planners would like development to take place. Second, the negative tool of regulatory controls as represented by development control which gives planners the power to accept, modify or refuse applications to develop or change the use of land. The right of private interests to develop land as they see fit has therefore been conceded in the public interest.

The negative power of development control is backed up by a second concept that a refusal or imposition of conditions in the public interest does not lead to any compensation being made for losses or profits foregone by private interests. This is quite different from the situation outlined in Chapters 2, 3 and 4, in which farmers and foresters have by and large been exempted from development control over their land-use operations, but, where they have had to concede some control over their operations in the public interest, for example in Sites of Special Scientific Interest, they have been paid very generous compensation.

A third concept is that the guiding and shaping approach to planning is widely accepted by all political parties, and even the Conservatives in their most radical period in the early 1980s argued:

> The planning system balances the protection of the natural and the built environment with the pressures of economic and social change. The need for the planning system is unquestioned and its workings have brought great and lasting benefits.
>
> (Environment, 1980) (1)

This means that planning for the built environment has remained remarkably unchanged in its key concepts since it was first set up in the 1947 Town and Country Planning Act.

Nonetheless, there have been many debates about the way in which planners should balance the competing pressures outlined above. One method favoured by right-wing parties is to take a reactive role and to arbitrate between developers both private and public and the rest of society in its various guises by acting as a neutral value-free umpire. The drawback to this approach is that if development is refused the planner is seen as the enemy of economic growth, job creation, and the freedom of individuals. If, however, development is granted, planners lose the rationale of their existence and may be seen as mere facilitators of development. A more fundamental drawback is that planners may not be able to act as value-free and all-knowledgeable individuals.

One solution to these drawbacks is to be more proactive and to attempt to resolve conflicts by bargaining and imposing conditions, or if possible to prevent them by forecasting demand and allocating land by zoning. One problem with this approach is that planners lose their value-free status and become involved whether they like it or not.

Whichever approach is adopted the planner's role is likely to continue to be, as it has been since the 1950s, as a quasi-legal arbitrator guiding and shaping the development/protection of the environment. This consensus does not mean, however, that there is not a considerable debate about the issues involved in executing this role.

Issues

What type of planning system? In terms of plan making the big issues include how many tiers there should be and how they should be linked. Ideally one can conceive of a four-tier system: National – Regional – Sub-Regional – and Local. Many people would argue, however, that four tiers are wasteful.

Related issues include the relationship between plans, which tier should have precedence, and whether plans should be created from the top down or the bottom up. First principles would suggest that plans should be top down, but only to provide a very general framework within which local plans could work out detailed plans suitable for their area, under the principles of subsidiarity and diversity within unity.

In terms of development control there are three big issues. First, should decisions be made on the merits of each case with reference only to the plan or should decisions be made almost entirely on the land-use zones set out by the plan? Second, how much control over design and/or socio-economic matters should be allowed in addition to the key decision, based on the environmental issues involved? Third, which developments or land-use changes should be covered and at what size of development should exceptions to control come in? Again, first principles suggest that development control needs to be set within a firm framework, but one that gives discretion when circumstances diverge from the norm.

What sort of environment? The big issue here is how much development should be planned for in an equation which attempts to balance the need for homes and jobs and their related developments with the case for conserving the countryside. The key argument centres around whether urban expansion should be focused as traditionally in and around towns or whether radically new models need to continue to be developed as in the New Towns of the 1950s and 1960s or the new villages pioneered in the 1970s and advocated for the future by the Town and Country Planning Association (1990). Traditionally, a certain amount of expansion has been allowed but only in existing villages, and then only in certain selected villages, and the key ethos has been one of preserving farmland and the countryside from urban growth.

Within the limited amount of development planned for, issues have focused on which villages should be selected for expansion, how many and what type of houses should be constructed, and whether there should be any attempt to favour local people or local building styles in these decisions. Other issues have focused on ideal patterns based on the economies of scale both internally and externally in terms of service provision, and attempts to provide so-called balanced communities by age, class or ethnicity.

Underlying all these issues are an enormous desire for individuals to live in the countyside, an economic imperative from employers to seek green-field sites near motorways and other transport links, a general trend for all developments to be serviced by motor transport, and a general raising of space standards so that even a static population needs more space. For example, household sizes have fallen constantly, thus requiring far more houses, for the same popluation, thus increasing pressure to build in the countryside. In an unplanned Britain, the possibility of a spread city based on telecommunications, and already found in many parts of the United States would soon become a reality. The challenge for planners is to balance the forces driving spread-city formation with the desires of those who wish to conserve the countryside and the delights of urbanity.

Who decides? Planners are the only group, however, involved in decision making, albeit a group central to giving advice about key decisions about land allocations for development. At the macro level, according to Adams (1994) three broad groups interact – the state; the finance industry; and the construction industry – to set the broad parameters of legislation and policy, while at the micro level, according to Marsden *et al.* (1993) a number of actors, for example farmers, financial institutions, builders, planners, councillors and residents, interact to make decisions about rural land-use changes. Collectively, the key actors in the process are developers, regulators, facilitators, and affected parties (Glasson *et al.* (1994). In theory this is a pluralist exercise in which every participant has an equal say, but in practice it is apparent that certain groups have become dominant. In particular, it is believed that farmers and middle-class groups have formed an unholy if unofficial alliance to maintain preservationist policies: farmers to maintain the base of their livelihood, and to keep development land prices artificially high so that they can cash in whenever any limited development is allowed, and middle-class groups so that they can according to Murdoch and Marsden (1994) keep the countryside exclusive and also cash in on rising house prices caused by the supply of new houses being artificially restrained by planning policies. In contrast, developers and private sector employers are desperate to develop on the same land, notably the countryside within 50 miles of major towns, and are keen to change policies towards a less restrictive attitude. The resulting conflict between the development wing of the Conservative Party (the paymasters) and the pragmatic wing of the Conservative Party (the voters who actually elect them) is particularly virulent in the south-east, where any wavering in the anti-development stance could lose crucial seats, and thus power.

It is clear that planning is not the neutral, value-free, technical exercise that it may appear, but is, in fact, dominated by the socio-economic and cultural milieu in which it operates. It is thus important to appreciate the different theoretical viewpoints that have attempted to conceptualise this milieu. Rydin (1993) has provided a useful attempt which is based on an evolutionary account. In the first stage, planning evolved through an 'environmental deterministic' view that many of society's ills derived from poor environments. The first planning systems were thus rooted in a system of 'naive public administration'. This developed into the method of planning based on 'procedural planning theory' which has dominated the practice of land-use planning in the United Kingdom since 1947 and has inspired most texts to take an 'institutional approach' to analysing it. In contrast, many theoreticians both on the right and left of British politics have argued that planning should evolve into a different form, not dominated by the classical tradition of plan formulation leading to plan implementation and plan revision as a technical procedure within a pluralist 'liberal political economy' framework, but should instead take on explicit aims favouring certain groups or issues. These issues will be returned to again in Section E, after the actual practice of planning for the built environment has been outlined and discussed in Sections B, C and D.

B: EVOLUTIONARY EVENTS

Events leading up to the Town and Country Planning Act 1947

The roots of planning (Cherry, 1979) lie in the public health and design professions and in concerns about the squalid conditions of the Victorian industrial city. These led to the Public Health Act 1848, which laid down minimum conditions for housing. Fifty years was to pass before the Housing Town Planning Act 1909 allowed voluntary town planning schemes to be drawn up. Real progress was made with the Housing Town Planning Act 1919 making it compulsory for towns above 20,000 to draw up town-planning schemes, which were linked to the parallel development of council housing and suburbia inspired by the New Towns movement and its driving organisation the Town and Country Planning Association. Further progress was made in the Town and Country Planning Act 1932, which extended planning to everywhere, but under this Act far too much land was zoned for development because of the fear that a refusal of planning permision would lead to the payment of compensation.

The 1930s also saw a good deal of concern about the loss of farmland and scenic areas like coastline to the rapid suburbanisation of the period. Surveys of these changes, notably by Dudley Stamp's Land Use Survey and the Council for the Preservation (later Protection) of Rural England (CPRE), and a growing belief in the professional skills of the young planning profession represented by the youthful Town Planning Institute (later the Royal Town Planning Institute (RTPI)) founded in 1914 encouraged politicians to call for real controls over land-use change.

These calls became more strident during the Second World War, which ushered in socialist thinking, the idea of centralised planning and three crucial reports. First, the Barlow Report (Cmd 6153, 1940), which advocated postwar industrial relocation based on the stick of planning controls in prosperous congested areas and the carrot of incentives to relocate in depressed areas or in planned new towns. Second, the Scott Report (Cmd 6378, 1942), which emphasised the need to protect farmland from urban growth in order for Britain to become less reliant on food imports with the related recommendation that agriculture and forestry should not be fettered by planning controls. Third, the Uthwatt Report (Cmd 6386, 1942), which argued that planning could only work if a refusal of planning permission did not lead to compensation.

All of these recommendations found their way into a series of postwar Acts, including the Town and Country Planning Act 1947, but unfortunately although they were passed by the most socialist government of the century, which created the welfare state, this Labour Government was so busy transforming Britain that it diluted its environmental legislation into several Acts over several years which in time led to certain problems of coordination as the postwar period unfolded. The logical solution of environmental problems was thus confounded at the outset by political expediency and to some extent a misreading of the problems and their possible solutions caused by the atypical conditions of the Second World War and the preceding conditions of the 1930s with their mixture of agricultural depression and explosive suburbanisation. As Cherry (1979), in his excellent summary of nineteenth century previous planning history, has so succinctly put it: 'Planning legislation is the story of the incremental adoption of measures imperfectly conceived in respect of problems only partially understood' (316).

In order to assess if Cherry's conclusion remains valid for this century it is best to divide planning up into its two disparate activities, plan making and development, and then examine some of the related policies that have an effect on the built environment before coming to some sort of preliminary finding in advance of the more detailed analysis in Section E.

Plan making

Plan making has experienced two major active periods between 1947 and 1995, with a third confused period emerging in the mid 1990s. The first period lasted till 1968 but it lingered on throughout the 1970s as it was replaced and the effects of its policies continued to be felt well into the 1980s. In this first period plan making was based on the classical Geddesian model of Survey–Plan Preparation–Approval–Implementation–Monitoring–Review, and also the idea of the local authority as developer, as conceived in socialist postwar Britain. Hence the term 'development plans'. These plans were intended to be end-state long terms showing in detail how land would be acquired and developed.

This was soon undermined by the change in 1951 to a Conservative Government committed to 'freeing' the people from socialism and replacing public development with private development. This meant that plans which were to have been the development plans for local authority development suddenly became 'guides for development control', a role to which they were totally unsuited. In addition massive changes in the 1950s, notably rapid economic and population growth, a swing away from socialism and a new academic emphasis in planning on socio-economic rather than physical planning, meant that by the 1960s the plan-making system set up in 1947 had become totally unsuitable.

Two influential reports in the 1960s led to a second system of plan making being set up in 1968 and 1974. First, the report of the Planning Advisory Group (1965) concluded that there would have to be a radical change in the form and content of development plans and thus proposed: a) a different type of plan based on a systems view of the world; b) statements of policy only at a 'structure' level; c) local plans for detail, to guide development control, and to form a basis for positive environmental planning; and d) more local control over plan making. Second, the Skeffington Report (Committee on Public Participation in Planning, 1969) advocated more public participation throughout the process.

The main proposals of the Planning Advisory Group were contained in the Town and Country Planning Act 1968 which rewrote the plan-making sections of the 1947 Act, and lip-service was paid to the participation proposals of Skeffington in the 1971 Town and Country Planning Act which consolidated the changes made in the 1968 Act. However, implementation of the 1868 and 1971 Acts, was severely hindered by changes to local government made in the 1972 and 1974 Local Government Acts. These Acts, instead of following the proposals for unitary city regions advocated by Redcliffe-Maud (Committee on the Management of Local Government, 1967) and accepted by the Labour Government (1964–70), were rejected overnight by the Conservative Government after their victory in 1970 and replaced by a hastily thought-out ideological reform based on two-tier authorities. The difficulty with these authorities, which were set up in 1974, was that the metropolitan areas were too small, the concept of counties was out of date, and the district boundaries were absurd. In Scotland, however, Regional Councils were set up in 1974 to produce Regional Reports and Structure Plans and provided a model of what could have been in England and Wales.

The new system was introduced area by area throughout the 1970s and in the early 1980s. Advice as to how to prepare plans under it have been contained in three documents: first, in the 1970s by the Ministry of Housing and Local Government's 'Manual on Form and Content' (Housing and Local Government, 1970); second, in the 1980s by the Department of the Environment's (1984a) Circular 22/84; and third in the 1990s by the Department of the Environment's (1992b and 1992c) Planning Policy Guidance Note 12 and Good Practice Guide. In essence two types of plan have to be prepared. A Structure Plan by county councils which sets the overall framework, and a Local Plan by district councils which sets out detailed policies within the context set by the Structure Plan.

Although there have been changes to the detailed process of plan preparation since the early 1970s the core features of the process remained the same till the mid 1990s. Key features of the process include: survey and review; initial consultation; pre-deposit publicity and statutory consultation; a public inquiry stage; a report on the inquiry; and the adoption process following the inquiry stage. The process is thus a lengthy one and can cause conflict rather than cooperation between authorities, notably because structure plans are prepared by county councils but are implemented by district councils via development control. The plans should include proposals for housing and employment land, transportation links, land to be protected from development, and proposals for other land uses such as mineral extraction and recreation.

In the late 1980s proposals for reform were made (Cm 569, 1989) which recommended: 1) the abolition of Structure Plans and their replacement with short statements of policy; and 2) that district councils and National Parks should prepare district-wide development plans. This was followed by the Planning and Compensation Act 1991 which: 1) retained Structure Plans but freed them from DOE approval in line with Local Plans which had been freed in the 1980s; and decreed that 2) district-wide Local Plans outside Metropolitan areas would have to be prepared by 1996; 3) Minerals Plans and Waste plans would have to be prepared by counties by 1996; 4) National Park plans would have to be prepared by National Park Authorities for both land use and minerals by 1996; and 5) plans would have more say in development control decisions as from 1992 (for more detail see the development control subsection below). A number of other measures also speeded up the process.

Further changes were introduced by a chaotic, piecemeal and apparently democratically driven reform of local goverment in England due to be implemented in 1997 which was expected to consist of 15 two-tier counties operating the former system, and the rest operating a system based on unitary authorities. Elsewhere, an imposed but consistent reform of local government in Scotland (Cm 2267, 1993) and Wales (Cm 2155, 1993) introduced unitary authorities, which were due to prepare Development Plans for their areas as from April 1996. In addition, under European Union legislation and the Rio 1992 agreement on sustainable development (see Chapter 6 for more detail), land-use plans in the 1990s have to carry out an Environmental Appraisal (Environment, 1993) of their plans preferably in the form of a Strategic Environmental Assessment.

Development control

Some initial points. The first and most important point is that development control has stood the test of time and its basic form has remained virtually unaltered since 1947, although there have been considerable changes in detailed practice. Indeed, there has only been one major government study of the system, the Dobry report of 1975 (Environment, 1975), which concluded along with the Government (Environment, 1976a) that the system was sound, although there could be problems depending on how it was operated. Second, it is the sharp end or coal face of planning, rather than the sexy glamorous area of plan making and it has thus been neglected and overlooked. Third, it is, however, where the impact of planning is felt and can be measured. The fate of millions of pounds and hundreds of jobs are decided every day, let alone the fate of both local neighbourhoods and much-loved environments. Fourth, like birth control its effects are hard to assess because of what has not happened, and also because behaviour has been changed because of its existence, in parallel with say sexual behaviour in the light of AIDS.

What is development control? The power of development control lies in the ability to permit, refuse or impose conditions on a planning application for the development of land. But what is development? The planning Acts have defined it as 'The carrying out of building, engineering, mining or other operations, in on or over the land, or the making of any material change in the use of any building or land.'

There are two distinct types of development: first, physical, when land is physically altered or buildings are constructed or altered; and second, a change of use when buildings may alter little except perhaps internally but the use may change markedly. The Acts are deliberately vague so that the Government can fine tune the definition according to the needs of different land uses, areas, and periods of economic growth. Physical development is defined in various General Development Orders (GDO) and advice about interpretation is given in circulars and Planning Policy Guidance notes (PPGs). The main exemptions to development control are given for agriculture and forestry.

The GDO has altered quite a lot over the years and is no longer uniform across the country. For example, the agricultural exemption and the favoured position of MAFF as a participant consultant altered quite a bit between 1986 and 1992. In 1988 the need to consult MAFF over applications including farmland was reduced to only those involving a loss of 20 hectares or more of prime (grades 1, 2, 3a) land. In 1986 in the National Parks and some surrounding areas, Local Planning Authorities (LPAs) were given the power to impose siting and design controls over farm and forest buildings – but not the principle of construction, and in 1992 this power was extended to all areas. Also in 1988 the need for full permission was imposed on farm buildings on holdings below five hectares.

Spatial differences date from 1981 when the limits set in 1977 were retained in National Parks, AONBs and Conservation Areas, but in other areas control was relaxed, notably by increasing the percentage enlargement of either living or work space that could be constructed without planning permission being needed. In the 1980s Enterprise Zones and Simplified Planning Zones were introduced in which development control can be either virtually withdrawn or much relaxed, but it was not intended that there would be many such zones in truly rural areas.

Assuming that planning permission is needed an application is made on a form to the Local Planning Authority. Before 1974 this was the county council, but since 1974 district councils have dealt with all planning applications except for those applications dealing with mineral matters, county matters raising issues of county-wide relevance, and those that straddle National Park boundaries. In National Parks applications are dealt with by the Park Committee.

Applications can be for outline or detail. The advantage of outline for developers is that they can test the waters without much expenditure on detailed plans, while for planners the stakes are low and there can be a genuine negotiation. The disadvantages for planners are that once the principle is conceded only the details of the application can be discussed, while for developers going for outline then detail slows down the process. Whichever route is chosen a number of other people/organisations are consulted either as a statutory obligation or as a matter of courtesy.

An application may also have to include an Environmental Assessment (EA). This became mandatory for certain applications, for example, oil refineries, and discretionary for other applications in 1988 under EC Directive 85/337. Advice as to when and how to ask for an EA is contained in DOE Circular 15/88 (Environment, 1988a). An EA must examine the direct and indirect effects of a project on: a) human beings, flora and fauna; b) soil, air, water, climate and the landscape; c) the interaction between a and b; and d) material assets and the cultural heritage, so that the project's environmental effects and the scope for modifying them may be properly evaluated (Glasson *et al.*, 1994). In the 1990s the Government has been under pressure to extend the scope of EAs to leisure developments like golf courses and to major changes to farmland by land drainage or afforestation, and in 1995 permitted development projects were made subject to an EA if there were likely to be significant environmental effects.

The planning officer then makes a recommendation to the planning committee after dialogues with all concerned. The planning officer's advice and the committee's decision are based on three background factors: 1) Government advice on how to interpret the legislation; 2) planning case law; and 3) planning case lore based on peer-group contacts and textbook knowledge. In particular advice has been given in DOE circulars 9/76, 22/80 and 16/87 and Planning Policy Guidance notes 1, 3 and 7 (Environment, 1976a, 1980, 1987a, 1992d, 1992e, and 1992f) as shown in the accompanying summary box.

Summary of advice contained in DOE circulars and Planning Policy Guidance notes on development control

Circular 9/76: 'Permission should be granted unless there is a sound and clear-cut reason for refusal. Onus is therefore on the local planning authority to show that development is unacceptable rather than on the applicant to show that it is.'
Circular 22/80: The Government's concern for positive attitudes and efficiency in development control does not mean that their commitment to conservation is any way weakened: in particular they remain committed to the need to conserve and improve the countryside, natural habitats, and areas of architectural, natural,

historical or scientific interest and listed buildings. There is no change in the policies on National Parks, Areas of Outstanding Natural Beauty or conservation areas. The Government continues to attach great importance to the use of green belts to contain the sprawl of built-up areas and to safeguard the neighbouring countryside from encroachment and there must continue to be a general presumption against any development within them. Nor will the Government allow more than the essential minimum of agricultural land to be diverted to development, nor land of a higher agricultural quality to be taken where land of a lower quality could reasonably be used instead.'
Circular 16/87: Introduced a new need to foster rural diversification and thus allow some farm diversification developments in light of surpluses. Another change was altering the justification for control to a continuing need to protect the countryside for its own sake rather than primarily for the productive value of the land.
PPG7 1988: Reiterated the need to protect the countryside for its own sake, but added the qualification: 'The priority now is to promote diversification of the rural economy so as to provide wide and varied employment opportunities . . . including those formerly employed in agriculture.'
PPG1 1992: Section 26 of the Planning and Compensation Act 1991 introduced a presumption in favour of development proposals which are in accordance with the development plan unless material considerations dictate otherwise.
PPG3 1992: put increased emphasis on re-using urban land as a means of relieving pressure on the countryside. The three main aims of countryside policy were stated to be:

1 to encourage economic activity;
2 to conserve and improve the landscape; and
3 to conserve the diversity of wildlife.

PPG7 1992: Building in the open countryside should be strictly controlled but elsewhere developments which benefit the rural economy are acceptable. The 1995 'Rural White Paper' Cm 3016, 1995) proposed a new rural business class and re-using rural buildings for business rather than residential purposes.

Decisions by the planning committee had to have regard to the provisions of the Development Plan before 1991, but since then as the summary box shows there has been a presumption in favour of development proposals which are in accordance with the Development Plan. Nonetheless discretion is still allowed, unlike zoning systems practised in almost every other country. Each application is thus effectively treated on its own merits. As the summary box also shows the onus is on the LPA to show that an application should not be allowed, using the principle of innocent until guilty. The decision will also need to take into account a checklist of factors including: a) the effect of the proposal on road safety, amenity and public services; b) possible conflict with other public proposals; c) the need to protect natural resources; and d) the public interest, that is, an unacceptable change in character of area, but not the loss of, say, a view or livelihood by individuals. Conversely deserving cases cannot be excepted from control, for example, the need to build an extension to house a dying granny.

Three types of decision can be made: approval; approval with conditions; and refusal. In more detail, few approvals are made without conditions, and even then there is a time limit of five years. Approvals with conditions are the most common decision. Between 1972 and 1992 conditions were often imposed under Section 52 of the 1971 Act, as long as they met the DOE test (Environment, 1985) of being: necessary; relevant to planning; relevant to the development to be permitted; enforceable; precise; and reasonable. For example, policies or conditions which attempt to restrict housing to 'locals only' were deemed inadmissable under planning law or *ultra vires* in 1981 (Clark, 1982) when the DOE in their decision letter ruled that the Lake District Structure Plan could not include such a policy to be implemented by development control, since: 'planning is concerned with the manner of the use of land not the identity or merits of the occupiers.' In 1991 conditions were replaced with 'obligations' under Section 12 of the Planning and Compensation Act 1991 and advice as to how to agree them was included in Circular 16/91 (Environment, 1991). Increasingly since 1979 and the advent of Thatcherism, planners have employed the motto of 'getting things done' and so most planners now seek to collaborate with developers in partnership schemes and planning departments have often been renamed Property Development Departments.

The fear has been expressed that this relationship may be too cosy, and even lead to corruption, or at the best, the 'sale' of planning permission in return for some public gain, for example, public recreational facilities provided by an out-of-town superstore. Indeed 'planning gain', the assumption that planning permission will only be given in return for some trade-off, is now a widespread, if not yet universal, feature of development control work (Bunnell, 1995). Other fears concern the extra costs imposed by landscaping and design conditions on housing in attractive landscapes, thus further excluding less wealthy people. In addition there is a widespread abuse of conditions restricting occupancy of a house to a farmer or farm worker (Gilg and Kelly, 1996c).

A refusal of permission is the least common, largely because of the deterrent effect of the planning system. Thus people never even think of applying or are deterred at the stage when they make an infomal inquiry.

However, once an application is made planners are reluctant to say no because of the mass of advice which encourages them to say yes and the fear of not only fury from refused applicants, but also from councillors seeking taxes and jobs from applications and thus 'brownie points' for their re-election campaigns.

If an application is refused, the imposition of conditions is felt to be unreasonable, or a decision has been too slow, an applicant can appeal to the Secretary of State for the Environment. Except for legal errors the decision of the Secretary of State or his/her agents is final. In the early 1990s most decision making was delegated to the Planning Inspectorate, an Executive Agency created at the same time from the corps of planning inspectors in the DOE. In 1995 the Agency employed 220 salaried inspectors and around 110 part-time inspectors (Cm 2807, 1995). The number of appeals rose rapidly in the 1980s, from 10 to 25 per cent, as applicants tested the system in the light of Thatcherite policies encouraging free enterprise, but in the 1990s the number of appeals fell back from 26,000 in 1990–1 to the long-term average of around 14,000 in 1993–4 (Cm 2807, 1995). Appeals can be made through three procedures: a Private Hearing (about 80 per cent of cases); a Public Inquiry (15 per cent); or Written Evidence (5 per cent). The evidence is collected by a Planning Inspector who writes a report and in combination with other colleagues will make a decision. In about 5 per cent of cases the decison will go to politicians, and in very contentious cases right to the Cabinet.

There is also a power of a 'call-in' by the Secretary of State over controversial national matters, large developments, or a substantial departure from development plans. Examples have included major airport proposals and coalfield developments in rural areas, notably Stansted airport and the Belvoir coalfield. There is also the possibility of Parliamentary Bills, either private or public, for example, the Channel Tunnel Rail Link through rural Kent.

Planning controls are backed up by a set of enforcement and special controls which include: Enforcement Notices; Stop Notices; and the power to revoke or modify a permission before development has started; and a Discontinuance Order, which can rescind a permission after a building has been put up. The first two powers provide the sanction of fines, or even prison sentences, and the ultimate sanction of demolition, while the last three involve the payment of compensation. All of these powers were tightened up in the Planning and Compensation Act 1991 to strengthen the real teeth that development control in general has. In addition to the general powers, planners also have a set of specific powers as follows.

Green Belts. These are derived from Circular 42/55 (Housing and Local Government, 1955) which gave them three roles: 1) to restrict the sprawl of large built-up areas; 2) to prevent the merging of adjacent towns; and 3) to preserve the setting of a town. In the 1980s an attempt was made to downgrade the importance of green belts but after much pressure DOE Circular 14/84 (Environment, 1984b) actually added two new roles: 4) to safeguard the surrounding countryside from further encroachment; and 5) to assist in urban regeneration. These roles were confirmed by PPG2 in both 1988 and 1992 (Environment, 1988b and 1992g). The basic implication is that permision will be much harder to obtain in areas designated as Green Belt, and indeed policy was further tightened when PPG2 was revised in 1995, to add a further objective, 6) providing access to the open countryside for the urban population (Environment, 1995a).

Green Belts derive their legality from approval in Development Plans and between 1979 and 1986 the area covered by green belts grew from 1.75 million hectares to 4.49 million hectares (*Journal of Planning and Environment Law*, 1988, 682–3) so that virtually every major town or city was ringed by a belt. They have, however, been accused of creating attractive but artificial countryside based on institutionalised use (such as hospitals, universities and schools) or leisure use (golf, horticulture and country parks), or encouraging hobby farming (Elson, 1986). Conversely many real farmers are driven out by vandalism or are seduced by high land prices to let land go.

Listed buildings and conservation areas derive from diverse legislation. The original 1947 Act only referred to listed buildings. The Civic Amenities Act 1967 introduced the concept of conservation areas to protect the wider environment of individual buildings, and this concept was strengthened in the Town and Country Amenities Act 1984. In 1990 listed buildings and conservation areas were given a separate Act, the Planning (Listed Buildings Conservation) Act 1990. Mynors (1995) has provided comprehensive information on how the systems work and the Department of the Environment (1994a) has provided advice on how the legislation should be implemented.

The core principle is the duty to compile a list of buildings of special architectural or historic interest. These may not necessarily be pretty or even old and listing even postwar buildings can now be considered. In common with green belts, the number of listed buildings grew rapidly in the 1980s. Between 1947 and 1992 listing was carried out first by the Ministry of Housing and Local Government, then by its sucessor the DOE. Since 1992 listing has been done by the Department of National Heritage (Cm 2811, 1995) and in 1994 they had some 500,000 listed buildings on their books, and 8,000 conservation areas. Controls over the fate of buildings however, remained in the hands of the DOE or LPAs, depending on the nature of the issue. This division of responsibility between the listing and controlling authorities could be seen as a fatal flaw.

Control over listed buildings takes the form of Listed Building Consent.

This means that a listed building or any building in a conservation area cannot be altered without consent. A non-listed building can be given six months protection via a Building Preservation Order. Churches were exempt from control till 1994 when some denominations were brought under control, but not the major Christian churches who have internal systems of control acceptable to the Government. This is a significant factor given the dominance of churches in the rural landscape, and the need to find alternative uses as congregations dwindle. However loopholes are provided by provisions which allow for the demolition or alteration of unsafe buildings which can arise cynically by neglect or even by accidents which occur in the middle of Sunday night! The fines for abusing the system are relatively small given the profits that can be made from redevelopment.

The key to preservation is in fact via positive conservation. This means finding a sustainable use for listed buildings and conservation areas via imaginative planning policies and grant aid. In addition, a flexible attitude to the conversion of old buildings, for example barn conversions, can provide an economically profitable use. This can, however, lead to charges of middle-class gentrification (Larkham, 1992) and once the initial renovations and conversions have been allowed, to over-zealous protection rather than organic growth, thus preventing the eclectic growth which created the charm in most villages.

Tree preservation and Article IV Direction Orders. Tree Preservation Orders prohibit the cutting down, lopping or wilful destruction of a tree, groups of trees or woodland. If trees have to be cut down a replanting condition can be imposed. However, like listed buildings, the damage is often wilful and replanting can probably never replace the trees lost. A more insiduous problem is neglect or allowing cattle to graze thus preventing regeneration (Watkins, 1984). Article IV Direction Orders give planners the power to bring normally exempt operations under the GDO, into planning control. Examples include so called leisure plots with gimcrack huts in fields or painting cottages in unsuitable colours.

Some related policy areas

Although rural areas are attractive places to live, and have thus attracted affluent migrants and more new jobs than urban areas, there are several social problems that planning needs to address. In particular, affordable housing is in short supply, as a result of in-migrants bidding up prices, and restrictive planning policies limiting the supply of new housing below the demand. In addition, service provision is poor, due to the scattered and low density of the population, and the widespread use of cars. In more detail, a Rural Development Commission survey (Cm 3016, 1995) found that in 1994 71 per cent of parishes did not have a daily bus service, 52 per cent did not have a school, 43 per cent did not have a post office, and 41 per cent did not have a shop.

Housing policy. In addition to the policy guidance for new housing set out in various DOE Circulars and PPGs, a number of other policies have an impact on rural housing development. The Rent Agriculture Act 1976 gave security of tenure to tenant farmers and their families and imposed a duty on local councils to rehouse farm workers displaced from tied cottages. This narrow piece of legislation, favouring rural housing for rural people, was however swamped by the Housing Act 1980 and subsequent amendments. The main thrust of this legislation was designed to sell off council houses to sitting tenants at substantial discounts depending on their length of tenure. Safeguards to prevent resale in protected scenic areas except to local people were included, but for most rural areas this legislation allowed rapid gentrification of villages in the 1980s (Phillips and Williams, 1983). In addition to the existing stock being sold off, new council housing building came to a halt in the 1980s under Thatcherism. This was partly offset by an 'exceptions' policy introduced in 1989, under which local planning authorities can give permission for affordable housing for local people, which would not normally be allowed (Cm 3016, 1995) and by the work of housing associations under the Housing Corporations Rural Housing Programme. But in the early 1990s only about 10,000 housing units (Cm 3016, 1995; and Gilg, 1995) were built compared to Clark and Dunmore's (1990) estimate that between 120,000 and 180,000 social housing units needed to be built between 1990 and 1995, if the depradations of the 1980s were to be made good. The 'Rural White Paper' did, however, forecast an increase in aid of about a quarter (Cm 3016, 1995), and ruled that houses in all communities below 3,000 people would not be eligible for the proposed Purchase Grant Scheme, which would allow housing association tenants to buy their homes. Even with these recent changes the policies have hastened the transition of rural areas to middle-class enclaves of privilege dedicated to anti-development attitudes, and have thus enhanced the lobby in favour of planning policies restricting houses in the countryside.

Transport policy. The Transport Acts 1980 and 1985 had the theme of replacing subsidised public transport with flexible less regulated free-market 'operators who could make a profit in a deregulated environment. Rural areas, however, suffer from an extreme version of the public transport curse of uneven demand during the day and the week. Those unable to drive a car or unable to afford one have either left the country areas or suffer extreme deprivation. These policies have again favoured the transition to a middle-class culture in rural areas and related policies geared to their specific wants rather than the needs of all.

Employment. The wartime Barlow report introduced the concept of regional policies designed to help disadvantaged areas, including poor usually upland or remote rural areas, and since then a combination of sticks (restrictive development control) and carrots (flexible development control and grant aid) have been employed. In the 1980s, however, these policies were much reduced under the Thatcherite revolution in public policy. Fortunately, they have been replaced or supplemented by the continuation of existing policies or the development of new policies.

For example, the European Social Fund and the European Regional Development Fund (Structural Funds) introduced grant aid specifically targetted at economic revival in disadvantaged rural areas, defined as Objective 5b regions, in the 1990s or at any area suffering acute economic difficulty under the Objective 1 programme. In addition the EU Leader programme offers funds for innovatory local schemes. British schemes include the Single Regeneration Budget which local authorities bid for, and Regional Selective Assistance targeted at areas facing rapid economic change, like Cornwall (Cm 3016, 1995). Much more long-standing aid has been provided by the Rural Development Commission (before 1987 the Development Commission) since the turn of the century in England. In 1994–5 the Commission employed 331 staff and spent £42 million (Cm 2807, 1995). They have targeted nearly 90 per cent of their financial aid at Rural Development Areas in the 1980s and 1990s, which in 1995 covered 35 per cent of the area and 6 per cent of the population, and was due to attract some £11 million in support for businesses. Though these areas have varied they have centred on the south-west, eastern England and the north. At the national level the Commission provides information and advice and encourages action by local communities by various pump-priming schemes, including 'Rural Challenge' which offers six prizes of £1 million each annually.

In Wales, the Development Board for Rural Wales and its predecessors have provided similar aid, and in 1994–5 the Board planned to spend £27.4 million (Cm 2815, 1995). In Scotland, Highlands and Islands Enterprise (between 1967 and the early 1990s the Highlands and Islands Development Board) has used a wide variety of policies including: stimulating business development; encouraging training and learning; fostering social development; and enhancing the environment, with a grant-in-aid of £58 million in 1994–5 (Cm 2814, 1995), to lift the economy so successfully that parts of the Highlands are outperforming south-east England, albeit against a background of low incomes and poor infrastructure.

Much has been written about teleworking in the 1900s, and notably, the telecottage, but so far the rhetoric is greater than the reality according to Richardson *et al.* (1995) but its potential not to create *new work*, but to move work to a *new setting* should not be ignored, in that it removes one of the obstacles to the growth of a possible spread city, namely, the high environmental cost of travelling to work in such a dispersed environment.

Mineral extraction. Although mineral extraction is normally covered by development control, a number of separate issues are specific to mineral extraction. This has been recognised in a number of ways, for example, the publication of a separate series of Mineral Planning Guidance notes (MPGs) and by the Planning and Compensation Act 1991 which introduced Minerals Plans at the county or National Park level. The main issues involve the restoration of quarries and other sites after extraction ends, the length of time that such extraction lasts, the conditions imposed on operations notably with regard to dust and noise, and the problem of long-standing permissions that do not meet modern standards. Many of these issues were addressed by the Town and Country Planning (Minerals) Act 1981. Since then aftercare, especially restoration to amenity wetlands or woodlands, or even productive farmland, has improved. However, proposals for new extraction, notably for aggregates in the south-east have become so controversial, as clearly demonstrated by Murdoch and Marsden (1994), that developers have had to seek permission to exploit so-called super quarries in Scotland, specifically for a mile wide and a third of a mile deep quarry in the Hebrides, with severe environmental and social implications (Johnson, 1994). Finally, the Environment Act 1995 brought in new powers to review and update mineral permissions so that modern standards could be continuously applied, initially for permissions dating from between 1948 and 1992, and then on a 15-year rolling programme (Cm 3106, 1995).

Water issues. The two editions of this book have both been written in severe summer droughts (1975 and 1995), but between the two editions water management has been transformed from a cosy monopoly dominated by local and central government quangoes to a fiercely political and controversial battleground (Cmnd 9734, 1986). The change came when the Water Act 1989 set in motion the privatisation of water supply to private water companies. Regulation of water resources remained with the state in the guise of a newly created National Rivers Authority (NRA). Although theoretically this removed the judge-and-jury cartel of the previous system, the poacher/gamekeeper relationship set up by the 1989 Act has been full of difficulties, notably because the NRA has been starved of resources to carry out its duties. In 1996 the NRA was due to be subsumed into the new Environment Agency discussed in Chapter 6.

More problematically, the water companies took control over major water bodies with existing and potential recreational use on the surrounding land covering some 180,000 hectares. Although 110,000 hectares of this land in protected or designated sites cannot be developed without covenants or management agreements, the remaining 70,000 hectares can be developed for residential or leisure uses subject only to normal planning control. Since the provision of water has always been a most powerful factor in developing the built environment and in deciding which areas should be favoured with expansionary policies in Development Plans, the new water companies are in a very strong position to negotiate planning permissions in return for water supplies delivered. There is thus a severe danger that one group of companies could dictate rural planning policies implicitly and secretly without any overt explicit evidence of their influence.

Pressure groups. The clash between development and preservation is mirrored in the pressure groups interested in planning. The conservation lobby is represented most notably by the Council for the Protection of Rural England, while the development lobby are represented by the House Builders Federation and other employer organisations like the British Roads Federation. A different sort of pressure group are those groups trying to look after social issues, notably the Rural Community Councils, and their umbrella organisation Action for Communities in Rural Areas (ACRE). Conventional wisdom has it that planners use pressure groups when they can help them, but otherwise largely ignore their representations, albeit politely.

Summary. Healey and Shaw (1994) have provided an excellent summary of how the planning process may thus be represented or seem to be, and which is a perfect yardstick by which to measure the various more critical assessments, including Healey and Shaw, that are contained in Sections D and E:

> The institutional arrangements and policy tools of the planning system appear, therefore, to have the flexiblity to accommodate the task of regulating the use and development of land in such a way as to achieve the objectives of sustainable development. The key challenge of this task is to work out how demands for development and resource exploitation can be combined with the environmental objectives of handing on to future generations an environmental inheritance equivalent to that received by the present. This task has both material considerations – the balancing of one set of objectives against another – and moral dimensions in terms of value placed on environmental qualities and relations with the natural environment. Thus the planning system appears to provide a flexible *regulatory regime*; a set of formal mechanisms and practices built around them through which the tensions between the potentially conflicting objectives of economic development, meeting social demands and needs, and conserving and enhancing environmental quality may be managed. Its traditions are dominated by an administrative-legal discourse within which judgements combining knowledge and value are typically intertwined.
>
> (426)

C: THE SITUATION IN 1995

Unlike the areas covered in the previous three chapters the annual report Command papers to Parliament are less use in town and country planning since the lead department, the DOE, delegates most powers to local authorities. In more detail, Regional Planning Guidance, in at least draft form, had been published for all the English regions (Council for the Protection of Rural England, 1995). Many county councils appeared to have been reprieved in 1995 after the reform exercise carried out by the Local Government Commission, so that by 1997 it was expected that there would be a chaotic mixture of two-tier counties and unitary county/districts, leading to the worst bureaucratic nightmare imaginable for all those involved in planning. In Scotland and Wales unitary authorities were imposed as from 1996. National Parks were due to be reformed under the provisions of the Environment Act 1995 (for details see Chapter 6). Most Structure Plans had reached the first or even second review stage, after undergoing several revisions in the interim. In contrast the introduction of area-wide Local Plans was still in progress in 1995 and 54 per cent of non-metropolitan districts and 28 per cent of National Parks had an adopted Local Plan, with 91 and 72 per cent expected to have an adopted plan by 1996 (Cm 2807, 1995). Likewise Minerals and Waste Plans were in their introductory stage with 54 per cent having a Minerals Plan and 18 per cent a Waste Plan in 1995 (Cm 2807, 1995).

The DOE has been the lead department for planning issues since it was created in the early 1970s. In 1995 (Cm 2807, 1995) it had 10 main responsibilities which included: planning; local government; the countryside and wildlife; and environmental protection. The last two of these are considered in Chapter 6. The DOE had eight priority aims in 1995 which included: promoting sustainable development; guiding the development, use and reclamation of land in a way which gives appropriate weight to economic, environmental and social factors; and improving the quality of life for people who live and work in rural areas. These policies were delivered through spending programmes, mainly through sponsored bodies and local authorities rather than directly. A big problem here is that LPAs are often of a different political hue from that of central Government. For example, after the 1995 local elections the Conservative Party only controlled four districts. Policy aims were also achieved through market-based instruments, through cooperation with other departments and bodies or through regulation and monitoring.

Turning to planning specifically, the general policy of the DOE in 1995 was to define the boundaries of planning control, by providing the legal and policy framework for the land-use planning system which they argued regulates the development and use of land in the interests of economy, efficiency and amenity. The DOE's aims and objectives were: to reconcile development which is necessary to provide homes, investment and jobs with conserving and enhancing the built and natural environment; and to operate the system effciently, at the lowest cost to central and local taxpayers, and to those proposing development, that is compatible with achieving these aims. The means of delivery was achieved largely through LPAs responsible for the preparation of Development Plans and the development control process, with the DOE being responsible for:

- the legislative framework;
- monitoring and evaluating the performance of the planning system;
- issuing national planning policy guidance and regional guidance for local authorities and other users of the system;
- commissioning research;
- determining appeals against LPA decisions (this work is largely undertaken by the Planning Inspectorate Agency); and
- the determination of planning applications called in for decision by the Secretary of State.

Responsibility for the preservation of the built environment rests with the Department of National Heritage (Cm 2811, 1995). One of its six aims was to preserve sites, monuments and historic buildings and to increase their accessibility for study and enjoyment, not just because these are aims worth pursuing in their own terms, for enjoyment, relaxation and stimulation, but also because they contribute to the cultural and economic achievements of the nation. These aims were achieved by a combination of legislation, policy-making and executive action. However, the core policy was maximising private-sector involvement by keeping historic buildings in active use. English Heritage, a grant-in-aid body funded by the Department, looks after buildings in the public sector and gives advice and grants concerning conservation to the private sector. Grants accounted for about a third of total expenditure by English Heritage of £116 million in 1993–4, of which about £100 million was provided by the Department.

It is of course easy to set the framework for planning, but delivery is another thing. Attention is thus now turned in Section D to an assessment of the effect of planning on the built environment of the countryside. This will concentrate on the two key themes of policy balancing the demands of economic growth with the preservation of the countryside.

D: ANALYSIS

Because planning is divided so starkly into its two key activities this section follows the same division and examines plan making briefly, before undertaking an in-depth analysis of development control, since this is where the effect of planning is seen and can be assessed.

Plan making

If we work down the hierarchy Regional Plannning Guidance is seen to have existed in two forms. First, the DOE-dominated Regional Planning Councils between 1964 and 1979 (for whom I once worked between 1969 and 1970) and second, the combination of regional groups of LPAs working with the DOE to produce DOE Regional Planning Guidance notes (RPGs) in the 1990s. These have provided useful frameworks but their impact has been limited by the fact that they are not rooted in any meaningful local administrative area and the Council for the Protection of Rural England (1994) has criticised them for their limited amount of public participation and environmental input. In the 1970s and 1980s Regional Reports in Scotland produced by the Regional Councils at the time showed what could be done, and not surprisingly the Labour and Liberal Democratic Parties have promised elected Regional Councils for the late 1990s. In the meantime the DOE has set up 10 regional centres in which other government activities like transport and employment planning are coordinated, but only by civil servants (Cm 2807, 1995). This exercise does not, however, satisfy Bruton and Nicholson's (1987a and 1987b) call for clear statements of national land-use policy *à la* Scottish planning guidelines so that the implications of the fundamental nature of the planning process – the management of change through the negotiation, agreement and promotion of policies, proposals and resources for change – could be fully recognised.

In Structure Plans, progress was at first slow and a complete coverage was not achieved till the early 1980s (Healey, 1983). Analyses of the first generation of plans found them to be a) too broad but also too detailed and contradictory; b) too physical with not enough attention to socio-economic matters; and c) found that public participation had taken too long, had been too expensive, and had had little effect. Thus Barrass (1979) concluded that the plans had taken too long to prepare and were out of date by the time they had been adopted. Bracken and Hume (1981) also made the serious criticism that forecasting had been undertaken with little regard to planning powers, resulting in a confusion of wishful thinking and muddled intervention which had not given clear guidance for private and public investment. Bruton and Nicholson (1985) criticised the lack of strategic planning at the regional level, and Bruton and Fisher (1980) also noted a lack of clear national guidance except in Scotland and so proposed a move to Regional and Structure Plans being prepared by unitary authorities.

Cloke and Shaw (1983) found that most Structure Plan policies contained restrictive policies, notably in the south-east. However, Elson (1981) argued that these restraint policies could not buck the forces for development, and that they would have to give way to economic pragmatism. Indeed, Cross and Bristow (1983) argued that structure planning had by the 1980s become an incremental process arbitrating between powerful interests over specific issues and had lost any pretence at being a long-term or comprehensive decision-taking process. Cloke and Shaw (1983) concurred and argued that policies of the LPAs had become intentions, desirable courses of action and a base line for negotiation.

Between the early 1980s and 1990s most Structure Plans entered a revision stage during which housing provisions were gradually scaled back in most counties, though a substantial number increased allocations (Coates, 1992). Some fierce battles were fought between local residents and housebuilders during this period, and even between Cabinet ministers, as a later section reveals. Academics, however, neglected Structure Plans in this period, since the findings concerning the first generation appeared to remain valid, namely that they tend to be out of date, and have little relevance to the real planning battle of day-to-day development control.

In contrast Local Plans, as one of their jobs, provide detailed guidance for development control, especially since the advent of so-called plan-led decisions following the Planning and Compensation Act 1991. Unfortunately, progress in Local Plan preparation was even slower than for Structure Plans, largely because they had to wait until a Structure Plan was in place, and also because many of the 400 or so districts set up in 1974 did not have the resources to prepare plans. Bruton (1983) estimated that the rate of progress between 1974 and 1982 would mean that Local Plan coverage would not be complete till well past 2001. By 1988, large parts of rural England away from the rural fringe were still without a Local Plan (Environment, 1988c). According to Bruton and Nicholson (1987c) 28 per cent of all local authorities were planless in 1985, but the situation was offset by the common use of non-statutory plans to avoid a complete policy vacuum for development control and other decisions (Bruton and Nicholson, 1983). This slow progress meant that throughout the 1980s and even into the 1990s (McClenaghan and Blatchford, 1993) LPAs either operated without a Local Plan, or used a pre-existing 1947-style plan written before 1974, or relied on the vague policies of the Structure Plan – hardly a formula for logical and consistent planning. In the 1990s, however, the short cut procedures introduced by the Planning and Compensation Act 1991 were, according to a survey of local authorities, leading to much quicker Local Plan preparation (Vickery, 1995). In addition, county councils have been producing Rural Strategies as a result of promotional work by the Rural Development Commision, the Countryside Commission and English Nature in an attempt to identify rural needs and respond with flexible and targetted policies.

Underwood (1982), from a survey of plans, concluded that there was a good deal of uncertainty about the role and function of Local Plans. Healey (1989) went further and argued that Local Plans were unable to provide an effective and comprehensive approach to the management of land-use change. This is largely because, as Bruton and Nicholson (1987a) pointed out, the broad aim of the planning process is to negotiate and agree, locally, policies, priorities and resources for change. In this process land-use plans offer a means of recording the results of local agreements and negotiations. In this respect the process may be more important than the product. In an era like the 1980s and 1990s when planning became the 'management of change/ getting things done', the day of the plan as a neutral technical document which ignores the struggle behind its preparation may have had its day.

However we need to go back to McLoughlin (1973) for real words of wisdom:

> Policies and plans of various kinds should, therefore, be seen as the outputs of a continous planning process. . . . The weakness of statutory physical planning is its inherited emphasis on the *plan* itself, rather than the process of strategic choice which produces it. Its overriding strength is that it forces the planning system to 'come clean', i.e. to be publicly committed about its intentions.
>
> (262)

The key point to emerge from this analysis is that plan making, although perceived as the creative side of planning, became largely discredited and ignored as a self-indulgent exercise in the pragmatic world of the 1980s and early 1990s and that attention should thus shift to the long-time Cinderella of planning, development control.

Development control

Traditional meta-narratives based on empirical observation

The main concept that has been tested by this approach is the link between policy and events on the ground as influenced by the development control process. In particular most work has focused on the efficacy of development control in preserving the countryside and guiding development to the locations selected by plans. A simple cause-and-effect hypothesis is thus set up, and this is then evaluated by the use of development control data.

However, Gilg and Kelly (1996a) have outlined seven main difficulties in using development control data for this purpose. First, policies are subject to change leading to a time-lag and overlapping of often contradictory policies. Second, policies can derive not just from different sections of the planning policy document arsenal, but also from different public or private bodies, for example, policies relating to infrastructure provision. Third, policies are only guidelines, even in the so-called plan-led 1990s, and can be interpreted by decision makers in many different ways. Councillors and planners are only human, with all the prejudices that this implies. Fourth, there are the 'birth control and displacement' effects already noted. Fifth, the lead-up to a decision may have been influenced more by some groups than others, and thus some decisions of most interest to influential groups will be different to those over which they have less interest. Sixth, there will be unforseen effects as the real world changes, and also because of the perverse reaction of some to ignore policy signals. Seventh, and finally, policy makers can only expect to achieve a certain success rate. For example, green belt designation does not mean that no development will be allowed. However, few such policies contain explicit statements of what sort of permission rate is admissable, preferring instead to hide behind global totals of housing provision. It is never stated what will happen if these allocations are too generous or too harsh, and that policy will change at a given rate of development. Nonetheless, the suspicion remains that subliminally development controllers will adjust their permission rates to some implicit target rate, like the apocryphal driving-test examiner.

In spite of these difficulties much good work has been done with the data and methodologies have been developed to circumvent them. For example, Brotherton (1992) has attempted to construct a model and general theory of planning control using the concept of application quality, defined as the extent to which applications conform to the material considerations that determine policy. Applications can then be categorised as weak or strong depending on the degree to which they conform to policy and, conversely, decisions can be categorised as weak or strong depending on the degree to which they conform to policy. Brotherton (1993a and 1993b) has further developed these ideas with reference to the interpretation of both planning appeals and refusals.

In a development of Brotherton's approach, but one based on a radically different methodology, Tewdwr-Jones (1995) has proposed a behaviourist/humanist approach in which 10 aspects of decision making at the local level could be evaluated in contrast to Brotherton's numerical analysis.

Statistical and cartographic analyses. The simplest thing to do with the data is, however, to subject them to classical empirical analysis, especially bearing in mind that over half of all houses have been constructed in the postwar period and have thus been processed by the system. Furthermore, the vast majority of houses built since 1980 have been constructed by the private sector under the so-called 'spec' house-building system in which builders build houses speculatively in the hope that someone will buy them.

The simplest analysis begins with the rate of refusal. In this regard about 10–20 per cent of applications are refused, of which around 10 to 25 per cent go to appeal, of which about 60–80 per cent are refused. A simple analysis of overall application rates, which average around 450,000 a year, demonstrates that applications rise with booms in the economy and fall with recessions, demonstrating the essentially speculative and unplanned nature of demand. The planning system does not seem to have influenced the erratic behaviour of the building industry.

It is easy but can be misleading to map planning applications and the decisions made. This is because published data do not weight planning applications, but as Curry and McNab (1986) have shown it is vital to do so, since one application may be for one house or 1,000 houses. The first study to weight data was by Gregory (1971), who used weighted 1957–66 data to show that green belts were being better protected than shown by official unweighted data, but that green belts were nonetheless being eroded: a) around existing clusters; b) by realignment of the boundary; and c) on one or two large sites.

Gregory's work inspired a series of studies in the 1970s and 1980s. For example, Curry and McNab (1986) were able to demonstrate very variable rates of demand across the country and a high rate of applications in the Cotswolds from people in the south-east of England. A similar effect was demonstrated by Brotherton (1982a), who showed that restraint areas, namely the National Parks, had most applications per head, while London had the least, thus demonstrating the latent desire to move out of big cities. The whole point of development control is of course to uphold policy and Pountney and Kingsbury (1983), in a study of whether decisions agreed with a plan or not, have shown that decisions in accordance with the plan are lowest in green belts and market towns, but nearly highest in rural areas, revealing perhaps a pragmatic reaction to severe development pressures in the semi-rural areas, but continued restraint in the open countryside.

Empirical studies have not, however, been confined to development control data and a number of studies have used a combination of data to study the impact of planning. A consistent theme has been the protection of farmland, and the control of urban sprawl as one of the main aims of postwar planning. This work was initiated by Coleman (1976), who used a comparison of data derived from the first land-use survey of the 1930s with data from the second land-use survey of the 1960s to argue that the rate of land loss had not slowed from the prewar average of around 25,000 hectares a year, and that planning was failing to maintain a clear division between town and country, leading to the creation of an urban fringe. In contrast Best (1977), using annual data from the MAFF census, argued that the rate had fallen significantly to around 15,000 hectares per year, and would continue to do so since much land had been lost in the 1960s due to the need to rehouse slum clearance areas at modern lower densities, and that in the future standard densities of around 30 houses per hectare, infilling, and the use of derelict land would aid the reduction in the rate of land loss.

The debate was rejoined in the 1990s when the CPRE disputed data from the DOE (Environment, 1995b) derived from annual returns from the Ordnance Survey which purported to show that the rate of land loss had in fact not only fallen, but had fallen to an amazing 5,000 hectares a year by 1989 (Environment, 1992f), partly because by 1992 47 per cent of new housing was being built on previously developed land compared with

38 per cent in 1985 (Cm 3016, 1995). Detailed work by Sinclair (1992) for the CPRE using a variety of data has, however, produced a consensus figure which shows a much greater loss. In more detail, the average loss was estimated to be 15,700 hectares per year between 1945 and 1990, with figures as high as 21,000 hectares per year in the 1960s, but as low as 11,000 hectares per year in the second half of the 1980s. However, this was still more than twice the DOE rate in the 1980s and between 1945 and 1989 Sinclair has estimated that total land losses were not the 525,000 hectares officially recorded, but instead 705,000 hectares. In addition Sinclair's data show a growing gap between reported agricultural census losses of land from agriculture, and unreported losses, derived from the total agricultural area. From these data Sinclair argued that DOE policies regarding a more lenient approach to land release were flawed, since they were based on complacent and unreliable data, and that if current trends continued, the urban area would increase from 15 to 25 per cent by 2050, or the equivalent of a town the size of Norwich each year.

Studies focusing on urban containment. The most classic study of containment remains that carried out by Hall *et al.* (1973) and Hall (1974). Their main thesis remains very valid in the 1990s, namely that decentralising forces in the population and the workplace are being opposed by containment policies. The Hall team came to three important conclusions. First, urban areas had been contained but private owners had decamped over green belts into villages or expanded smaller towns, with council tenants being rehoused in tower blocks. It was argued that this was a civilised British form of Apartheid with Tory shires being protected by planning from working-class incursions and Labour voters. Second, there had been a growing separation of work from residence. Third, there had been a severe inflation of land and property values and like the other trends this continued, so much so that land in the 1990s represented over 34 per cent of the cost of a house (Meikle, 1991). Hall *et al.* (1973) thus concluded:

> It certainly was not the intentions of the founders of planning that people should live cramped lives in houses destined for premature slumdom far from urban services or jobs, or that city dwellers should live in blank cliffs of flats far from the ground without access to play space for their children. Somewhere along the way a great deal was lost, a system distorted and the great mass of people betrayed.
>
> (433)

In a partial update from secondary data Hall (1988) pointed out four great changes which were causing profound changes in the geography of contemporary Britain: 1) zero population growth but a marked redistribution of population into the so called Golden Belt and the Golden Horn (see next paragraph for explanation), leading to a voracious demand for conversion of rural land to urban purposes; 2) a shift of land out of agriculture so that the paradox exists of substantial areas of wasteland near the hearts of the cities, while demand for land is buoyant at the peripheries; 3) a shift away from manufacturing to high-technology services; and 4) a demand for out-of-town warehousing/retailing. Moss (1978) in an earlier analysis has graphically portrayed the processes at work as shown in Figure 5.1 and though the detail has changed since then the same broad structural processes are still in operation, although there has been an attempt to divert growth from expanding and new towns to the dying inner area.

These trends can be verified by macro-studies based on the population censuses of 1971, 1981 and 1991.These show that rural areas experienced a population increase of 1.8 million between 1971 and 1991, and a percentage rise of 16.9 per cent compared to 3.9 per cent for England as a whole. Employment figures were also up by 13 per cent between 1981 and 1991, compared to a rise of only 1.2 per cent for England as a whole (Cm 3016, 1995). For example, as Figure 5.2a shows, population has been growing rapidly in the arc from south-west England to East Anglia (Hall's Golden Belt and Golden Horn), returning the population pattern slowly to its pre-Industrial Revolution pattern. Furthermore, rural areas have grown at the expense of old industrial areas and Breheny (1993) and Townsend (1993)

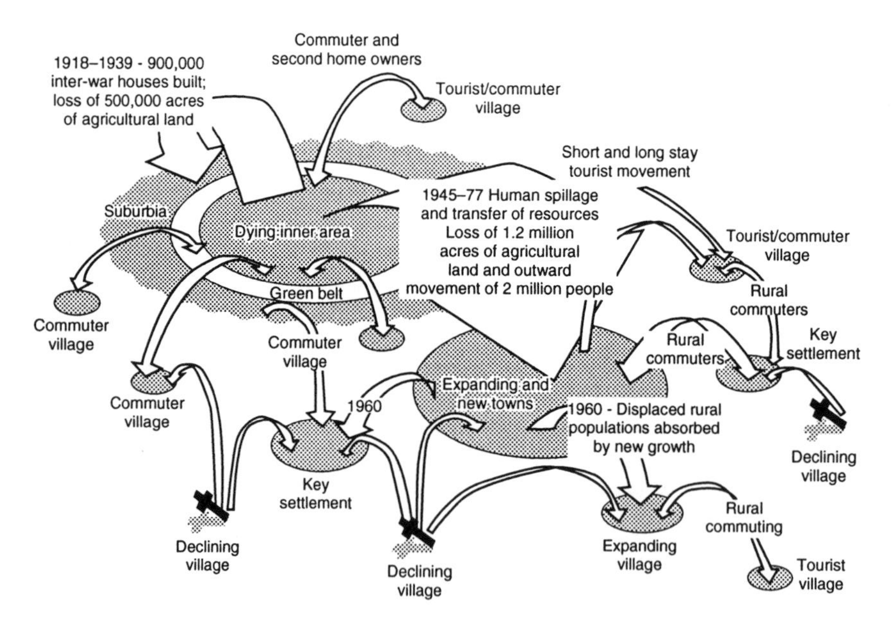

Figure 5.1 Processes of population and settlement change in the countryside around major cities

Source: Moss, 1978

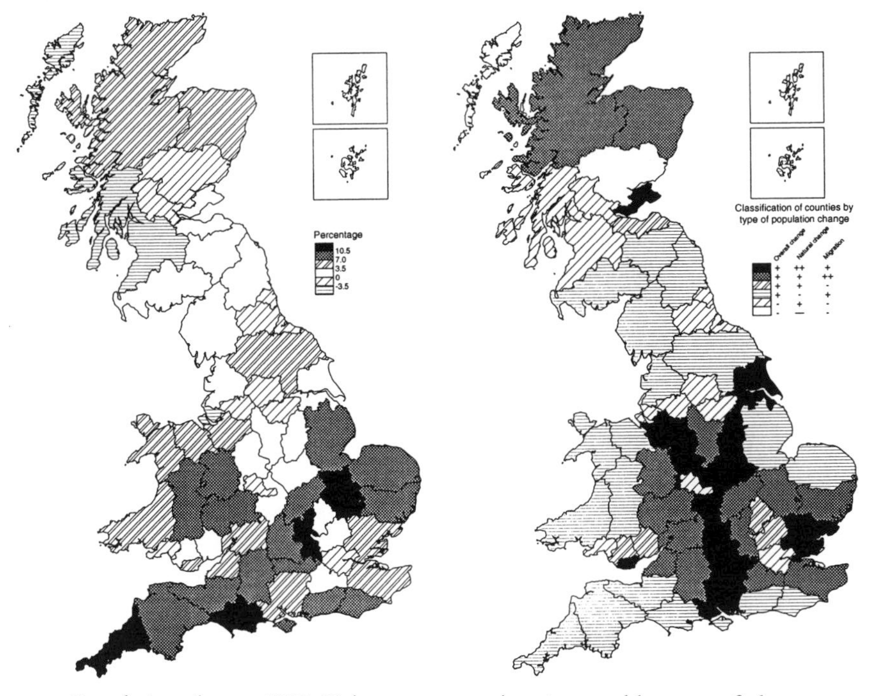

Population change 1981–91 by country and region, and by type of change

Figure 5.2a Population change, 1981–91, by county and Scottish region
Source: Champion, 1993

Figure 5.2b Type of population change in counties and Scottish regions, 1981–91
Source: Champion, 1993

have shown that these changes encompass employment as well. D. Owen (1995) has also shown how this is related to the development of a high-tech Britain represented by a 'Golden Crescent' curving between Oxford and Cambridge, with a young dynamic population which grew overall by 12 per cent between 1981 and 1991, via the population changes shown in Figure 5.2b. From these and other data (Champion *et al.*, 1995; Gordon and Forrest, 1995) it is clear that we are witnessing a fundamental shift in the structure of housing and employment patterns, as well as retailing and leisure, which planning has not been able to prevent, but only to ameliorate.

In more detail, the most intense pressure has, however, been felt in the south-east of England, where Munton (1983) has shown that much land has been lost in the metropolitan Green Belt, notably around Heathrow airport and in Essex. Keyes (1986) has also shown how a change of policy can have an effect. Between 1980 and 1984, 83 per cent of applications were refused inside the Green Belt, but when the policy was relaxed in 1982 the rate of refusal fell from 96 per cent in 1980–82, to only 55 per cent in 1982–84. Outside the Green Belt the refusal rate remained exactly the same at 25 per cent.

Such changes in policy allied to attempts by Mrs Thatcher to weaken Green Belt policy in the mid 1980s, encouraged a group of developers calling themselves Consortium Developments to test the waters with several planning applications for totally new settlements of around 10,000 people in or around the London Green Belt. The first such case, at Tillingham Hall in Essex on totally open arable land, was turned down by the Secretary of State in 1987 on Green Belt grounds. A related application to the south-west of London was also turned down, but one nearby was referred back to the developers, before a change in policy occasioned by the semi-greening of Mrs Thatcher and the appointment of a more conservation-minded Secretary of State in the late 1980s ended the hope of developers that they could breach the long-standing policy against major incursions into the Green Belt, but as Herington (1984) has shown major population growth has leapfrogged the Green Belt into the countryside beyond. In the mid 1990s the DOE gradually turned its face against the further exurbanisation of Britain on energy-saving grounds and several PPGs now advise against developments that are not self-contained, like commuting developments and out-of-town shopping centres.

The problem is that, as Towse (1988) has shown, the London Green Belt, and in particular the golden triangle cornered by Gatwick–Heathrow–Farnborough, contains 395,000 manufacturing jobs, 25 per cent of the south-east total and more than half as many as south-east Wales. Faced with severe restraint policies the response by firms has been to indulge in *in situ* development and to seek planning permission on former institutional land, like hospitals, by offering planning gain. This only offers a short-term solution to the century-old problem that the south-east contains the driving force of the British (now pan-European economy) but that it also by the same token contains the intelligentsia most dedicated to preserving their high-quality living environments. Not surprisingly, Elson (1993) has found that the Green Belts have achieved their aim of preventing development, but that some more development, especially if it were recreation based and improved the environment, could be allowed.

A related problem is high house prices caused by the restrictive planning policies, and a study for the Department of the Environment (1992h) estimated that restrictive planning policies may have increased house prices by as much as 40 per cent in the south-east. However, the study also estimated that there would have to be significant relaxations of policy for there to be any effect on house prices, which could lead to unacceptable losses of countryside. In any case, to a large extent the problem seemed to have been solved by 1995, when the effects of the over-touted negative equity of the mid 1990s and a general lack of confidence in employment prospects meant that house prices had fallen dramatically since 1990 and houses were more affordable in 1995 than at almost any time since the Second World War.

A related issue is the degree to which LPAs have allocated enough land for development in their plans. Planners have always argued that they have followed DOE advice in various circulars and PPGs which have called for between 5 and 10 years of building land to be available and that it is builders who are at fault if they do not make long-term plans to acquire land banks of developable land. They point to studies by, for example, the DOE (Environment, 1978) which show that about 20 per cent of permissions are not taken up, with 60 per cent of these owned by *non*-house-builders, thus implying that speculators are using building land to make a 'quick buck'. The DOE concluded that planning permissions did not give a good indication of either demand or land availability and that planning could not be blamed for any shortfall in building land. Hooper *et al.* (1988) in another study concluded that a majority of counties had enough land to last for more than five years based on 1982–6 completion rates. Nonetheless in the housing boom of the late 1980s many counties exceeded their Structure Plan provisions though this confirms the long-term suspicion that builders only find fault with planning when their own short-sightedness is to blame, and that planners are in fact encouraging them to take a longer-term perspective by indentifying land where they may be able to develop for at least 10 years into the future.

Studies focused on landscapes and villages. Studies of protected landscapes which are under severe pressure because of their high-quality environments have been carried out for National Parks and AONBs by Blacksell and Gilg (1981) and Anderson (1981) in addition to the work already discussed by Curry and McNab (1986) and Brotherton (1982). In general they conclude that the policies appear to have been successful, urban growth has been contained, and the built environment of the National Parks and AONBs has changed remarkably little. But visually the limited numbers of houses that have been allowed have either employed banal suburban styles, or have had pseudo-vernacular details superimposed, not only rendering such buildings architecturally eclectic, but also often putting their price above the reach of local people, thus encouraging the creation of weekend ghettoes of long-distance visitors.

Another specific set of studies have examined villages. A central issue has been the long-standing policy to select a few villages for public investment and private development. The theory behind this was that economies of scale could be achieved, since there were natural thresholds for certain services. For example, a small school needed a population of say 500, and the next size of school, 2,000 people and so on. This application of Christaller's Central Place Theory was thus developed to select villages for growth, no change, or even decline, both theoretically in the South Atcham study (Housing and Local Government, 1969) and actually in Durham and Norfolk, for example (Clout, 1972). Most plans have thus contained so-called 'key settlement' policies, and though the term has fallen out of favour in the 1990s, the number of 'selected' villages in 1990s terminology has if anything been an even smaller percentage of the total settlements than ever, falling from around 15 to maybe 5 per cent of rural settlements.

This is in spite of several studies showing problems with such policies. For example Gilder (1979) found that demographic and distance factors, varying capital factors, differing standards of provision and other variations eliminated any internal economies of scale thus invalidating the core argument for key settlements. Gilder therefore concluded that the cost of future growth would be less costly if it was dispersed since the marginal costs of making use of existing fixed assets was likely to outweigh the benefits of concentrated development. Powe and Whitby (1994), in a reworking of these types of data, found a U-shaped curve which suggested that the cost of providing services could be reduced if small settlements were allowed to grow until their costs of service provision fell to the bottom of the U-shaped curve. Accordingly Powe and Whitby argue that the cost of providing sevices should be an important element in rural settlement planning, and that the current trends in migration towards rural areas may be beneficial in reducing the costs of service provision. This was recognised by the Government in 1995 when they announced their intention to undertake research into the cost of rural service provision (Cm 3016, 1995).

Other studies have examined whether key settlement policies were achieved on the ground. For example, Gilg and Blacksell (1977) found that in some instances key settlements had lower planning permission rates than non-key settlements. Overall, however, there was no difference between the permission rates of all villages, thus suggesting that the policy had had little impact. Nonetheless, development had been restricted to villages. Cloke (1979) in another study found that private housing had a poorish fit with village development policies, but that not surprisingly council housing had a better fit. Cloke also found that the skilled working class had tended to move to key villages, but that the professional and managerial classes had tended to move to non-key villages. Cloke concluded that planners had been unable to adhere closely to their policy strategies in the day-to-day decisions which represent the basic task of land-use control, and that the scope of local planning was not sufficiently broad to provide effective answers to the wider rural problems of transport, housing, employment and services.

In a later overview Cloke (1983) argued that settlement planning was failing because of too much emphasis on physical rather than social planning. Nonetheless, this has not prevented a major transformation of the built environment of many villages, with a bijou olde worlde vernacular style in village centres and standard suburbia in new estates on the fringes (Bowers and Cheshire, 1983) in spite of long-standing design guides based on the pioneering 'Essex Design Guide' of the early 1970s (Gilg, 1978). S. Owen (1995) has also pointed out that villages have become less distinctive this century as a result of forces including planning which are inducing standardisation. However, from the results of a survey of eight structure and six Local Plans he found a more positive, if patchy, attitude to developing local architectural styles, and he thus concluded that more flexible and discriminating policies would be needed if villages are to regain their distinct identities. This should come about since the 'Rural White Paper' (Cm 3016, 1995) announced the intention to revise PPG1 to encourage a more regionally and locally based approach to design and planning.

Sillince (1986) in a slightly different finding found from a study of 34 villages (17 key and 17 non-key) that the key settlements had attracted growth until a change in policy in 1977 had allowed the non-key villages to grow again. Herington (1985) in another longitudinal study found that between 1951 and 1981 the biggest absolute growth had been in key settlements, with albeit a big percentage growth in non-key villages between 1961 and 1971, but that this had slowed down in the 1970s, so that between 1971 and 1979 development in key villages had grown to 75 per cent of the total.

Nonetheless, a lot of villages have registered losses of population in the postwar era, due to a combination of restrictive housing policies, lower household sizes, and the conversion of two or three cottages into one house. Weekley (1988) in a study of the Midlands found that 45 per cent of villages had lost population in this way in a remarkably even pattern.

In a national study of development control for the DOE, Martin and Voorhees (1980) found that both key settlement policies in particular and development control in general work better in pressured areas than in areas of slow growth or decline. This finding of course reflects the weak position of planners as reactive agents, in that they can only plan when there are development proposals to react to, in an economy which since 1979 has removed so many public investment opportunities via the privatisations of the 1980s and 1990s.

All the empirical analyses so far discussed also suffer from the same problem, in that they can only react to data-derived evidence. Accordingly, in the 1980s as the main impacts of development control began to be described, but not adequately explained, researchers began to look for more meaningful forms of analysis. Two schools developed, a structural school and a behavioural school. These are now discussed in turn.

Structural analyses focused on political economy and regulation theory (or: planning as a struggle between groups for power)

All the studies in this sub-section embrace the concept that planning is a struggle between different groups for power, and some of them also contain the explicit assumption that the eventual winner will be the group that represents the interests of capital accumulation. The view of planning as a power struggle first strongly emerged in the mid 1980s and this has continued to be a powerful concept in the mid 1990s.

The interactions between the various key groups were used by Short *et al.* (1986) to make certain observations about the development process from a study conducted in one of the most pressured rural environments in Britain, the county of Berkshire, and in particular the M4 corridor south of Reading. This revealed that planners were seen as a barrier to be 'got past', but that different groups adopted very different tactics. Four tactics were identified: naive; cautious; aggressive; and negotiating. Large builders tended to be negotiators and also tended to be most successful because they had found out what planners wanted. Protestors were drawn largely from the owner-occupier set, but significantly they were much more likely to be involved in development control issues rather than plan-making matters.

In conclusion Short *et al.* (1986) hypothesised the following process: first, the anti-growth opinion of locals is expressed by refusals; second, developers react to this by making frequent appeals to the DOE; third, the DOE justifies imposing higher growth on appeal by adding planning gain. For example, in the above example, the DOE added 8,600 houses to the submitted Structure Plan over the original plan for 31,000 houses, bringing the total to 39,600. From this and other evidence, for example the *Journal of Planning and Environment Law* (1988, 682) which showed that the housing allocations in 13 out of 18 Structure Plans had been increased by the DOE, Short *et al.* concluded that planning was only a tool of capitalism and that it was in fact merely 'legitimising capital accumulation' by giving the impression of local democracy and power over development, while central Government reserved the right to restore the capital imperative if it was threatened.

There is thus a clash between the central and the local state which has been developed by Cloke and Little (1990) using evidence from Gloucestershire. Cloke (1994) had begun to explore this relationship after he had been frustrated by the lack of explanation provided by the logical positive analyses he had conducted in the late 1970s and early 1980s (see the previous sub-section) and he and Little have concluded that the form and function of the central state reflect both the power of dominant capital interests and the maintenance of the basic structures of capitalism. The central state is committed to the role of initiating policy which facilitates the reproduction of suitable environments for capital accumulation. The central state is thus subject to considerable and definitive constraints on action. However, from the evidence they collected Cloke and Little then argue that the state cannot be viewed merely as a puppet of capital interests. Its role is much more than that of parasite or neo-government. In terms of constraints on policy and action, there remains at the central state level some considerable autonomy and discretion. However, at the local state level, the constraints on action become more firmly prescribed, and the political art of the possible permits less autonomy and less discretion than in the central arena. Not only do the structures and requirements of internationally dominant capital and nationally dominant class interests press in on the local scale, but action in the locality is also constrained by the relationship between central and local states, and between the governments working within them.

These conclusions raise all sorts of problems, notably whether planning is a genuinely participatory exercise or merely a cosmetic camouflage giving the impression of democracy but not the substance. In addition, there is the problem already raised of different political parties being in power at the central and local state level. Finally, in the 1980s and 1990s Conservative Secretaries of State have rhetorically claimed that all their reforms, notably in local government finance, were devolving power from Westminster, whereas many people claim they were in fact circumscribing freedom of decision making at the local level. Nonetheless, Brotherton (1992), from a rigorous analysis of development control data using the concept of weak and strong applications and decisions outlined earlier, has concluded that LPAs take the lead in determining policy tightness and that the central planning authority (the Planning Inspectorate and DOE ministers) partially at least take the lead given by LPAs, so that planning control is locally led, with circulars influencing rather than directing policy.

In addition to the studies by Short *et al.* and Cloke and Little which have examined both the central and the local state, a number of studies have been made at the local level. One of the first was carried out by Rydin (1985) in her study of two contrasting areas in Essex: Epping Forest and Colchester. Her methodolgy involved: a) interviews with planners; b) a study of plans; c) an analysis of development control data; and d) a survey of applicants. Rydin found that Epping Forest, a leafy green middle-class rural area, had an anti-development ethos, while Colchester, a prosperous town, had a pro-development ethos. This was not, however, reflected in the permission rates and Rydin, using Lukes' concepts of power discussed in Chapter 1, argued that those involved in the planning process may be using covert arguments to get their way by manipulating perceptions.

In particular, the profits to be made from the developments being discussed were rarely mentioned, and Rydin concluded that the professional ideology of planners, with its broad rhetoric, had obscured the distributive aspects of the planning system, and that in this sense planning had an ideological function in legitimising the benefits received by these economically powerful groups. In a later paper Rydin and Myerson (1989) developed the theme of rhetorical discourse to take a hybrid political economy and post-modern stance when they concluded that the rhetoric used by actors should be used to complement rather than replace explanation, since, language not only constructs reality, but also reflects and helps to create the interests and ideas of dominant social groups.

This theme was developed by Pacione (1990 and 1991) in a study of planning in western Scotland which concluded that planning decisions highlight the way in which the state adjudicates between divergent interests and achieves its dual imperatives of capital accumulation and socio-political legitimation.

This line of thinking was also espoused by the work of the University College London team (already outlined in Chapters 2 and 3) in the late 1980s and early 1990s. As part of their work the team, notably Murdoch and Marsden (1994), examined the land development process in Buckinghamshire using the concept of middle-class domination and employing a case-study methodology. They demonstrated how the middle classes in the area have created a vision of 'traditional village communities' to create a strong anti-development stance among village residents, and thus have constructed an image of exclusivity which is mutually reinforcing, particularly in the way in which the planning system operates. This has prevented farmers from diversifying into certain activities, and has funnelled them into activities which, for example, re-use old buildings in middle-class ways, or develop land into socially acceptable golf courses. In common with Rydin, Murdoch and Marsden argue that the middle classes have not used the real arguments to advance their case, namely, the desire to exclude less well-off people or people who do not conform to being white, English, and family oriented, but have instead used arguments based on 'acceptable' concepts of peace and quiet, and landscape preservation.

Human agency, behaviouralist and post-modern perspectives

To some extent studies of these types are interchangeable with those in the previous sub-sections, and selection has been made on where the key theme lies. For example, Rowswell's (1989) study of planning in East Sussex between 1971 and 1981 could be classed as empirical in that it compared growth in the key villages of 9.8 per cent with growth in the non-key villages of 12.3 per cent. It is Rowswell's explanation, however, that marks it as behavioural, since he attributed the differences not to planning policies, but to the differential attitudes of landowners to development, with some landowners failing to cash in their land for development even in key villages, thus frustrating the ambitions of both planners and developers.

In a similar study, this time of farmers' influence on planning committees in particular, and in general, the exception afforded to houses with an 'agricultural occupancy condition' in the open countryside, Gilg and Kelly (1996b and 1996c) found in common with Rydin that farmers used all sorts of covert arguments to gain planning permission in North Devon. When these failed they then adopted a range of behavioural tactics which undermined both central and local state policies and cut across all sorts of other divisions like social class or political party. In so doing they managed to gain planning permission in many cases against the wishes of the planning officers and the rest of the planning committee. This case study reveals that in certain circumstances local human agency can still be very important.

A key feature of the tactics employed by farmers was setting the agenda in which the debate was conducted. This is one of the central tenets of post-modernism and has been used by Buller and Hoggart (1986) in their study of planning in Sussex which emphasised the importance of non-decision making, or keeping certain items off the agenda. This could also be interpreted as a political economy approach if the agenda setting emphasised middle-class or capitalist values. The degree to which rural areas have become dominated by middle-class values is indeed a long-standing topic in rural studies dating back to Pahl's (1965) classic work in the 1960s which was later extended by Pahl (1966 and 1967) to include the concept of two distinct types of middle class in-migrant: first, wealthy retired people who bought old houses in the village centre and then zealously preserved them and their neighbourhood; and second, spiralists, a group of young ambitious professional people who bought houses on the fringe of the village in modern estates as a temporary measure in their upward climb to success.

Since then a number of studies have shown how planning has been an accomplice in this process. For example, Pacione (1980) in a study of a large village in the west of Scotland has shown not only its rapid growth but a significant spatial division of house types between middle-class private housing and lower-class public housing. There were also major differences in the source of migrants with the middle-class areas attracting people from all over Britain, while the public housing areas were restricted almost entirely to migrants from Glasgow. In addition, council house tenants exhibited low social linkages limited to their own area, while middle-class residents had strong social linkages not only in their own areas but also into the council house area. Pacione (1991) in a later article has demonstrated how this middle-class pressure on the rural areas around Glasgow has continued at the edge of villages in a long-standing gentrification of such areas.

In the more traditionally rural county of Hampshire, once dominated by the squirearchy social hierarchy of squire, parson, teacher, shopkeeper and worker, Harper (1987) has shown how three types of village have emerged: first, established farming villages still dominated by the squirearchy and of the type so vividly described from personal experience by Gilg (1992b); second, villages in transition; and third, metropolitan villages now totally urbanised by middle-class commuters. Harper partly explained the spatial pattern with reference to distance from Southampton, with exceptions being explained by the reluctance of some landowners to give up their social and economic power by selling off land in a thought process well documented by Newby *et al.* (1978) in the 1970s.

The process of gentrification can have unfortunate side-effects on service provision and social life for the diminishing number of traditional rural residents as they are surrounded by the surging surf of middle-class culture and two-car ownerships. Cloke and Little (1987) from a study of six key service issues – the post office, minibuses, buses, doctors, housing, and industry – concluded that: a) positive impacts on these services had largely resulted from the private sector; b) more housing which might bring house prices down to levels affordable by traditional locals was strongly resisted by a strong middle-class lobby; c) the more deprived groups, the poor and the elderly, suffered most from the lack of service provision; but d) there was, however, a high level of adaptability by rural comunities to lack of services and therefore that the ideology of making do had serious implications for rural planning.

Indeed the study of rural service provision and deprivation which flowered briefly in the late 1970s (Moseley, 1979 and 1980; Shaw, 1979) was cut down in the 1980s by the powerful ideology of making do and 'self-help' popularised by Mrs Thatcher's homespun philosophy based on the economics of the village shop. Accordingly, work on rural communities either had to take this philosophy on (McLaughlin, 1986a, 1986b and 1987) or couch work on deprivation in different terms (Cloke and Thrift, 1987; and Cloke and Millbourne, 1992) in order to get funding. When such work was funded and allowed to be published (Shucksmith *et al.* (1994), it revealed growing divisions of wealth between rural incomers, or as Cloke and Thrift have termed them the 'service class' (or Pahl's 'spiralists'), and traditional rural inhabitants. The 'underclass' so patently visible in urban areas to those who dare travel through them, is, in contrast, patently invisible to those who look at the countryside through cottage-tinted spectacles or who ignore the 'ecological fallacy' in which average scores for deprivation pick out urban deprivation because it is concentrated, but fail to pick out rural deprivation because it is spread thinly (Walker, 1978).

Returning to the wider canvas, Healey and Shaw (1994) have argued that the development of discursive, communicative approaches to exploring the form and content of development plans is more likely to reflect the breadth of contemporary understanding than the calculative approaches used by the Government and currently being developed (PIEDA, 1992). Thus Healey and Shaw (1994) use a textual analysis of the discourse contained in postwar planning documents to identify five strands in which the environment has been treated by planning: 1) a welfarist-utilitarianism, combined with a moral landscape aesthetic (1940s on); 2) growth management, servicing and containing growth while conserving open land (1960 on); 3) active environmental care and management (1970s on); 4) a marketised utilitarianism, combined with a conservation of nationally important heritage (1980s on); 5) and finally, in the 1990s, sustainable development. These strands have contained three types of approach. First, viewing the environment as a functional resource, a reserve of non-renewable resources and amenities for human enjoyment, which Whatmore and Boucher (1993) have shown treats the environment as a commodity in the process of negotiating environmental 'planning gains'. Second, the moral and aesthetic notion of the environment as a backcloth or setting to human life. Third, the environment as a constraining factor. However, although the five strands and three approaches have co-existed with each other and other planning policies, it is the modernist rational discourse of economic growth, instrumental rationality, and the ability of the human species to control and manage nature, which has won out. Therefore, the tendency has been to restrict the planning agenda to a narrow remit focused on land use and development and to a discourse dominated by forms of utilitarian functionalism. The challenge for the new environmental agenda is thus to gain real leverage over economic discourse, not by a reinforcement of traditional strategies and policies but by a fundamental re-thinking of the form and content of the system in terms of conceptions, technical methods and policy processes.

In conclusion, planning is a creature of the society it finds itself in, and currently British society is dominated by middle-class values based on self help, capital accumulation through life, and hard work. Much work has been done on the impact of the middle class, but Murdoch (1995) would like to see attention now being directed not on concepts of class grounded in production, but by the patterns of consumption, and how classes form and re-form themselves over time. He argues that we must identify how actors come together and act collectively, how they use resources and assets in such collective action, how they seek to effect further redistributions of these resources and assets, how these collective actors distinguish themselves from others, and how they stabilise such distinctions.

For example, planners may start out as socially concerned graduates, but those whose careers take them to rural areas soon find themselves sucked into a strong rural culture dominated in a few areas still by farmers, as shown by Gilg and Kelly's (1986b) work in North Devon, or increasingly by incomers who live but do not work in the area, and so have a strong interest in creating a pleasant rural environment based on some mythical view of an ideal countryside (Cloke, Lapping and Phillips, 1995). There is perhaps no great harm in this, when one considers the alternatives, or visits areas in North America or Mediterranean Europe where similar restraints have not operated. There is, however, still room in the generally protective environment to allow a greater case to be made for some socially orientated planning of the built environment than is at present the case.

E: COMMENTARY AND SOME THOUGHTS ON THE FUTURE

Alternative theoretical and evaluative perspectives

In addition to the three perspectives already used, a number of other approaches have been advocated for assessing the impact of planning. For example, Brindley *et al.* (1989) have developed a typology of planning styles based on a six-cell matrix. The two columns are divided into either market-critical or market-led attitudes to market processes, while the three rows are divided into buoyant, marginal and declining areas. The resulting six cells produce a six-fold typology: regulative planning; trend planning; popular planning; leverage planning; public-investment planning; and private-management planning. Alhough the matrix was developed for urban areas it can easily be applied to rural areas as well, and an interesting exercise for the reader would be to try to restructure the studies outlined above into the matrix. A similar perspective has been provided by Rydin (1993) in a matrix which involves four columns: an ideal market model (new right); market failure (liberal political economy); a Marxist model (new left); and institutional economics (institutional approach), and three rows: the state; planners; and the public. Once again, the reader is invited to recast section 5d into this format.

A quite different assessment method has been provided by consultants working for the DOE (PIEDA *et al.*, 1992). They argued that the methodology would need to reflect the function and character of planning which seeks to balance competing interests in land. For example,

- land is both a factor of production and an environmental resource;
- the demand for and use of land is a reflection of complex social and economic processes involving a multiplicity of private interests;
- planning draws its objectives from a wide variety of interests, and seeks to strike a balance in the 'public interest';
- planning seeks to achieve its aims through a variety of instruments which operate in parallel with other strands of public policy;
- planning operates across a number of scales of government often involving the *need* to balance local and wider interests; and
- in comparison with most other areas, planning involves long timescales and substantial time lags.

Set against these characteristics the consultants argued that evaluation must take into account *inter alia*: the need to reveal the performance of planning in balancing multi-faceted sets of objectives and interests underlying the use and development of land; the need to take into account the variety of instruments available; the need to highlight the interrelationships existing between all the different bodies involved; and the need to consider the time dimension.

A core feature of the methodology is the division of planning objectives into: objectives aimed at meeting the needs of economic growth and development; objectives aimed at protecting and enhancing the quality of life for the present generation; and objectives aimed at conserving environmental resources for future generations. The methodology was tested by the consultants in three case studies, and was found to work quite well, though it would be more effective if it was built into the plan-making process at the outset, and was backed up by a plan-rich environment with good development control and land-use change information systems.

However, this mechanistic approach may offer only a partial evaluation since it tends to leave out post-modern and behavioural critiques which stress the nature of power and the use of language. For example, Forester (1993) has attempted to apply Habermas's overarching theory and the concept of communicative interaction to try to assess how groups involved with planning interact, how they come to have the sets of beliefs they have, and how those beliefs might change. In addition, Fischer and Forester (1993) have argued that language is not neutral, but that instead language shapes political and social understanding, and not only describes, but creates the world. Both these contributions stress the need to probe underneath the surface, and not to accept language as it stands, but to assess how planners have created a language of their own in order to give plans greater power. For example, words like 'the public interest' imply that planners have found out what that is, and are speaking on the public's behalf, when in fact they are basing such concepts on a limited agenda.

It is thus very clear that evaluating the effectiveness of planning is an almost impossible task, not only because of the multi-objectives outlined by PIEDA and the problem of overlapping time scales, but also because planning is not the neutral, technical exercise it is sometimes claimed to be. Nonetheless, some sort of evaluation must be attempted, and a few brave souls have put their head above the parapet.

Overviews of planning

The first point that should be made, however, is that public policy is a messy and incremental process as described by Delafons (1995) writing from the experience of working for 40 years in the DOE, and thus that any overview by definition will impose an apparent rationality or meta-narrative on what was in reality a pretty uncoordinated and ad hoc set of events. In spite of these cautionary comments, Healey's summary of the postwar development of planning instruments, as shown in Figure 5.3, provides a useful framework with which to read the overview attempts that have been made and also refers back to the Gilg/Selman spectrum of planning powers outlined in Chapter 1. Healey concludes that the strength of the system rests on its attention to detail and treating each case on its merits, but that its

Type of land policy measure	Specific Instruments	Periods of government: 1945-51 Labour; 1951-9 Conservative; 1960-4 Conservative; 1964-70 Labour; 1970-4 Conservative; 1974-9 Labour; 1979— Conservative
REGULATORY MEASURES	Control of development	1947 •
	General development order	1947 •
	Use classes order	1947 • } major revisions
	Rights of consultation	1947 • public agencies and government departments — 1971 • publicity for some types of development
	Rights of objection	1947 • direct property
	Public inquiries	1947 •
	Regional policy controls	1945 • Industrial Development Certificates — 1965 • Office Development Permits
		1949 • national parks and areas of natural beauty — 1967 • conservation areas
	Special area provisions	1947 • comprehensive development areas — • 1968 • general improving areas (housing) (industry) 1980 • enterprise zones; 1986 • simplified planning zones
DEVELOPMENTAL MEASURES	Land acquisitions for planning purposes	1947 • — 1963 • by agreement for any purpose — 1972 • for some planning purposes; 1975 • Community Land Act (for public purposes) — •
	Compulsory purchase Powers for planning purposes	1947 • for planning purposes
	Special development agencies	1946 • New Town Corporations — Most NTC's being 'wound up'
		1947 • Central Land Board — 1967 Land Commission • — • 1980 Urban Development Corp.
	Compulsory disposal of public land	1980
FINANCIAL MEASURES	Taxation of development gain	1947 • Development charge • (1952)-1953 — 1967 • Betterment Levy • 1974 Development gain tax 1976; 1976 • Development land tax — • 1985
	Public purchase of land and property at a privileged price	1947 • — • (1959)
	Planning agreements	1971 Section 52
	Subsidies for land reclamation; for development	1966 •Derelict Land Grant; 1980 • Urban Development Grants; 1980 • Enterprise Zones (financial powers); 1986 • Urban Regeneration Grants
INFORMATION AND GUIDANCE	Development plans	1947 • 'One-tier' plans — until revoked (major revisions discussed); 1968 •structure and local plans; 1980 •registers of public land holdings; 1985 •unitary development plans
	Public inquiries	1947 • — (local plans only); 1972 •examination-in -public for structure plans
	Consultation procedures	1947 • for government departments and public agencies; 1968 • with the public at large in respect of plans

Figure 5.3 Instruments of the English planning system 1945–88

Source: Healey, 1988

failure has been in not seeing the wider picture, not assessing the impact that it is having; and in being too reactive.

This has been confirmed by the Audit Commission's (1992) study of development control which found that, in practice: most authorities lack systems to ensure that their decisions are faithfully implemented, and that most planning departments, having given attention to the decision-making process, do little to monitor and thereby demonstrate the success or otherwise of their activities. In an attempt to remedy this gap Pearce (1992), in a general overview begins by making five key observations:

1 There are problems in assessing effectiveness, in particular:
   a) whose goals should be evaluated?
   b) if we try to limit goals to official goals only, these are difficult to trace and they have changed over time;
   c) what would have happened without planning? (the 'birth control' effect);

   d) what about unintended consequences; and
   e) there are limited data and statistical methods to compare goals with results.

2 However, the postwar goals of planning are generally agreed to have been:
   a) facilitating and encouraging private enterprise;
   b) ensuring the supply of land for private housing;
   c) containing urban areas;
   d) re-using derelict land;
   e) protecting residential amenities; and
   f) conserving historic or heritage buildings, and the landscape;

   The list used to be longer but in recent years some previous goals have been downgraded, notably:
   a) attempts to overcome regional imbalances by redistribution;
   b) planned decentralisation;
   c) securing self-contained and balanced communities;
   d) public expenditure on the provision of public housing and new settlements;
   e) segregating employment and residential uses.

3 The main means for achieving policy goals has been development control.

4 The main achievements have been:
   a) protection of countryside and residential areas;
   b) increased re-use of derelict land;
   c) more efficient use of public infrastructure.

5 The main unachieved goals or unintended costs have been:
   a) reduced economic growth due to delay plus constraints and bureaucracy;
   b) inflation of land values; and
   c) separation of home and workplace.

6 The possible sources of failure have been:
   a) information imperfections;
   b) bureaucracy and lack of coordination;
   c) use of multiple and conflicting goals;
   d) lack of competition between monopoly suppliers of planning services;
   e) limited administrative capacity and resources; and
   f) government externalities, for example, 'public bads' located in marginalised areas.

In 1995 Pearce was commissioned by the DOE to put some empirical flesh on his observations, and the results which were due out in 1997 are awaited with interest.

Thornley (1990) in an examination of the shorter period covered by Mrs Thatcher's time as premier has made the following points. First, the period witnessed a reorientation of planning purposes to: a) a greater acceptance of market criteria; b) a selective application of environmental criteria; and c) the removal of social criteria. Second, a reorientation of procedures away from local democracy towards centralised government supervision. And third, a division of planning into three areas: areas with strong controls and good environments; areas where economic criteria dominate; and areas where planning powers have been relaxed to encourage economic recovery.

Thornley concludes by asking if the changes will last. He finds three weaknesses in Thatcherite philosophy: first, the market is not all-efficient; second, green issues will grow in importance; and third, it represents an attack on democracy. For the future, Thornley argues that since the system is basically still sound, if advice on how to operate it changes then all will be all right. Therefore he advocates a need to restablish community control over land and take it back from capitalists.

The importance of capitalism to planning is taken up by Hobbs (1992) in a detailed analysis of the postwar period which adds much flesh to the bones of Figure 5.3 (see page 178). He finds that planning policies responded to changes in the economy, so that they became weaker in times of recession, and stronger in boom times. Ironically in times of recession, planning emphasised how it could improve environments at the local level. Overall, however, despite the formal emphasis being placed on the management of land-use change in the public interest, the overriding priority lay with support for production. Unfortunately, despite this underlying concern for supporting economic growth, planning has been poorly integrated with other economic strategies and has lacked powers to direct economic activity. There are thus fundamental limitations on the ability of planning to shape land use change. Nonetheless, at the local level Hobbs concludes that the system does give discretion to reflect uneven economic performance.

This discretion will be enhanced by the move to a more plan-led system following the Planning and Compensation Act 1991 and MacGregor and Ross (1995) have concluded that the market may have to take planning more seriously in the future, but only if up-to-date and unambiguous plans are in operation. In spite of these reservations, Diamond (1995) has been able to credit postwar planning with five successes and only two failures. The five successes are: achieving significant improvements in urban form; achieving the virtual abolition of slum housing; effective management of land-use conflicts; undertaking countryside planning for city dwellers; and considerable success in containing regional disparities. The two failures are: limited examples of successful inner-city regeneration; and a failure effectively to relate transport and land-use planning. Although this analysis is urban-centred its relevance to the countryside is that improvements to towns reduce the desire to relocate in the countryside and to seek recreation there, as discussed in the next chapter.

Finally, Cullingworth and Nadin (1994) have reflected on postwar planning and concluded that 'comprehensive planning based on firm predictions of the future course of events is now clearly impossible. Incrementalism is the order of the day' (19). In particular they point out the widening gap between a capitalist system driven by global forces able to make decisions almost instantly which can make or break whole regions economically, and a planning system which is driven by the needs of people for homes and jobs and the need to protect the environment. This gap between land-use development and 'needs' throws considerable doubt on the adequacy of a planning system which is based on the assumption that land uses can be predicted and appropriate amounts of land 'allocated' for specific uses. It also implies that planning will have to embrace the agents of the market, and adapt its regulatory mechanisms to a system of negotiation or even partnership with developers.

The future

As Cullingworth and Nadin have pointed out, planning has had to learn to deal with planning for uncertainty, especially with regard to the economy. Fortunately population movements are less volatile, and take effect over shorter periods, people now being less mobile than capital, and the overall population is not growing. Nonetheless, a predicted continuation (Breheny, 1995a) in the long-term trend to much smaller households will mean that even a static population will require millions more houses according to current OPCS projections.These projections include an extra 4.4 million households by 2016, and a population rise of 12 per cent by 2025 in rural areas (Cm 3016, 1995). In theory this could lead to the vigorous debate about land release in the south-east becoming ever more heated, but there are other factors that could throw either petrol or water on the conflagration.

First, the explosive development of the Internet in particular, and information technology in general, could lead to any one of the three scenarios outlined by Miles (1995) coming to pass, namely: more home-based work, more divisions between those in or out of work, or more part-time/shared work. To some extent any one of the scenarios reawakens the dream of sustainable and self-contained new settlements envisaged at the turn of the turn of the century by Ebenezer Howard. In the 1990s the Local Government Management Board (1995) has produced an illustrated technical manual for planners, designers and developers to convert the rhetoric of sustainability into practical action at the level of settlements, and Rudlin and Falk (1995) have shown how to build houses to last in existing built-up areas in order to reduce commuting. The need for sustainability, as Chapter 6 will discuss, arises largely from the Rio agreements and the associated Agenda 21. Planners are now beginning to learn how to reconcile the imperatives of Agenda 21 with strategic policies (Connell, 1995) although in the short term the Government's response, notably in PPG13, has been to attempt to force or encourage more and more development back into the cities by stricter controls, and by making cities more attractive, to integrate developments into a few selected sites in order to reduce the need to travel, and to make more use of existing housing and developed land (Cm 3016, 1995).

Breheny (1995b) has questioned the efficacy of this policy and has calculated that higher urban densities will have little impact on the emission of greenhouse gases from transport, and that planning policies in urban form offer fewer savings than other policies. Furthermore, the momentum towards rural living and employment may be politically impossible to reverse, and current policies may not survive. In addition, the European dimension will grow in importance, and though nobody is suggesting the imposition of a pan-European planning system, individual countries can no longer operate a planning system that merely drives employment or people to other parts of Europe. Indeed, conversely the British Government's Maastricht opt-out from EU social legislation is claimed to have given the United Kingdom a huge competitive advantage in the mid 1990s. It is hard to see any Government imposing planning policies that undermine this type of advantage.

At the end of the day, however, the future will depend on the vision that we have of the countryside we want, what sort of electoral priority this is accorded, and what sort of principles are needed. Accordingly, the seven principles set out by the Countryside Commission (1989b) for the planning system are an extremely valuable way not only to end this chapter, but also to introduce the wider concepts to be discussed in the final two chapters. The seven principles are:

1 Natural beauty and landscape diversity should be conserved.
2 'New' countryside should be created wherever possible.
3 Green belts should serve a wider purpose.
4 Maximum environmental benefits should be secured from development that has to take place in the countryside.
5 New housing in the countryside must make a positive contribution to the rural scene.
6 Rural enterprise is welcome if it is developed harmoniously with the countryside.
7 Major developments in the countryside should be strictly controlled and of the highest standard of design and landscaping.

Finally, the accompanying summary provides a précis of this Chapter.

Policy evaluation checklist derived from Chapter 1.

1 *Were the problems understood and which value system dominated?* In 1947 the problems were ones of food shortages and fears of urban sprawl. The value system was thus one of protecting farmland. This has now changed to protecting the countryside for the middle classes who can afford the house prices.

2 *Which goals were to be achieved and were the policy objectives clearly spelt out?* The goals were initially those of protecting food production resources, but these were only spelt out later as the policies came under review, and the DOE and LPAs had to justify policies of restraint.

3 *Were the policy responses accretionary, pragmatic and incremental?* Plan making has seen one major non-accretionary change in the 1970s, and an incremental change in the 1990s. Development control has, however, evolved incrementally since 1979.

4 *Which powers in the Gilg/Selman spectrum were mainly used?* Exhortation and advice have been the main weapons in plan making, once the socialist system was abandoned in the 1950s. Development control is a classical regulatory device. Financial measures are used at the margins.

5 *Were planners given sufficient powers/resources and sufficient flexibility for implementation?* At the outset, yes, but plan making soon became devalued into exhortation. However, development control allows considerable flexibility and 'treating each case on its merits' is the great strength of the system.

6 *Were performance indicators built into the policies?* Not for many years, and even now the main indicator is the speed, not the quality, of decision making. However, in such a complex and subjective area performance indicators may not be very useful, except as crude measures.

7 *Have the policies been underresearched?* The overall impact of the policies has been woefully underresearched. But, individual aspects, for example, the containment of urban growth, and the regressive social and economic impacts, have attracted some good, but not comprehensive research.

8 *Have the policies been effective overall?* Urban growth has been contained, the pattern of villages has been maintained, and agricultural production has been protected overall. There have been, however, a number of individual breaches and the resulting built environment has often been characterless.

9 *Have there been uneven policy impacts, by group or by area?* Yes. Planning works best in areas where there is pressure it can react to and so its biggest impact has been around major cities. Its

restrictive policies have favoured better-off-people but harmed the less wealthy.

10 *Have the policies been cost-effective and are they environmentally sustainable?* Planning is a cheap public service. However, restrictive policies have slowed down or prevented maximum economic growth, but at the same time have prevented an even faster slide into an unsustainable Megalopolis.

11 *What have been the side-effects?* Creating an increasingly middle-class England based around the new jobs of a post-industrial society, raising land prices, slowing economic growth, and aiding the creation of uniform committee-style environments.

12 *Were the policies modified by unforeseen events/policy changes elsewhere?* Yes, the potentially good reforms of the 1968 and 1971 Acts were undermined by the local government reforms of 1974, by the move to a mobile society and politically since 1979 by the need to 'manage change' pragmatically.

13 *Were the policies overtly sectoral and were they compatible with other policy areas?* No, the original aim was to provide a comprehensive development plan. This was then modified to coordinating other sectors in the 1970s, which has evolved into making competing land uses compatible.

14 *What have been the effects of 16 years of Tory rule?* According to Thornley (1990) they have been a greater acceptance of market criteria, a selective use of environmental criteria, the removal of social criteria, and the creation of three tiers of planning control ranging from strong to weak.

15 *Should the policies be modified?* Yes, but only in detail, since the system is inherently sound. The main weakness is the lack of a sensible administrative structure at the regional and local levels which will be made worse by the 1996 and 1997 changes.

16 *What alternatives are there?* In addition to the creation of city regions as the core administrative structure, policies need to be more pro-active in terms of land assembly. Crucially, a decision has to be made over continuing towards the unsustainable 'spread city' or returning to urbanity, and restricting new rural settlement to only those who need to use the extensive environment.

17 *Overall verdict.* Britain would be a far worse place without planning and thus its main impact has been in preventing bad things. What it has allowed, however, has too often been mediocre and could have been more sensibly restricted to an urban location.

6

PLANNING FOR THE 'NATURAL' ENVIRONMENT

A: INTRODUCTORY CONCEPTS

Unlike planning for forestry and agriculture or for the built environment, which produce physical goods for sale and consumption, planning for the 'natural' environment produces intangible results with few economic benefits. We therefore need to produce a philosophy of why planning for the 'natural' environment should be built into planning for productive activities, or become a productive planning activity in its own right by charging *directly* for the consumption of environmental goods instead of *indirectly* by taxation. Alternatively we could attempt to produce some monetary measures by which to value the 'natural' environment, since traditionally it has been seen as a free good to pollute or to enjoy recreational activity in.

Traditionally also, however, the 'natural' environment has been seen as an adjunct to economic activity, and planning for it has involved ameliorating the deleterious effect of either internal or external forces. For example, a farm may have a fine wildflower meadow particularly along a streamside public footpath. Internal financial forces may push the farmer to plough it up, while external forces may damage it by excessive recreational visits and picking of the flowers, thus preventing regeneration by seeding. The planning response would normally be to advise the farmer not to plough up and to play on his public spiritedness. If that failed, then a management agreement could be offered, and then if that failed, a Site of Special Scientific Interest could be declared. Likewise, walkers could be asked not to pick or trample the flowers, and if that failed they could be banned at certain times or even totally. Therefore, the traditional approach has been dominated by working through the Gilg/Selman spectrum outlined in Chapter 1 from the voluntary end and has tried to keep management as far as possible in the private sector.

If progress is to be made with this softly softly approach then arguments for managing wildlife and landscapes for conservation and recreation need to be convincing and easily understandable by both land managers and the public. In essence three key arguments can be used: first, the practical argument that conservation provides a genetic reserve and preserves the balance of nature and helps our survival; second, the spiritual argument that nature acts as a safety valve for modern living in the face of declining religious belief and the growth of a secular society; and third, the amazing damage caused over the past, and notably recent past, with the extinction of individual species and habitats by reciting the arithmetic of woe. All of these arguments are widely used, notably by natural history and travel programmes which command prime times on national TV.

In more detail, Gilg (1981) found three main philosophies used to justify planning for conservation. First, an ethical/aesthetic argument view which includes 'man' as part of nature, the apparent dominion given to 'man' over nature by the Bible in the book of Genesis (Passmore, 1974), and nature as a replacement for religion. Subsequent opinion polling by MORI (1987) found that this view was supported by 64 per cent of people. Second, a scientific and educational view which includes the concepts of Darwinism, the need to preserve a gene pool, and the use of nature trails to educate children in scientific observation. This view was supported by only 7 per cent of MORI respondents. Third, a preservationist and disaster argument which includes the ecosystem idea and working with nature instead of against it, and thus the core concept of the survival of life on the planet Earth. This view was the most supported, by 75 per cent.

Although the arguments so far considered may be eminently reasonable, as soon as they are related to the real world, severe problems are encountered. First, there is not in fact a 'natural' environment to conserve, since in most of the world, and certainly in the United Kingdom, virtually all the natural ecosystems were destroyed long ago. The arguments thus centre on a continuum ranging from a re-creation of what are thought to be natural environments, or the management of arrested plagio-climaxes such as semi-natural woodland, to allowing nature to co-exist with farming as in the NAL landscapes of Chapter 3. These arguments then raise the questions of whether we plan for what we actually have, what we might have, or what we did have in history.

A second argument, then, refers to the main purpose being sought, and here there is a big division between planning for a) Landscape Conservation which aims to maintain and manage the large features of the landscape, and b) Nature Conservation which aims to conserve species and ecosystems. There is also a difference in scale and time horizon: a landscape can survive for decades, even centuries, even though the ecosystems it depends on are in decline, while a key habitat can be destroyed in a matter of minutes. A further problem is that a landscape may be beautiful and attractive for recreation but not ecologically 'correct' or sustainable.

A third question, then, arises as to what habitat or species we should give priority to. For example, rare species because of their rarity, common species since they contribute most to wider ecosystem survival, or attractive species because they are popular with the public? This in turn raises the question of who should decide, an elite scientific community, a patrician set of recreation planners dedicated to improving people through quiet contemplation of the countryside, or the lowest common denominator of popular demand satisfied by a Lions of Longleat or a Walt Disney world? Nature, of course, does not in fact exist as an entity, it is a social construct, and as such those who have the power to define it and construct ideologies around it, notably the scientific community, have been able to exercise considerable power, as Toogood (1995) has shown. Recently calls have thus been made for other issues, including ethical issues, to be given more prominence and to downgrade the mechanistic view of nature as something to be measured and managed via the scientific method which has so dominated much twentieth century thinking (Clarke, 1993; and Redclift and Benton, 1994).

A fourth question refers to how planners might achieve their goals, assuming they had found answers to the previous questions. Should there be development control for nature conservation? This is beguilingly attractive but landscapes are made up of living organisms and their control would be a bureaucratic nightmare if detailed features were to be covered. In addition, calls for development control over natural habitats misunderstand the nature of the resource, which is basically a man-created habitat in which most areas are only kept going by whole farming practices, and thus conserving one habitat in isolation is a short-term palliative. The voluntary approach, in contrast, may also do little more than slow down or ameliorate the rate of loss.

In more detail, most nature plans have developed along a continuum similar to the Gilg/Selman spectrum, and a pyramid of conservation priority has been created as shown in Figure 6.1. The basic idea is that rare species and habitats are the most important and should be protected by negative means like legal bans on killing or picking species, and by positive powers like land acquisition or management agreements. Conversely, common species and habitats are regarded as being unworthy and not in need of help. The core concepts are thus rarity and protection by legal controls and site designation.

There are, however, severe philosophical and practical difficulties with these concepts. First, philosophically, the concept of rarity, although used as a measure of value elsewhere, for example, in the idiosyncratic art world, is deeply flawed since it leads to priority being accorded to things just because they are rare, not because of their intrinsic importance or wider value to the broader scheme of things. Second, it is too static and deals with what we have, or have inherited, rather than what we might have or what we did have a long time ago. In many cases this has led to policies which have attempted to conserve the very poor moorland habitats that we inherited in

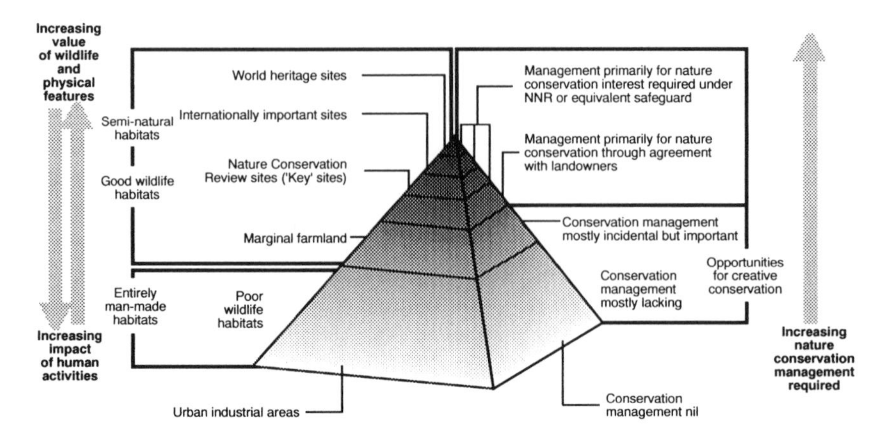

Figure 6.1 Concept of a pyramid of nature conservation priority

Source: Nature Conservancy Council, 1984

the twentieth century, rather than imaginatively trying to re-create the native woodland that could be there. Third, there is the 'Island dilemma' which argues that wildlife species reduce with the size of habitat, and with the distance between habitats (Diamond, 1975). The concepts outlined above have led to an approach based on site protection or a 'zoo mentality' which is ultimately self-defeating, since wildlife in zoo-like islands amidst a sea of modern farming cannot survive very long. One of the themes of this chapter will be the gradual realisation that policies based on rarity and site preservation may once have been a useful and necessary 'finger in the dyke', but that in the long term the pyramid approach to conservation shown in Figure 6.1 must be replaced with a more holistic approach based on total environmental management and habitat re-creation.

Such an approach of course lies at the heart of the sustainability/multiple-use approach much talked about, but little practised, in the 1990s. This approach itself is not without its difficulties, for example, the problem of too many goals, and how to rank and interconnect them. For example, in the farming context, how does one reconcile the demands of: production; nature conservation; ecology; landscape; and recreation, and then put the result into practice?

One long-standing method at the macro-level may offer a way forward, namely the use of the range of monetary evaluation methods shown in Figure 6.2. These argue that since the 'natural' environment does not normally have a price tag, people can be asked to estimate what they might pay. These so-called 'willingness to pay' or 'opportunity cost' methods have been widely used not only to prioritise different types of environment, but also to evaluate the cost-effectiveness of existing policies. They also became

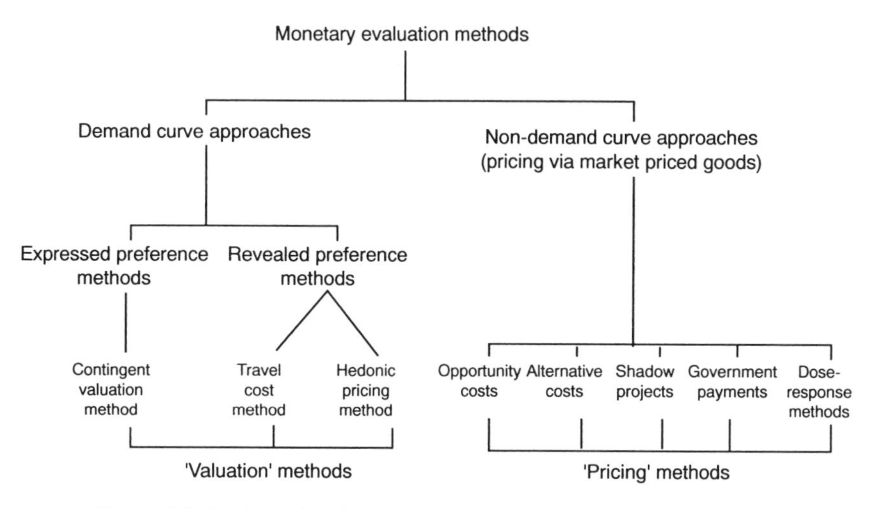

Figure 6.2 Methods for the valuation and 'pricing' of environmental goods and services

Source: Bateman, 1995

valuable because they fitted with the Conservative ideology of market- and accountancy-inspired 'bottom-line' evaluation procedures for public policy, and as such these methods for 'costing the countryside' have been subjected to a good deal of attention (Flynn, 1993). Nonetheless, it is easy to query the degree to which people can disentangle the different values they really put on one set of goods which are in reality subsets of others (Hutchinson *et al.*, 1995), and whether they provide a better way of managing the countryside than other methods (Flynn and Pratt, 1993).

People's landscape and nature preferences are also revealed by the type of recreation they indulge in, and indeed studies of actual recreation travel and expenditure form the basis of several of the methods shown in Figure 6.2. In addition, recreational use is one of the main arguments advanced for planning 'natural' environments, both because of the benefit society is supposed to gain from refreshed individuals, and because of the profits that can be made from such recreation, both directly in the form of entrance fees and so on, but also indirectly by the sale of, for example, walking attire and associated goods.

Recreation as a crucial subset of planning for 'natural' environments

The type of recreation benefit depends largely on the type of recreation experienced and here the big divide is between 'formal' and 'informal' recreation. 'Formal' recreation takes place on managed sites, often owned or leased by profit-seeking organisations. Management normally takes the form of: a) providing special areas; b) zoning sites within these areas; and c) rationing demand by access controls including an entrance or membership fee, or by imposing maximum numbers. Golf courses provide a prime example and Murdoch and Marsden (1992 and 1994) have clearly shown how golf courses have been used by an unholy alliance of developers and middle-class groups to create exclusive countryside which is 'formally' for their recreation or profit only and from which both parties gain benefits.

In contrast, 'informal' recreation refers to the practice of driving round the countryside on a Sunday afternoon looking for somewhere to picnic, read the papers, and probably sleep. It may also include a strenuous walk of as far as a mile to a notable viewpoint or other attraction. This type of activity is likely to be undertaken by the skilled working class, and the countryside or environment provides the backdrop, but is not an essential feature, since by definition 'formal' access away from public roads and footpaths is not attempted. Little benefit is gained by anyone from this experience, but there has been a probably erroneous feeling that it somehow threatens the countryside with a wave of ill-informed over-use.

Whichever type of recreation is engaged in, a number of key variables are thought to influence the degree of participation. First is the concept of the time budget which works out how much time an individual has left after working, sleeping and other essential activities, for recreation. Second is the amount of disposable income, again after essential spending has been accounted for. Third, and highly related to income, is car ownership, and especially in 'informal' recreation the car is the key variable which acts as a focal point, and an extension of the house allowing a visual rather than a physical contact with the countryside, because most of the people who act in this way have a fear of farmers and the unseen dangers of the countryside. The fourth concept is of 'intervening opportunity', that is, how many attractive sites are there between home and a possible end destination, and how is the choice made between visiting one site or set of sites rather than alternatives. A key sub-concept here is that there appears to be a minimum distance that has to be travelled before recreation is thought to have taken place. Once this threshold has been passed there is a distance-decay function. However, this will vary depending on the physical nature of the recreation and its need for either an ordinary or a specialised site. The fifth and final concept is of a round trip, and recreation patterns follow a cyclical process both in time and space as shown below:

Anticipation – Travel to Site – On-site experience – Travel back from site – Recollection

These concepts can be used to work out the likely demand for recreation based on three types of demand identified first by Lavery (1971): 1) effective demand – the actual numbers taking part; 2) deferred demand – the numbers who would like to but cannot because there are no facilities or they are ignorant of them; and 3) potential demand – the numbers who would participate if they had more cash and/or spare time. Examples of how demand can then be assessed have been provided by Smith (1983) and Pigram (1983). These types of model have for a long time been able to predict rural recreation patterns very accurately (Wall, 1972) but although age, social group, education and so on are key predictors there are still the wild cards of 'personal taste' and 'personal predisposition' based on lifetime experiences which can cut across these variables. In addition these models tell us little about *why* people feel the need to go to the countryside, and *why* they do some things rather than others, and *why* some landscapes are liked more than others.

Different types of model based on the so called Clawson method have partially provided some answers to these difficult questions, by concentrating on the monetary cost of recreation and by developing the range of methods shown in Figure 6.2 which can be used for recreational studies as well as wider environmental studies. These methods relying heavily on monetary assessments also rely heavily on distances travelled, and thus sites close to major centres of population are valued more highly. These apparently quantitative but in reality largely speculative methods may not thus be the best way to base decisions about land use.

The physical capacity of the countryside to absorb numbers is also complicated by the concept of carrying capacity, which is the level of recreation use an area can sustain without an unacceptable degree of deterioration of the resource or of the recreation experience (Countryside Commission, 1970). Brotherton (1973) has defined four types of carrying capacity. First, economic capacity, the number of visitors which will yield the maximum economic benefit for a multiple land-use site. Second, physical capacity, the number of users capable of using the site, for example, the number of car parking places, or fishing positions, or pitches on a rock climb. Third, ecological capacity, the amount of use before damage takes place, for example, taking more fish than the annual breeding increment, or disturbing nesting birds more than, say, ten times an hour. Fourth, perceptual capacity, the level at which enjoyment begins to fall, for example, an overcrowded ski run, or one other red anorak in a wild landscape, when one is pretending to be Scott of the Sahara preparing to do battle with an African tiger! (Tickner, 1978).

In addition, the limited natural environments suitable for rural recreation are further limited by problems of access to private property. Harrison (1991) and Marsden *et al.* (1993) have pointed out that the English and Welsh legal systems give enormous weight to private property rights in contrast to the very limited access they give to individuals, largely through laws governing access to fishing and other field sports, as well as to casual rambling. In the uplands, however, there has been a tradition of *de facto* if not *de jure* access, and in Scotland, the law regarding rambling access is far less restrictive.

In more detail, a threefold set of consequences tends to unfurl given the powerful position of private property. First, landowners control most positional goods in the countryside and thus wish to restrict access even in National Parks. Many of them have a paranoia about access and certainly do not want ramblers invading their territory. Second, planners seek to defuse the conflict by providing controlled access to key sites whose main *raison d'être* is to foster *healthy outdoor quiet recreation at low densities* and thus educate the public into 'proper' use of the countryside. Third, the general public have low expectations, little knowledge and poor information sources. Therefore, they tend to congregate at known beauty spots where car parks, toilets and refreshment facilities are provided. Planners and landowners thus form an unholy alliance and encourage this pattern by 'herding' such people into 'honeypots'. The justification for this is that it helps to protect fragile sites from over-use, using the concept of carrying capacity already outlined above, but many commentators are suspicious that this is a conspiracy between middle-class planners and farmers to keep so called 'fragile' sites for the exclusive use of the cognoscenti and to keep the 'riff-raff' at bay. Harrison (1991) has written at length about the way in which a *'countryside aesthetic'* has been used to justify discriminatory types of recreational access in the countryside.

Such a perspective, however, may be a little narrow, and yet over-elaborate in that such a conspiracy, if it did exist, runs counter to one of the main themes of this book, namely, that ill-informed incrementalism has been the order of the day, rather than masterminded plotting. Thus taking events as they appeared to be, rather than as they have been perceived by those who are wise after the event, planners have tended to become involved in recreational issues when an imbalance between supply and demand has occurred, and especiallly when this imbalance has threatened the resource qualitatively or even its very future physically. Another cause for intervention might be a conflict between two or more types of recreation, for example, angling, boating, and sub-aqua activities.

Having decided to intervene, planners have to decide what to do to prevent either damage or conflicts. The possible solution depends on the degree of conflict, and whether the great recreational planning cliché has been broached, namely:

> Are visitors by their sheer weight of numbers destroying the beauty they have come to visit?

The possible solutions are as follows. First, to ration access, for example, by price or by a queuing or booking system. Second, to divert use away from the threatened site by encouraging the use of spoiled or already urbanised sites (honeypots). Third, to create new and different attractions, for example, theme parks, or country parks. Fourth, by providing a better home environment, thus taking away some of the need to visit the countryside. Fifth, by encouraging different leisure uses in different areas.

For example, instead of saying, 'Don't do this', say, 'Do this, it will be better.' Such an approach can also use education and example, based on the Reithian belief which dominated the BBC until recently, namely that public servants have a duty to educate as well as to entertain, and that a basic duty of public life was to improve people to live more rewarding lives.

Cynics could say that this approach was similar to the conspiracy outlined above in which recreational sites, like conservation sites, are allocated by the pyramid concept. And to some extent one can see common features in that both approaches in essence have three big themes:

The conspiracy theory

1 'Big brother' (farmers and environmentalists) knows best;
2 middle-class recreation areas should be protected;
3 the 'rest' should be encouraged to enjoy passive (sic) recreation in honeypots.

The Reithian theory

1 The scientific elite knows best;
2 fragile natural environments should be protected from recreational over-use; but elsewhere
3 education and positive planning should encourage everybody to indulge in a much wider range of recreation spread across much wider tracts of the countryside.

The Reithian theory thus provides a link back to managing the 'natural' environment which has been reckoned by Glyptis (1991) to involve two related approaches. The first is to recognise recreation as an important rural land use, and an important contributor to an integrated social, economic and conservation strategy for the countryside. The second is to keep human needs to the fore, not to the detriment of the natural environment, but as the prime *raison d'être* for wishing to conserve it.

The complex and ever-changing strands between conservation and recreation ideologies have thus informed and will continue to inform the wider debate about planning for the 'natural' environment, as the next section, which traces the evolution of legislation will clearly show.

B: EVOLUTIONARY EVENTS

Events within the United Kingdom

Evolution to 1949. As the box below shows, the traditional approach to planning for the 'natural' environment has been to protect sites of high conservation value and encourage recreation elsewhere. This attitude can be traced to the initial growth of rural recreation in Victorian Britain which romanticised wild landscapes and the practice of field sports in such environments, notably hunting and shooting. Such pursuits were carried out by wealthy people on their own land. In contrast, the growing number of middle-class people created by the Industrial Revolution wished to ape the pursuits of the upper classes and the *nouveaux riches*. In time this desire also filtered through to the working class, and by the mid 1930s a mass movement of ramblers and cyclists had been formed who desired access to the wild landscapes owned by the upper classes. This led to an inevitable clash between the landowning classes and the recreationalists. Matters were brought to a head by stage-managed 'mass trespasses' of Kinder Scout, a part of the Peak District theoretically highly accessible from the major cities of Manchester and Sheffield, which has been graphically described by Hill (1980), one of the key activists in the struggle to gain access to the countryside.

Public land management spectrum: The traditional approach to reconciling conservation and recreation

High conservation value area	Low conservation value area
Low public access to protected site	High public access to Country Parks

Many of these activists were also keen naturalists and they realised that the best way to gain access to the countryside would be to assert their environmental credentials and call for National Parks on the American model as a long-term goal. In the short term they would seek to gain access by demonstrating their love of natural countryside and concern over its future. Thus was born the alliance between: a) countryside conservation by designation; and b) recreational access for healthy outdoor types interested in quiet enjoyment of the natural beauty of the countryside, which has so dominated postwar thinking.

The first fruit of this alliance was a weak piece of legislation, the 1939 Access to the Mountains Act. Progress was then made in the socialist atmosphere of wartime with a key report by John Dower (Cmd 6628, 1945) whose son became not only the National Park Officer for the Peak District in the 1980s, but also the Director of the Countryside Commission in the 1990s. Crucially the Dower report set the tone for the postwar alliance between conservation and quiet informal recreation in certain landscapes when it said that a National Park should be:

> An extensive area of beautiful and relatively wild country, in which for the nation's benefit and by appropriate national decision and action:
> a) the characteristic landscape beauty is preserved;
> b) access and facilities for public open air enjoyment are amply provided;
> c) wildlife and buildings and places of historical and architectural interest are suitably protected; while
> d) established farming use is effectively maintained.

The Dower report was followed by a report from the Hobhouse Committee (Cmd 7121, 1947) which largely confirmed Dower and both recommended that most of the designated areas should be in the uplands, although some areas in the lowlands were proposed, for example, the Cotswolds, Chilterns, the Downs and the Norfolk Broads, but not as National Parks. In contrast, Scotland with its large expanses of remote upland was not thought to need National Parks beacuse of a more lenient trespass law, less pressure of population, but also because of the huge power of the Scottish landowning lobby (Cmd 6631, 1945 and Cmd 7814, 1947). Instead National Scenic Areas were eventually created in which extra development control powers became available. Other reports examined footpaths and access (Cmd 7207, 1947) and the conservation of nature (Cmd 7122, 1947).

The 1949 Act. Although all these reports did lead to legislation it came at the end of the Labour Party's radical transformation of Britain, and when the Labour Party was running out of steam. In addition, many of the potential powers that an environmental act could have had, had already been included in the 1947 Town and Country Planning Act or were precluded by the emphasis given to agricultural production in the 1947 Agriculture Act. Thus the opportunity to create an integrated rural planning system was lost, and instead a rather limited, but at the same time contradictory, set of goals were legislated for in the 1949 National Parks and Access to the Countryside Act. This Act set out:

> To make provision for National Parks and the establishment of a National Parks Commission, to confer on the Nature Conservancy and local authorities powers for the establishment and maintenance of nature reserves, to make further provision for the recording, creation, maintenance and improvement of public paths, and for securing access to open country, to amend the law relating to rights of way, and to confer further powers for preserving and enhancing natural beauty.

A number of themes can be seen here, notably the creation of National Parks and nature reserves, followed by the preservation and enhancement of their natural beauty, and developing public footpaths and access to open country. These are now considered in turn until the next period of legislation in the 1960s.

First, a National Parks Commission and 10 National Parks were set up in the 1950s and a start was made on the designation of Areas of Outstanding Natural Beauty. However, both the National Parks Commission and the local government National Park Committeees set up to administer the Parks were given conflicting roles by the Act, namely:

1 to preserve and enhance natural beauty (Sections 1 and 5); and
2 to encourage the provision or improvement of facilities for the enjoyment of National Parks by open-air recreation and the study of nature (Section 1); but
3 to have due regard to the needs of agriculture and forestry (Section 84).

At the time these didn't seem to be in conflict since most recreational users were dedicated walkers who knew and respected the countryside, and the 'dog and stick' farming techniques then in vogue didn't pose much of a threat to the traditional landscapes of the uplands, and indeed had created the so called 'natural' beauty of the Parks. Thus it didn't seem to matter that no priority was given to the two key aims of beauty and recreation.

More problematic at the time was that there were inadequate resources of cash and staff. Most staff were seconded on a short-term basis, and thus no National Park culture or career structure could develop. Not surprisingly apart from the fact that 20 per cent of England and Wales was designated as either National Park or AONB in the 1950s, little else was done. By the same token nature reserves began to be designated in the 1950s by the Nature Conservancy but little thought went into their management. It was perhaps naively thought that designation would be enough and that the Nature Conservancy should concentrate on scientific research.

Second, under the Act public footpaths had to be shown on definitive maps to be collated by county councils at a large scale. They then had to·be transferred to the new 2½-inch (1/25,000) maps newly created by the Ordnance Survey for use by ramblers. In addition long-distance footpaths were to be created by negotiation. In practice little work of this nature was accomplished in the 1950s, and similarly very few access agreements were negotiated.

Thus, by the time the 1960s arrived little progress apart from designation had been made. Crucially, though, the ideology of 'quiet enjoyment had become entrenched in policy, and the concept of nature had been appropriated by the NCC as a scientific one which could be used as a value-free and neutral arbiter of conflicting demands on the countryside. However, the real world had moved on. In particular the demand for recreation had grown rapidly with greater affluence and more leisure time, and in the medium term a rapidly rising population drew forecasts leading to the spectre/promise of a so-called fourth wave of change crashing over the countryside in the form of a mass recreational movement. This wave would not only be large, but it would be different in kind, composed no longer of informed caring walkers, but instead of car-borne people, ignorant of country ways and in search of often noisy pursuits based on motor engines. Interestingly enough the man who gave most prominence to this fourth wave was the son of John Dower, Michael Dower (1965) though he did not necessarily subscribe to the doom-laden scenario predicted above by those who saw the fourth wave as a threat, rather than the dawn of a new utopia.

The 1968 Act. The newly elected Labour Government, buoyed up by its vision of a meritocratic society, saw the fourth wave as an opportunity, but one tempered by the still-powerful farming lobby when in a 1966 White Paper (Cmnd 2928, 1966) entitled *Leisure in the Countryside* they wrote:

> Given that townspeople ought to be able to spend their leisure in the countryside if they want to, that they will have more leisure and that in future they will be able to buy more cars and boats and otherwise spend money on their weekends and holidays, the problem is to enable them to enjoy this leisure without harm to those who live and work in the country and without spoiling what they go to the countryside to seek.
>
> (3–4)

Two Acts followed, the Countryside (Scotland) Act 1967 and the Countryside Act 1968. In both cases Countryside Commissions were set up with a number of recreation and conservation duties, but since the conservation duties infringed on the work of the Nature Conservancy Council they were seen to be covert recreation agencies. Curry (1994) has described the Acts' recreation and access provisions as slender, and has argued that as a result countryside management had to invent innovatory mechanisms outside the statutory framework. Nonetheless, the apparent thrust of the Acts was a series of duties designed to deal with the explosion of new forms of leisure in the 1960s, in particular, duties for the Commissions to:

1 provide and improve facilities for the enjoyment of the countryside mainly through grant-aiding the public and private sectors; and
2 to secure public access to the countryside for the purpose of open-air recreation; and for all public bodies under Section 11 of the Act to have regard to the provision of recreation, thus potentially opening up much land in public ownership, notably forests and land around upland reservoirs. However, Section 37 enjoined public authorities to have due regard to the needs of agriculture and forestry and to the economic and social interests of rural areas, thus, as in the 1949 Act, effectively hamstringing them.

Nonetheless, by replacing the National Parks Commission in England and Wales with the Countryside Commission, and by setting up the Countryside

Commission for Scotland, with general powers to review countryside conservation and recreation the Government did set up agencies which could prick its conscience from time to time.

The Acts are mainly remembered, though, for the new areas they created, Country Parks. These must be: a) accessible and have the capacity to absorb large numbers; b) capable of deflecting pressure from protected/fragile areas; and c) able to prevent pressure on farmland notably in the urban fringe. They can be run by either the public or private sectors and in both cases grant aid is available. The concept was extended in Scotland with powers to create Regional Parks under the Countryside (Scotland) Act 1981. Returning to the 1960s, another new type of recreational area, Picnic Sites were created as a rather poor relation of Country Parks. In all these cases the conspiracy theory of keeping the countryside clean for countryside cognoscentii as outlined in the preceding section A can be seen to be behind the legislation. The main thrust of the 1968 Act was thus one of controlled access and restrictive control-based policies (Curry, 1994), and a tacit acceptance that the radical hopes of both Dowers and the watered-down 1949 Act had failed, and that a more limited agenda had to be followed.

If the 1960s was the decade of recreational excess, the hangover came with a vengeance in the 1970s. First, the postwar boom came to an abrupt end, and second, people began to believe that there was an environmental crisis. In the field of 'natural' environment planning two major problems had also emerged by the 1970s. First, a serious conflict between recreation and conservation, and second, the fact that National Park administration was local not national with two-thirds of the Committee composed of local councillors and only one-third being appointed by the DOE (Brotherton, 1985). This meant that local economic interests often took precedence over conservation in terms of the built environment, while the large number of local farmers on the Committees (MacEwen, 1982a) meant that recreational access was hardly a priority.

National Park issues in the 1970s. A major change for National Parks came, however, with the Local Government Acts of 1972 and 1974, which ushered in the following changes from 1974: a) each National Park to have a permanent staff and senior officer (a National Park Officer) in charge. However, there was no change in the committee structure; b) each park to produce a five-year management plan. Typically these have employed zoning to divide parks into a spectrum from extensive recreation areas to honeypot areas; c) 75 per cent of the budget to be paid for nationally and the budget to be doubled. This was a great improvement and has been sustained; however, the sums involved are very small.

Ironically at the same time as the reorganisation of National Parks was taking place the Report of a National Parks Policy Review Committee, popularly known as the Sandford Report (Environment, 1974) was published. This made a number of findings and recommendations. The main one was that in some places the conflicts between recreation and conservation had become irreconcilable. Where this was the case the report recommended that conservation should take precedence. The Government reply (Environment, 1976) endorsed the precedence of conservation over recreation but only as a last resort.

Organisational changes. The early 1970s also saw changes to the Nature Conservancy, which had already begun life in 1949 prior to the 1949 Act but was given a big boost by it. It had suffered a setback in the 1960s when it became part of the newly formed Natural Environment Research Council (NERC) but in 1973 it was restored but renamed as the Nature Conservancy Council (NCC) in the Nature Conservancy Council Act, 1973. It then became subdivided into two related bodies: a) the NCC, to oversee trends and manage sites; and b) the Institute of Terrestrial Ecology (ITE), to carry out and fund research in addition to research carried out by NCC and NERC.

In more detail the NCC was given more powers than the Countryside Commissions created in the late 1960s in that it had executive powers to establish and maintain National Nature Reserves (NNRs), and to designate and protect Sites of Special Scientific Interest (SSSIs). In addition the NCC had a much bigger budget, more staff, and also covered Great Britain. In common with the Countryside Commissions the NCC was also given powers to: a) provide information and advice on nature conservation; b) grant aid research and practical projects; and c) advise the Government on nature conservation.

Further changes came in the early 1980s when both Countryside Commissions were made independent grant-in-aid bodies and staff were no longer civil servants. Significantly the Countryside Commission was relocated from London to Cheltenham, while its Scottish counterpart remained in Perth. The Commissions' three main functions had by then clearly evolved to be: conserving landscape beauty in the countryside; developing and improving facilities for access and recreation; and advising Government on matters of countryside interest. Problems with these functions included the lack of priority between the various roles and thus the problem of reconciling clashes of interest. Three main areas of work were carried out: 1) acting as a catalyst since neither commission had any landowning power, no executive authority, and no real money; 2) providing advice on the ground and by reports; and 3) commissioning research and publishing documents. During the 1970s and 1980s the Countryside Commision had about 100 staff, a budget of around £20 million and was run by 13 Commissioners nominated by the Government. The main criticism made of the Commissioners was that they were weighted towards farming interests (Lowe, 1982 and Brotherton and Lowe, 1984) and also towards the Conservative voluntary/financially induced approach to conservation.

The NCC also became a grant-in-aid body in the 1980s with new headquarters in Peterborough, run by 15 nominated councillors and a staff of around 800 spread around the country. Typical NCC expenditure in the 1980s was about £36 million, of which about half would have been spent on staff and support costs. Of the remainder, £1.9 million was typically spent on the direct maintenance of owned NNRs, £5.1 million on scientific support, £5.2 million on management payments, lump sums and leases, and £2.4 million on grants.

The 1981 Act. The big change of the 1980s, however, was the implementation of the 1981 Wildlife and Countryside Act (Denyer-Green, 1983), which brought in major changes to the philosophy and practice of conservation and also introduced the disastrous principle of paying compensation to those people denied private profit in the public interest. The MacEwens (1982b) called it an unprincipled Act in that the payment of compensation flew against the practice in all other areas of land use planning, and also because, practically, it was a 'dead end' idea because it would be pouring small amounts of money into a bottomless pit.

The Act relied heavily on the 'voluntary' principle backed up by compensation and the threat of compulsory purchase if persuasion failed, and thus employed progress along the Gilg/Selman spectrum as its basic principle. However, the first part of the Act, Sections 1 to 27, dealt with the negative protection of flora and fauna. This is the area of the last chance saloon and a reactive response to wildlife preservation when a species has nearly reached the extinction stage. The Act thus included a revised list of endangered species, which is reviewed regularly by Statutory Instruments. The basic protection thus afforded by this part of the Act is a list of species and 'thou shalt not' rules, for example, take eggs, pick flowers, and kill species, backed up by fines and even prison sentences.

Part II of the Act took on the much more complex task of managing habitats so that species would no longer become endangered. It initially concentrated on SSSIs, but as the Bill went through Parliament the principles were extended to virtually all habitats. Nonetheless, the Act still has a number of sections referring to SSSIs. Some 2,600 SSSIs had been designated under the 1949 Act, but they were notified only to LPAs so that they could take the nature interest into account when they considered a planning application that might affect them. Farmers were often unaware that they had an SSSI on their land, let alone how to manage it. Thus the first innovation of the 1981 Act was to redesignate existing SSSIs and inform landowners of the fact. This process of re-notification was one of the most valuable features of the Act. However, it was what happened once re-notification or a new notification had taken place that caused controversy.

The basic procedure (Splash and Simpson, 1994) is that under Section 28 farmers have to give three months notice to the NCC of any intended change to an SSSI. Section 29 then provides a back-up power delaying the notified change for up to a year or in extreme cases compulsory purchase. More normally, however, Section 30 is invoked, which allows the payment of compensation if the operation of Section 28 and 29 has reduced the value of the land. In addition under Section 32 the NCC had to offer a management agreement with compensation where grant aid to improve (sic) the site with MAFF money has been turned down. The whole procedure was backed up by a Code of Practice and by a detailed DOE circular (Environment, 1983). Finally, Section 31 provided powers to restore a site that had been damaged illegally. However, such restoration is relatively meaningless, if, for example, a medieval woodland has been destroyed.

Apart from the objectionable principle of paying compensation the Act also opened up the possibility of spurious or even fraudulent claims by landowners who had no real intention of damaging an SSSI, but who knew that the NCC would have to call their bluff, and thus incur not only huge, but never-ending expenditure. At the time it was estimated by Adams (1984) that this could total £850 million over 20 years, with no guarantee that the site would be preserved if the compensation payments ceased. Furthermore, the cash for compensation came not from the mighty MAFF, the body causing the momentum behind improvement plans, but from the tiny budget of the NCC. Finally, the compensation included not only the profits foregone, and any loss of land value, but also the grant that might have been paid. This can be likened to the case of a potential university student who, on being denied a place at university, would not only be paid compensation for not being offered a place and the potential extra earnings that a degree should bring, but also the free fees and the loan/grant they would have been paid or been eligible for. This analogy shows the devastating fallacy at the heart of the 1981 Act.

Worse still, this principle was extended to other designated areas during the passage of the Act as the farming lobby saw that the Act was not a threat, but a way of seeming to be environmentally friendly, while at the same time earning money (Cox and Lowe, 1983). Thus under Section 39 all LPAs can make management agreements, and under Section 41 if an LPA or National Park Authority object to a farm capital grant, and grant aid is refused, they must offer an agreement with compensation (this obligation was however withdrawn in the 1990s). Though this was only to some extent an extension of management agreements which could already be negotiated under the Countryside Act 1968 and the Town and Country Planning Acts, it was the new principles of how compensation should be calculated that caused so much adverse comment and counter-comment at the time, notably by Ball (1985).

The procedures were also criticised by the Environment Committee of the House of Commons (1985) who called for an internal reappraisal of MAFF attitudes to give a greater priority to conservation, and to direct financial aid away from environmentally damaging operations to conservation-conscious methods, thus slowing the engines of destruction. They did not, however, challenge the voluntary principle, and in their reply (Cmd 9522, 1985) the Government welcomed this and argued that their policies were moving towards conservation already and that the amendments made by the 1985 Act, discussed next, had dealt with any problems.

The Act was amended in 1985 by the Wildlife and Countryside Act 1985, which gave the NCC more time to protect SSSIs and closed some loopholes. It also extended the duty of National Park Authorities under Section 43 to make maps of those areas that should be conserved from just moor and heathland to many other areas of natural beauty.

Recreational issues between 1981 and 1995. Returning to the 1981 Act, Part III of the Act dealt with recreation, and in particular Section 53 put county councils under a duty to keep the definitive footpath map under continuous review (instead of reviewing it at specific intervals). This recognised the lack of zeal in setting up designated maps, especially in Tory councils dominated by landowners. So it was an acceptance of a status quo. Other sections dealt with bulls and ploughing up fields with paths. In general the Act weakened the power of planners and irritated pressure groups like the Ramblers Association, who coordinated opposition to the Act and notably the problem of bulls and path ploughing up which they felt gave farmers too much leeway. This part of the Act prompted more letters to *The Times* newspaper than any other issue the staff could recall, but all to no avail.

Some ground was recouped by the Rights of Way Act 1990 (Garner, 1993), which placed more onus on farmers to keep paths clear and gave greater powers for LAs to de-obstruct and/or restore them and to send the bill to farmers. The Act was prompted by much evidence of paths being obstructed and LAs failing to sign paths (Harrison, 1991). In 1994, however, the Criminal Justice and Public Order Act 1994 introduced the new offence of 'aggravated trespass' designed to prevent illegal music festivals, hunt saboteurs, and new age travellers trespassing onto land. This was seen by some as a retrograde step, but legitimate ramblers should not be affected (Cm 3016, 1995). Another factor reducing access has been the privatisation of public bodies since 1979. These bodies, which control or controlled large areas of attractive countryside, notably the Forestry Commission and the Regional Water Authorities, had been increasingly welcoming, and although safeguards have been written into the transfer of public sector land to the private sector, many commentators remain sceptical about the nature of such access, fearing the imposition of high entrance charges (Groome, 1993). In the longer term the Countryside Commission had already decided to make the 1990s the decade when the promises of the 1949 Act would be put into practice, with regard to the footpath network. They began by setting out the problems, which included: poor information about facilities;

poor access; and conflicts of interest (Countryside Commission, 1987a). The Commission then set out to achieve the objective of improving and extending opportunities for the public to enjoy the countryside by: making the public more aware of facilities; providing well signposted and maintained networks of footpaths; and by providing access to wider areas of the countryside. The Commission (1987b) then set out their priorities:

1 by the year 1990 all LAs and public bodies to have been made aware of their duties and the need to establish the rights of way network by the year 2000;
2 by the year 1992 all farmers to have been made aware of their rights and responsibilities; and
3 by the year 2000 major media networks to provide information about countryside recreation so that by then all people are aware of their rights to use the footpath network.

In 1992 and 1993 the Commission gave further impetus to improving access by setting up the Parish Paths Partnership initiative and the Milestones initiative designed to boost interest at the local level in both setting up schemes and assessing overall progress (Cm 3016, 1995).

The Uplands and the Broads. Returning to the 1980s, another big issue was the future of the uplands, the United Kingdom's largest wildlife area, albeit one consisting of wet desert, rather than the rich natural woodland it could be. Unfortunately, no national policy had been evolved for these regions, and they had become subject to many conflicting policies, notably forestry, moorland reclamation, recreational access and so on. Accordingly, the Countryside Commission (1984c) called on the government to state the role the uplands should play and in particular argued that the economic base of the uplands should be widened, and their environment should be better conserved by modifying the grant system to favour land management rather than farming and by bringing major afforestation under development control.

In their reply the Government rejected most of these proposals but floated the idea of Landscape Conservation Orders. These were dropped in 1988. However, a similar idea surfaced in the same year in the legislation setting up the Broads National Park. The idea was that key areas would be designated in which harmful operations would have to be notified. An order could then be made prohibiting them in return for compensation. In essence it is a parallel idea to the ESAs already discussed in Chapter 3, and represented a shift from reactive thinking to a more positive position in which farmers are made aware of their responsibilities.

This more positive guidance was also enshrined in the 1988 legislation creating the Norfolk and Suffolk Broads National Park, Britain's 11th and the first for 30 years. It was also the first genuine lowland park, except perhaps for the Pembroke Park. The area had witnessed a history of clashes between recreation and conservation, and farming and pollution/landscape change since the 1960s, all of which came to a head in proposals to drain the Halvergate marshes in the early 1980s (O'Riordan, 1985b). The area is indeed a microcosm of the issues involved in planning so called 'natural' environments, since the Broads evolved from medieval peat diggings which then flooded, leading to a finely balanced ecosystem based on certain land uses none of which can change without threatening the others. There was therefore a need for someone to hold the ring and provide a lead for consensus.

So, in 1988 after much consultation by the Countryside Commission (1984d) the Norfolk and Suffolk Broads Act 1988 set up the Broads Authority as a new type of National Park Authority but one which nonetheless did repeat many of the mistakes of the existing Parks. Its duties are to: a) conserve and enhance the natural beauty of the Broads; b) promote the enjoyment of the Broads by the public; and c) protect the interests of navigation.

In order to do this the Broads Authority prepares a plan and a list of habitats it is particularly important to conserve, and a list of operations in these areas which it might wish to control by notification leading to the offer of a management agreement. The Act thus endorsed but extended the principles of the 1981 Act and set up a full-scale experiment into whether this type of approach could work. At the time most commentaries, for example O'Riordan (1990), gave the approach a guarded welcome as moving away from limited ad hoc approaches to long-term strategies, albeit one still based on compensation for its success in the short term.

The Edwards Report. The advent of the Broads Authority prompted a review of National Parks (Countryside Commission, 1991d) in the Edwards Report (after its chairman), which began by reporting on good progress, notably in expenditure, which had risen fivefold in real terms since 1974. It then made a number of incremental rather than revolutionary recommendations:

1 a new National Parks Act to redefine the two aims of:
   a) conserving the environment, and
   b) promoting quiet enjoyment;
2 a new farm-support system based on managing the landscape via individual farm plans, with Landscape Conservation Orders as a last resort; and
3 independent National Park Authorities with more planning powers should be set up following the precedent of special National Park Boards in the Lake District and Peak District National Parks.

In their reply the Countryside Commission (1991b) gave a virtually full endorsement of the report and its proposals and proposed to pursue the introduction of a new statutory statement of National Park purposes, restated their intention to press for the creation of independent National Park Authorities, and called for the New Forest to become a National Park. Although reluctant at first, the Government eventually met many of the calls of the report, and the Commission, in the Environment Act 1995, notably: the creation of independent National Park Authorities to be constituted as follows. One half plus one to be appointed by local authorities, one quarter to be appointed by ministers from parish councillors, and one quarter by ministers from anywhere. The national interest could thus further be diluted and the farming lobby be even more over-represented. The authorities were given two new duties: first, to foster the economic and social well-being of their local communities; and second, Government departments and other public bodies were to have regard to National Park purposes in carrying out their functions in the Parks. In addition the two main purposes of conservation and recreation were redefined, in the expectation that they could be reconciled, but that in the few cases where they could not be, that conservation would have precedence (Cm 3016, 1995). The New Forest was not, however, made a National Park, instead in 1994 it was decided that the same planning policies would apply there as if it were a National Park (Cm 2807, 1995).

Further organisational changes and the role of other groups. The 1990s also witnessed changes in the agencies responsible for planning the natural environment at the national level and the Environmental Protection Act 1990 set in train the following changes. First, the NCC was broken up, largely because it had been too critical of the Government and had overplayed its devil's advocate role. Thus in 1991 a new organisation, English Nature, was created with responsibility for England. In Wales the existing NCC staff were merged with the Welsh staff of the Countryside Commission to form the Countryside Council for Wales. Similarly in Scotland, former NCC staff merged with the Countryside Commission for Scotland to form Scottish Natural Heritage, under the provisions of the Natural Heritage (Scotland) Act 1991. In order to allay criticisms that the national and international perspective would be diluted by these changes a Joint Nature Conservation Committee was set up to take a British view, and to represent Britain at the increasingly important international level.

Many people pointed out the inconsistency in combining the conservation bodies in Scotland and Wales, but not in England, and so the Government examined the merger of English Nature and the Countryside Commission in 1994, but rejected it at least *pro tem*, in spite of evidence from Northern Ireland where the conservation division of the Northern Ireland DOE has been argued by Wilcock (1995) to provide a possible model for Great Britain. In addition, the need for a holistic view of the environment has been one of the themes in postwar planning, and ironically

DOE advice to LPAs in two circulars on nature conservation (Environment, 1977a and 1987b) stressed that nature conservation must be taken into account in all activities which affect rural land use and in the planning process, since the survival of the nation's wildlife cannot be achieved solely by site protection.

This theme was continued in PPG9 (Environment 1994b) but was specifically linked to countryside features which provide wildlife corridors, links or stepping stones from one habitat to another, since these help to form a network necessary to ensure the maintenance of the current range and diversity of our flora, fauna, geological and landform features. PPG9 goes on to advocate for the first time that LPAs should include policies for managing features of the landscape which are important for wild flora and fauna. This represents a significant departure according to Southgate (1995) because previous Government policies emphasised that management issues had not been a land-use planning matter.

PPG9 also reminds us that environmental planning is not just the concern of selected government agencies, but also of LPAs and others. For example, LPAs include plans for nature, recreation and the landscape in their Structure and Local Plans, reactively through the operation of development control and TPOs, and pro-actively by seeking management agreements and access agreements, as when forestry land is sold off. Other powers include seeking Capital Tax Relief for landowners who agree to maintain outstanding scenery (Countryside Commission, 1986a); grants to purchase heritage countryside under the National Heritage Memorial Fund set up by the National Heritage Act 1980; the use of Countryside Commission, MAFF, NCC/English Nature, and Forestry Commission tree-planting and similar landscape improvement grants (Countryside Commission, 1986b); grants for countryside conservation and recreation (Countryside Commission, 1986c); and hedgerow grants (Countryside Commission, 1993b). These typically take the form of 50 per cent grants to farmers, public bodies and voluntary groups.

Voluntary groups are also an important factor, as pressure groups, but increasingly as major landowners. The expansion of both activities has been allowed by an explosion in membership, which doubled overall in the 1980s (McCormick, 1991), with some groups showing explosive growth. The pattern of membership is not, however, even, and Cowell and Jehlicka (1995) have demonstrated sharp regional differences in group membership, providing qualified support for the frequently made assertion that concern for environmental issues is felt predominantly in southern England. They also highlight very low membership densities in the former industrial heartland and ex-metropolitan counties. Ironically, those with the most favoured environments appear to be keener on environmental improvement, than those in poor environments, although the explanation is probably one of disposable income, rather than differential desires for good environments.

The pre-eminent group is the National Trust, which derives its power from the National Trust Act 1907 and three other main sources. First, it has well over 2 million members and although many of them see membership as a cheap form of access to historic buildings and landscapes rather than as membership of a lobby they still represent a considerable culture dedicated to preservation. Moreover, it is a rapidly growing constituency with membership rising from a mere 100,000 in 1960, to 1 million in 1980, and doubling to over 2 million by 1991 (Groome, 1993). These figures also illustrate the way in which countryside conservation and recreation have become a major feature of British culture in the last 20 years. Second, the Trust has become one of Britain's largest landowners, with over 240,000 hectares and 550 miles of coastline, and although some of this land has been given to it as a source of income, for example, farming land, increasingly the Trust is pioneering holistic land-management policies on all its land. This is a change from a former policy which placed preservation on a pedestal and largely ignored nature conservation. In total the Trust spent 86 per cent of its income in 1994–5, or £109 million, on its primary purposes of preservation of the coast, countryside gardens and houses, more than the entire Government conservation budget. Third, the Trust has the power to declare land inalienable and thus free from any development, even by the Government. Unfortunately, the Government has attempted to break this rule four times, has succeeded twice, and tried again in 1994–5 when it proposed to build a road through Trust land in Dorset. It is important that inalienability is only rarely challenged because giving land and property to the Trust in the belief that it will always be preserved in good hands is a very important factor when people face their mortality. In conclusion the Trust is a powerful if elitist and traditional force for natural land management.

The Council for the Protection of Rural England, founded in 1926, is in contrast a small organisation which works almost entirely as a pressure group. Small it may be but it is a vocal group with influential spokespeople and chairmen often drawn from the media world. The Royal Society for Nature Conservation is a more scientific and gentlemanly organisation which acts as the umbrella organisation for county Wildlife Trusts who are potentially powerful landowners as they slowly buy up or are given key sites to manage, often with the help of dedicated volunteers. Likewise the Woodland Trust is a small but effective landowner of important woods. In contrast the Royal Society for the Protection of Birds (RSPB) has over 1 million members and like the National Trust is exerting a growing influence via land ownership as membership rises and with money from the National Lottery (Lawson, 1995). Finally, a relatively new group, SUSTRANS was awarded £42 million by the National Lottery in 1995 to develop a nationwide cycle network mainly along redundant railway lines, which are not only flat but have become important wildlife habitats since they were closed mainly by the Beeching axe of the 1960s.

In conclusion, this section has shown that there are many organisations involved in conservation and recreation. Indeed, the following bodies listed by Curry (1994) could have been added, if space had allowed: the Rural Development Commission responsible for the development of tourism in rural areas; the Forestry Commission responsible for recreation in state and private forests; the British Waterways Board responsible for recreation on inland waterways; the English Tourist Board responsible for encouraging tourism; the British Tourist Authority respnsible for encouraging visitors from abroad; the Sports Council responsible for providing facilities for sport including walking; district and county councils responsible for recreation provision via plans and development control; the National Rivers Authority responsible for recreation on rivers; and ADAS responsible for farm-based recreation. In addition, Curry notes that several government departments had responsibilities in this area over and above the DOE, notably: the Department of Education; the Department of Employment; MAFF; the Department of National Heritage; and the Department of Trade and Industry in the early 1990s. Groome (1993) has also pointed out the lack of coordination between local, regional and national recreational planning, and the need for strong regional guidance. This paragraph thus confirms one of the two main propositions of this book, that countryside planning is fatally flawed by sectoral fragmentation.

The impact of the international community

Nature and pollution are no respecters of international frontiers; for example, acid rain falls far away from its source (Tickle, 1994). Britain stands at a very busy natural crossroads which enhances its wildlife, but also places certain responsibilities on Britain to safeguard sites for migratory species. The international community has had a long history of wildlife and environmental conferences, but until the 1960s most of them were talking shops which achieved little practical progress (Gilg, 1981). However, since then major advances have been made, although noble words still dominate over effective action, and site protection or controls over endangered species remain the favoured weapon, in contrast to measures which address the underlying causes rather than the symptoms.

For example, the Ramsar convention, first agreed in the Turkish town of the same name in 1971 allows signatories to designate important wetlands, and the CITES convention controls the trade in endangered species, notably in their furs or skins. Useful as these are they represent the bottom line in conservation, and need to be backed up by wider strategies, notably at the global level.

Global initiatives began in earnest with the 1972 Stockholm conference on the environment. The conference outlined the basic concept of what was later to become known as 'sustainable development' and set out a first principle to follow: 'Man has the fundamental right to freedom, equality and adequate conditions of life in an environment of a quality that permits a life of dignity and well-being.' The start made at Stockholm was followed by the World Conservation Strategy (WCS) which was launched in 1980 with three main objectives: 1) to maintain essential ecological processes and life-support systems; 2) to preserve genetic diversity; and 3) to ensure the sustainable use by us and our children of species and ecosystems.

The WCS was then further developed by the World Commission on Environment and Development (1987), a United Nations organisation which set out a series of principles for the year 2000 in a report entitled *Our Common Future* or the 'Brundtland Report' after its chairperson, the Norwegian Prime Minister. These were based on the fundamental concept of sustainable development (commonly defined as *development that meets the needs of the present without compromising the ability of future generations to meet their own needs*) and the first principle of the 1972 Stockholm Declaration from which a number of measures were recommended which included: maintaining ecosystems and biological diversity; observing the principle of optimum sustainable yield; and preventing or abating pollution by establishing environmental protection standards.

Major progress was made in 1992 when the impetus built up by the Stockholm conference, the WCS and the Brundtland Report was followed by the United Nations Conference on the Environment and Development (UNCED) held in Rio de Janiero 1992, and attended by world leaders. Three main agreements emerged from Rio: first, a biodiversity agreement on protecting plants and species; second, a treaty on avoiding global warming; third, the so-called Agenda 21, a 800-page blueprint for action to lead development into environmentally sound areas, to be carried out by member states in their own countries.

The UK response to Rio was published in 1994 in four White Papers, one of which, on forestry, was discussed in Chapter 4. The first White Paper (Cm 2426, 1994) built on existing policies, and aimed to work through the Government's flagship mechanism, the market. In particular, energy prices were to be raised above inflation to 'increase people's awareness of the part their personal choices play in delivering sustainable development' and thus to create a counter-car culture, notably in terms of long-distance commuting and out-of-town shopping. Energy use, indeed, formed the focus of the second White Paper (Cm 2427, 1994) on 'Climate Change.' This proposed a long-term plan to raise petrol prices by 5 per cent more than the rate of inflation in order to cut emissions of carbon dioxide to 1990 levels by the year 2000. Plans to raise VAT on domestic fuel from 8 to 17 per cent had, however, been defeated by a backbench revolt in 1993, demonstrating how difficult it would be to achieve sufficient changes to a culture now so dependent on energy use. The third White Paper, on biodiversity (Cmd

2428, 1994), pledged to maintain habitats and native species, and was probably the most conservation-committed document of the four.

In spite of much apparent progress the real nettle of environmental politics has yet to be grasped, namely the gross disparity between consumption in the developed world and the rest of the world. In particular, the United States refused to yield at Rio, on such key issues as who should compensate Amazonian rain forest people, if they were forbidden from cutting down these forests. Nearer to home, it is hard to foresee any political party being elected on a policy that would cut deep into energy use by restricting car use over and above cosmetically. Nonetheless, Rio finally placed the environment on the political agenda. The next task is to accompany the fine sentiments with implementable policies that will have real effects.

Realistically, such measures will have to be put into place by nation states. The United Kingdom's response to Stockholm, the WCS and Rio has been to welcome them, and to claim that it was *de facto* putting most of their recommendations into practice already. Similarly, the UK response to European-scale environmental planning has been to claim the moral high ground, and to modify policies slightly to fit in with international imperatives. In particular, international commitments have been grafted onto the town and country planning system (Rydin, 1995). A prime example was provided in Chapter 5: the introduction of Environmental Assessments, as an adjunct to town and country planning procedures. Elsewhere, the EC/EU has in fact been instrumental in creating legislation that would not otherwise have been passed. For example, Part I of the 1981 Act was forced on the United Kingdom by the 1979 Birds Directive, and the National Rivers Authority was set up in the late 1980s because the EC deemed that it was illegal to place pollution control in the hands of private companies (McCormick, 1991). In addition pressure groups can bypass the British Government and appeal to a 'higher authority' in Europe. However, UK environmental agencies are often precluded from operating outside Britain, and Ward *et al.* (1995) have argued that this has much reduced their effectiveness, since this is where much environmental legislation is derived, notably from Brussels. They conclude that this problem has been compounded by restricting the remits of the Environment Agency and the Scottish Environment Protection Agency, set up by the Environment Act of 1995, to the United Kingdom.

A first attempt at a European policy was provided by a Council of Europe plan of action produced in the 1970s and from which maps of areas of ecological importance were produced (Gilg, 1981). The first European Community wildlife legislation, as opposed to the wider Council of Europe, was provided by the Birds Directive 79/409. The Directive provided basic protection by listing birds not to be shot at at all, or not in certain closed seasons, and by allowing Member States to create Special Protection Areas.

However, there have been severe problems, notably with interpretation and enforcement. There is thus little point in Britain and Scandinavia rigidly enforcing protection if the same birds are to be blasted from the sky in the different cultures of Mediterranean Europe.

Nonetheless, the limited success of the Birds Directive encouraged the EC to propose a wider directive on nature conservation in the mid 1980s. This was issued as a draft called Habitats Naturopa 2000 in 1988. It made slow progress, but was eventually agreed at Maaastricht on 12 December 1991. It became a directive on 21 May 1992 as the Habitats Directive 92/43 (OJ 92/L 206/1992) and came into force in June 1994. It should be fully implemented by the year 2004 (Sundseth, 1994).

In more detail, around 600 species and 200 habitats were due to be protected in 'Special Areas of Conservation' (SACs) (or NATURA sites) which will be similar to the 'Special Protection Areas' set up under the 1979 Birds Directive. SACs will be either habitats of community importance or the habitats of species of community importance. It is Member States that designate the SACs and then notify the Commission. Member States must provide management plans, take steps to avoid deterioration of habitats, and stop any significant disturbance. Proposals to alter the sites will only be given permission if the project is necessary for reasons of overriding public interest. A first list of 136 possible British land-site SACs was sent to the EU in June 1995; as expected most of them duplicated existing protected sites (Southgate, 1995).

The Habitats Directive was accompanied by Regulation 1973/92 OJ L206/1992, or LIFE (Financial Instrument for the Environment). This allocated 140 million ECU in 1991–2, and 400 million ECU in 1991–5 for environmental improvement. Within this total, 45 per cent was earmarked for the protection of habitats and nature, and 40 per cent for the promotion of sustainable development and quality of environment (Sundseth, 1994).

The European Community/Union has indeed had a much longer concern with environmental quality, and has had five Environmental Action Programmes. The first four of these, between 1973 and 1993, focused on pollution and ecological balance. However, as Briggs (1991) has demonstrated limited action took place during the first (1973–6), second (1977–81), third (1982–6) and fourth (1987–1992) Action Programmes. Also the Programmes concentrated on urban and industrial environments and on basic controls via emission standards. Implementation was also weak according to the Select Committee on the European Communities of the House of Lords (1992), with few inspectors and low fines, and though the fourth Action Programme called for effective implementation of directives, notably the 1979 Birds Directive (COM(91)321 final) it was generally agreed that the 200-plus pieces of legislation by then in place (they had grown rapidly, notably in the 1980s), were widely ignored and that subsidiarity was not working in practice, being seen as a means for ignoring legislation rather than adapting it to local circumstances.

Accordingly the Fifth Action Programme (1993–2000) *Towards Sustainability* (COM(92)23) has two big themes: 1) to shift the pattern away from regulatory instruments to collective responsibility; and 2) away from reactive measure to proactive proposals. The Fifth Action Programme has five target sectors: industry; energy; transport; tourism; and agriculture and forestry. In agriculture and forestry the three main aims are to: reduce pollution via controls, for example, Nitrate Sensitive Areas; improve financial incentives for environmentally friendly farming; and introduce new afforestation projects to soak up surplus farmland and pollution. An interim progress report on the implementation of the programme in the United Kingdom was made by the Department of the Environment (1995c).

The Action Programme is reinforced by the Directive on Integrated Pollution Control, which has the basic concept that controls over one pollutant only lead to another equally damaging pollutant being substituted. Thus overall control is needed. From this concept stem the principles of BEPO (Best Practicable Option), the polluter pays principle and the precautionary principle. Controls are, however, only seen as one measure while others like the 1993 Regulations on Voluntary Eco-Audit promote a culture of responsible environmental management, which will then complement the traditional 'command and control' approach based on the regulation of emissions.

Finally, information remains a key requirement. In this regard the European Environment Agency which was agreed in 1990 but only set up in 1995 in Copenhagen will be a useful assembly point for environmental information. Nonetheless, those who approach the Agency mistakenly thinking that it has any executive powers will be disappointed, since its remit is restricted to the collation and provision of information (European Environmental Agency, 1995). However, according to some activists, information is power and so the Environmental Information Regulations which came into force in 1994 under Regulation (90/313) will be useful (Poustie, 1993) since they allow anybody to obtain information, and thus pressure groups will be able to use much more informed data when pressing for environmental controls.

Turning to the future, the lack of any mention of environmental protection in the 1957 Treaty of Rome was remedied by the Single European Act of 1986 (Cole and Cole, 1993) and the Maastricht Treaty of 1992 which (under Title XVI/Environment Article 130r) dealt with the environment by setting four objectives and five principles which included all the measures outlined above, and the crucial principle that environmental protection should be integrated into all EU policies.

In conclusion, although the European Union is exerting a growing influence on British environmental policy, as are the Rio agenda and other international agreements, these tend at worst only to reinforce existing British practice, and at best, push Britain a little faster down the environmental road (Morphet, 1992).

Summary

Sheail (1995) has argued that the evolution of environmental planning in the twentieth century has progressed through three windows of opportunity. First, at the turn of the century, the voluntary movement put the issues on the agenda; second, the official movement put the basic legislation in place in the 1940s; and third, the emerging coalition between farming and conservation that is emerging in the 1990s presents a potentially fruitful partnership. Thus the core arguments about the need for conservation, and the right to recreation seem to have been won, but the mechanisms for putting these into place, as shown in Table 6.1 have been over negative in the face of entrenched opposition from landowners and other vested interests in maintaining the status quo. Accordingly, conservation and recreation policies largely work through the actions of others as the next section will show.

THE SITUATION IN 1995

Overall strategy

The overall strategy, which was set out in Cm 2426 (1994), Britain's response to Rio, sought to reconcile two fundamental aspirations: economic development and the protection and enhancement of the environment. This strategy was rooted in responses to previous UN conferences, for example, by the Department of the Environment (1986 and 1988), which claimed that the basic principles were already being followed in Britain, and in particular by the 1990 'Environmental Strategy' which set in train an annual series of White Papers on the environment under the title of 'This Common Inheritance.'

The fifth White Paper in the series (Cm 2822, 1995) reported on progress made in implementing both the 1990 and 1994 strategies and set out the Government's main priorities. In particular, the Government wished to increase the use of performance targets, so that progress could be more effectively measured by the various monitoring groups set up in the 1990s, notably the Panel on Sustainable Development. The Government also attached great importance to action by local government in implementing Agenda 21 via the Local Agenda 21 Initiative of the EU's Fifth Environmental Action Programme. In addition, the Government was committed to using the market, notably by levying taxes on undesirable uses. More generally, the approach was an incremental one, of reporting on progress made in implementing the 650 or so commitments, made in 1990, or since then in each White Paper. In the 1995 'Rural White Paper' the Government also announced that it would not introduce further protected areas by statutory designation, since the current structure was reasonably comprehensive and already had some overlaps (Cm 3016, 1995).

Department of the Environment

In addition to the work it carries out in the field of town and country planning (discussed in Chapter 5) the DOE also provides overall guidance for implementing the Government's countryside policies which were set out in 1992 in the policy document *Action for the Countryside*. The main thrust of this policy is to create a coherent framework for protecting and enhancing the beauty and diversity of the countryside and conserving its wildlife, while also encouraging the growth of a healthy rural economy and improving opportunities for public enjoyment of the countryside. In 1995 (Cm 2807, 1995) the DOE's main policy aims and objectives included: safeguarding and enhancing the landscape; promoting access to the countryside; conserving and enhancing wildlife; promoting tree planting; ensuring that agricultural and forestry policies take full account of the need to conserve and enhance the rural environment; and providing good social and economic conditions for rural people.

According to Cm 2807 (1995) the DOE achieved its objectives by persuasion, partnership and advice, and through a mix of policy instruments comprising legislation, regulation, grants and other financial support, as shown in Table 6.1. Many of its policies were implemented through other organisations, but it did have the final say on several things, notably: designation of Special Protection Areas (SPAs) under the Birds Directive; appeals against various orders made by local authorities over the creation, diversion or extinguishment of public paths; and the protection of trees and common land.

The DOE also deals with environmental protection. In 1994/5 the DOE spent on these aims an estimated £221 million the bulk of which was spent by the National Rivers Authority (£68 million). In addition, the British Waterways Board, whose job is to encourage the use of inland waterways (mainly canals) for leisure and freight and to conserve their heritage and environment spent £49 million, environmental research and monitoring consumed £39 million, waste and recycling initiatives spent £27 million, and water services £10 million. The main means of delivery was through setting targets and standards which were monitored by various enforcement agencies.The newly created Environment Agencies were expected to carry out much of this work from 1996, backed up by economic instruments, under the provisions of the Environment Act, 1995. In England and Wales as from 1 April 1996 the Environment Agency brought together the functions of

Table 6.1 UK and EU legislation, and international conventions, on nature conservation

Legislation	*Aims and aspects*
Town and Country Planning Act 1947	County council control over development.
National Parks and Access to the Countryside Act 1949	Formation of NCC, National Parks Commission (NPC), National Parks, Areas of Outstanding Natural Beauty (AONBs), NNRs and Local Nature Reserves (LNRs), rights of way and access to open country.
Countryside Act 1968	NPC replaced by Countryside Commission; provisions on access to the countryside and regard for conservation.
Ramsar Convention on Wetlands of International Importance 1971	Conservation of Wetlands of International Importance (WIIs), especially as waterfowl habitats.
UNESCO Convention for the Protection of World Cultural and Natural Heritage 1972	Identification and protection of natural and cultural areas of outstanding international value.
CITES (Convention on International Trade in Endangered Species of wild fauna and flora) 1973	Control of trade in endangered species of plants and animals.
Wild Birds Directive 79/409/EEC	Protection of wild bird species and their habitats in Special Protection Areas (SPAs).
Bonn Convention 1979	Conservation of migratory species of wild animals.
Wildlife and Countryside Act 1981	Designation of SSSIs, NNRs, Marine Nature Reserves (MNRs), and Areas of Special Protection for Birds (AOSPs); Nature Conservation Orders, Limestone Pavement Order, protected species.
Berne Convention on the Conservation of European Wildlife and Habitats 1982 (protected plant species list revised in 1991)	To conserve wild flora and fauna, including migratory species, and their natural habitats; contains lists of protected species; used as the basis for UK wildlife legislation and EC Habitats Directive 92/43/EEC.
Wildlife and Countryside (Amendment) Acts 1985, 1991	Revision of protected species designations; SSSI designation operative immediately on notification by NCC.
Nature Conservation and Amenity Lands (NI) Order (Northern Ireland) 1985	Duties of public bodies; declaration of NNRs, MNRs, ASSIs, and District Council Nature Reserves.
Regulation 797/85/EEC and Agriculture Act 1986	Environmentally Sensitive Areas (ESAs) for protection of wildlife by adoption of suitable agricultural methods.
Environmental Protection Act 1990	Replacement of NCC by regional NCAs and JNCC; various provisions for environmental protection, including additional protection for SSSIs, a prescribed list of polluting activities and substances and Integrated Pollution Control.
Town and Country Planning Act 1991	Requirements for planning permission.
Planning and Compensation Act 1991	Provision for additions to classes of project requiring EIA; LA powers to safeguard conservation areas strengthened.

Table 6.1 Continued

Legislation	*Aims and aspects*
National Heritage (Scotland) Act 1991	National Heritage Areas (NHAs), affordable special protection for both wildlife and landscape.
Habitats Directive 92/43/EEC	Special Areas of Conservation (SACs) for protection of habitats of wild fauna and flora; lists priority habitat types and species for SAC designations.

Source: Morris and Therivel, 1995

HM Inspectorate of Pollution, the National Rivers Authority, and local waste regulation authorities. In Scotland, the Scottish Environment Protection Agency took over the functions of the river purification bodies, HM Industrial Pollution Inspectorate, and local council controls over waste regulation and air-pollution control.

English Nature

English Nature published its third annual report (for 1993–4) in 1995. This provided a commentary on how English Nature spent its first two years on redefining its goals and methods, and which resulted in the publication of *A Strategy for the 1990s* in 1993. This set out strategic nature conservation goals for English Nature which included: setting the context for nature conservation through the Natural Areas approach; securing the integration of nature conservation objectives into all Government policies; achieving the sustainable management of all SSSIs; and developing the practice of sustainability including the promotion of personal stewardship. More specifically, the Report set out 8 objectives, which included:

1 Continuing to develop a more positive approach to the protection and management of designated sites, notably through the expansion of the Wildlife Enhancement Scheme. In 1994 there were 3,794 SSSIs covering about 870,000 hectares and looked after by around 23,000 owners or managers. There were also 150 National Nature Reserves (NNRs) covering nearly 60,000 hectares, of which 4,439 hectares were owned by English Nature. The Wildlife Enhancement Scheme in its four pilot areas had funded 175 management agreements covering 5,500 hectares of SSSI land, and the Reserves Enhancement Scheme designed to help voluntary organisations manage their SSSI land (13 per cent of the total) had included 5,522 hectares on 149 SSSIs. Procedures for negotiating management agreements had also been simplified, and 85 positive management agreements and 120 compensatory agreements were achieved in 1993–4.

2 Developing and promoting the concept of Natural Areas. These are based on the underlying rocks and landforms, the typical plants and animals of the area, and traditional patterns of land use. Six pilot areas have been chosen to take the concept forward.
3 Extending the Species Recovery Programme in order to enhance and widen the distribution of some of England's most vulnerable species.
4 Developing the NNRs as places for people to enjoy and to use them better as a means of getting the conservation message across to their three million visitors.
5 Advising on the implementation of the Habitats Directive, and submitting for designation 30 new sites as SPAs under the Birds Directive or as Ramsar sites. In 1994 there were 43 Ramsar sites covering 193,029 hectares and 45 SPAs covering 232,876 hectares. 33 sites were covered by both designations. In addition, 18 proposed SPA sites and 16 proposed Ramsar sites were under consideration.
6 Encouraging more public involvement with nature through the promotion of Local Nature Reserves (LNRs), participation in Rural Action, Community Action for Wildlife, and the School Grants Scheme. In 1994 393 LNRs were in existence. The three other programmes provide grant aid to allow communities or schools to fund environmental projects.

In order to carry out its work, English Nature had a budget of around £40 million, of which just over half was spent on running costs and its staff of 754. The remainder was spent on: maintaining nature reserves (£1.3 million); rents, leases and compensation payments (£8.6 million); conservation support (£2.0 million); grants (£1.8 million); capital expenditure (£1.2 million); the Joint Nature Conservation Committee (£2.7 million: in total the JNCC had a staff of 106 and a budget of £5 million (Cm, 2807, 1995b)) and publicity (£0.8 million).

In summary, the strategy was based on much more positive action than before, and a greater acceptance that the public should have access to protected sites. In particular the mid 1990s was witnessing a move away from the compensatory approach designed to prevent damage, which had so disfigured the 1980s, towards a move to positive agreements, designed to build on existing conservation value, for example, the Wildlife Enhancement Scheme, and on a broader canvas the MAFF-funded ESA scheme.

Countryside Commission

The Countryside Commission, in contrast to English Nature, acts mainly as a catalyst and an innovative initiator through pilot schemes. Their mission in 1995 (Countryside Commission, 1995b) was:

- to conserve and enhance the scenic, natural, historic and cultural qualities of the whole countryside for this and future generations;

- to secure opportunities for people to enjoy and appreciate the countryside for open air recreation; and
- to promote understanding of the countryside, its life and work.

Their vision was of a sustainable, multi-purpose countryside, beautiful, environmentally healthy, diverse, accessible, and thriving. In order to pursue this vision, the Commission: devised and promoted strong national policies; provided innovative and practical demonstrations of what can be done; and provided advice and grant aid, in partnership with a wide range of people and organisations to achieve practical action. In 1994/5 (Countryside Commission, 1995b) the 320 staff had a budget of £47 million, of which more than 70 per cent went towards supporting action on the ground.

The priority in 1995 was to prepare, with English Nature and English Heritage, a 'New "Character" Map of England' (Cm 3016, 1995) to reflect its historic, cultural, wildlife, natural features and landscape character as an aid to its future management. In more detail, in 1995 the Commission was working towards the transfer of the Countryside Stewardship and Hedgerow Incentive Schemes, which had been combined in 1994, to MAFF in 1996. Although this could be seen as a retrograde step, it transferred the budget to the much larger MAFF account, and in addition the Environment Act 1995, which brought it about, also included a section which requires MAFF to consult the DOE and countryside agencies on the development of environmental land-management schemes generally.

The Commission was also working towards the legal redefinition of the entire 140,000 mile rights of way network by the year 2000, the development of the 12 Community Forests shown in Figure 6.3, and the implementation of the Environment Act 1995 which enabled important hedgerows to be protected. In more detail, in 1994/5 the Commission: supported 171 countryside management services; had let 4,937 countryside stewardship and hedgerow incentive scheme contracts; had advised on 205 development control cases, but was moving away from this kind of work to commentaries on Strategic Plans; had funded 800 farm conservation adviser visits of which 65 per cent had resulted in positive action by farmers; and had funded 2,395 countryside staff training places. The Commission also has responsibility for National Parks, AONBs, heritage coasts, the 4,500 mile-long system of national trails (formerly Long Distance Footpaths) and the other areas shown in Figure 6.3. In 1994/5 the National Parks managed 300,000 hectares of land, gave 342 conservation grants, hosted 84 million visits, and had 33,000 hectares of land open to the public.

The situation in Scotland and Wales

Scottish Natural Heritage (SNH) was formed from the Countryside Commission for Scotland and the Scottish parts of the NCC in 1992 under

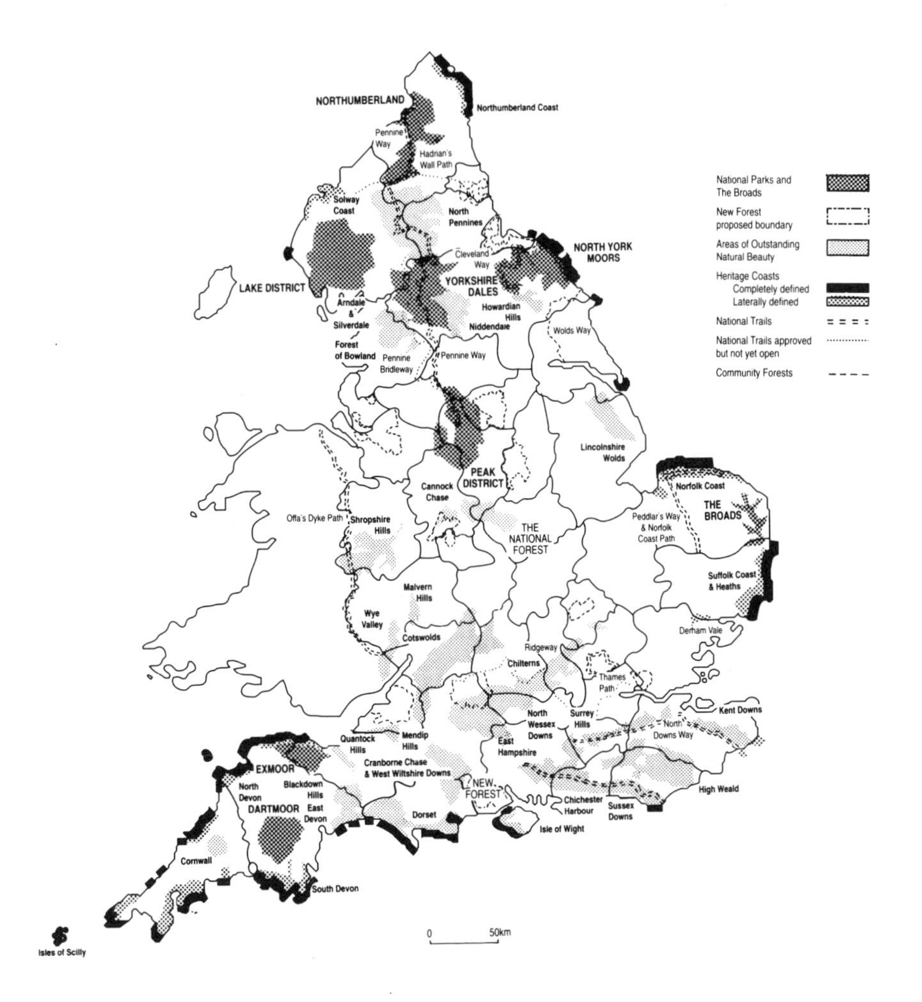

Figure 6.3 Protected areas and national countryside initiatives in England

Source: Countryside Commission, 1995

the Natural Heritage (Scotland) Act 1991 (Cm 2814, 1995). By 1994 it had evolved six external objectives and four internal objectives. Externally, SNH: placed great stress on influencing the actions and policies of others; developing thinking on sustainability; increasing awareness and understanding of the natural heritage and its wise use; promoting public enjoyment; protecting species, habitats and landscapes; and developing active partnerships. In 1992/3 SNH disbursed a budget of £39.6 million, of which over £7 million went on 757 grants. A further £2 million was spent on

Countryside Ranger services, £940,000 on the Central Scotland Woodland Countryside Trust, and £1.3 million on research.

In terms of designated sites SNH in 1993 had control or influence over 70 NNRs covering 114,478 hectares, 10 Local Nature Reserves covering 3,150 hectares, 1,359 SSSIs covering 835,083 hectares (of which 83,102 hectares were covered by 442 management agreements), 22 Ramsar sites covering 47,729 hectares, and 26 SPAs covering 52,249 hectares. If these figures are compared with the English totals it can be seen that Scotland is very significant in nature conservation terms. In addition, Scotland has three types of site not found in England, namely: nine Biosphere Reserves covering 28,768 hectares designated by Unesco; two Biogenetic Reserves covering 2,388 hectares designated by the Council of Europe; and one World Heritage Site, the 853-hectare island of St Kilda. Scotland still does not have any National Parks, but instead has 40 National Scenic Areas covering 1,001,800 hectares in which planning controls are much stricter. Many Scots now believe that they are better off without National Parks, believing that like the M25 these are the victims of their own success. There are, however, 36 Country Parks, and four Regional Parks covering 6,426 and 86,235 hectares respectively. Although access is less of a problem in Scotland, there are also four long-distance footpaths totalling 550 kilometres.

In Wales, the objectives of the Countryside Council for Wales are:

> to promote and keep under review the conservation and enhancement of the natural beauty of Wales; to conserve its flora and fauna, and its geographical and physiographical features; and to improve and extend opportunities for the public to enjoy, appreciate and have access to the countryside in Wales.
>
> (Cm 2815, 1995, 84)

The Council is also the Government's statutory adviser on all nature conservation and countryside matters affecting the principality. In 1994–5 (Cm 2815, 1995) the Council's 227 permanent staff planned to spend £20.6 million on various schemes including the conclusion of 140 SSSI management agreements, and 200 Tir Cymen (Countryside Stewardship) schemes. In addition the three National Parks in Wales attracted some £5.5 million in National Parks Supplementary Grant from the Welsh Office.

D: ANALYSIS

Although this chapter has so far treated conservation and recreation as linked themes, most analyses, much to their detriment, have concentrated on one or the other. Thus *faute de mieux* this section is divided into recreation and conservation sub-sections, but wherever possible the two issues are linked together.

Recreational studies

Traditional meta-narratives based on empirical observation

These approaches dominated the study of rural recreation in the period between 1950 and the mid 1980s when it concentrated on the explosive growth in numbers as the postwar boom ran its course. During this period survey work dominated, divided into *demand* surveys (what people did or wanted to do) and *site* surveys (what they did when they got there). These studies (Coppock and Duffield, 1975) produced a useful baseline for further work, which used statistical methods to recognise very strong links between some of the key variables, for example, rates of participation and car ownership. These links, however, were permissive rather than causal, and so the research effort began to move into the perception and behaviour of individuals, and thus the development of a more conceptual and theoretically informed explanation of rural recreation patterns (Owens, 1984). However, this work added little to the very strong relationships already found, and few new researchers wished to enter a field where apparently so much was known. Instead interest was refocused on the other growth industry of the postwar years, tourism. Accordingly, rural recreation research has tended to stay in the mode well developed by the mid 1980s, and divided by Smith (1983) into a four-stage continuum from description, to explanation, to prediction, to policy formulation, or by Morgan (1991) into: survey; analysis; plan; monitoring; and review.

Description and explanation studies have continued to focus on demand and site surveys, and have continued to find that a number of key variables have remained important, notably: disposable income, free time, ownership of a car, age and family status, and social class and education, and that countryside recreation is remarkably concentrated as shown in Table 6.2. All the factors explaining recreation demand have grown enormously since 1945, as Britain has become an affluent and basically middle-class nation, and recreation has come to be seen as a right, rather than as a privilege. Recreation and tourism now figure at the top of people's expectations, and countryside recreation is now enjoyed by an estimated 18 million people on a typical summer Sunday, and even 2 million people on a typical winter weekday (Countryside Commission, 1984e). In addition there has been an explosive increase in the number of people indulging in country sports (Harrison, 1991). Nonetheless, there is a significant number of people who are unable to participate, notably inner city or tower-block residents, because they do not have the financial or physical resources to do so.

An empirical study of the role of planning in providing for countryside recreation by Groome (1993) has concluded that: provision has tended to be grant-led; provision has tended to be opportunistic rather than the result of rigorous research; and that where techniques have been used they have

Table 6.2 Concentration of recreation: a summary of the findings of recreation surveys

A summary of the findings of recreation surveys	
By area	Within 50 miles of home a few accessible locations take most of the brunt
By distance	Over a third of trips are less than 10 miles, but another third are more than 40 miles
By car	Majority arrive by car and do not stray more than 100 yards from it
By site	Sites with an attraction/water/car park are favoured
By management	Unmanaged countryside is preferred, perhaps because it is free
By route	A round trip is favoured centred on car parks/refreshment/ shopping stops
By time	Notably between 1.30 and 5.30 pm
By season	Notably the summer months, but winter weekend breaks have grown
By inactivity	About half only stroll or walk around, about a third sight-see/eat/ sit
By activity	About a fifth walk over 2 miles, and less than 10 % watch nature
By motive	Majority enjoy the pleasure of the drive, the destination being an excuse
By day	Major peak on Sunday, minor peak on Saturday
By class	Middle class and skilled working classes are biggest pro rata participants
By age	Middle aged are biggest participants
By family	Families with young children are big participants

Source: Author

focused on forecasting future demands and searching for optimal patterns of land use. The main management measures have been zoning and channelling activities to selected areas and corridors.

Empirical studies have also focused on the impact of recreation on the sites visited. Stemming from the Dower report, this has traditionally been seen as detrimental, both aesthetically and physically, because visitors by their nature change the canvas on which the recreational experience is played. The impact is moreover concentrated. There is a marked tendency, first, for visitors to congregate at easily accessible beauty spots or 'honeypots', and second, not to move very far from this site. Accordingly, these sites are subject to extreme environmental erosion. This tendency for people to congregate has been exacerbated by planners who have seen it as an easy way to protect the wider countryside, most notably remote and/or more fragile habitats, from the 'hordes'. The long-standing aim of the Countryside Commission, to secure public access to the countryside for the purpose of open-air recreation, can thus be seen to be pretty much a failure, since very few access agreements have been negotiated. The main reasons have been the lack of money and unwillingness of landowners to concede widespread access which is less controllable than, say, a footpath.

Instead, a good deal of emphasis was put on Country Parks. Initially, these grew very rapidly (Groome and Tarrant, 1985) as many existing parks took advantage of the grant aid that became available after the 1968 Act. About 40 per cent are in the urban fringe and thus meet the objective of protecting this environment, and filtering of people from travelling further. However, there are far fewer around National Parks, thus leaving the National Parks less protected from mass visits. More generally Country Parks, most of which are local authority-run are more likely to be found in counties near big population centres. A typical example of such a park is provided by Cannock Chase, which has been the subject of long-term study (Countryside Commission, 1985). The plan for the park provides a perfect microcosm for recreational planning in general, since it divides the park into three zones: one where recreation objectives are given the highest priority; one where recreational use could be encouraged; and one where nature conservation is the priority and recreational use should be discouraged by removing or relocating parking. The plan developed the concept of carrying capacity in a 1930s study which proposed dividing up the area for different needs based on socio-economic class (Burton, 1974).

Since the mid-1980s, however, there have been moves to change these policies, and to encourage wider accessibility, largely based on new studies which have shown that the impact of recreation is not as damaging as first thought. For example, recreation has been demonstrated to be less damaging than afforestation, overgrazing and moorland reclamation (Sidaway, 1990), and the public perceive recreation to be less of a threat to the countryside than other factors like urban growth and pollution (Harrison, 1991). In addition, the Environment Select Committee of the House of Commons (1995) found that the number of visits to the countryside was steady, and that recreation did not cause significant widespread ecological damage. In fact they found that cultural conflicts were important and were often the root cause of many of the tensions. This was especially so in the case of active recreation, which seems to be growing, in particular, conflicts between noisy, dangerous or active sports and holidaymakers seeking quiet enjoyment. The way to resolve conflicts over noisy activities was to force manufacturers to develop quieter motors, and to develop special sites near urban centres. More generally, the Committee favoured high charges for car parking to ease pressure on popular 'honeypot' destinations and divert tourist traffic to less well known locations. Nonetheless, most recreational activity continues to follow the pattern as already set set out in Table 6.2. In essence this is very predictable, if little understood. It also reveals a very limited ambition, which is in some ways a great blessing, but also a great indictment of the failure of the postwar welfare state to raise intellectual levels to the same level as physical standards of life.

This theme is taken up by Glyptis (1991) in her text book on *Countryside*

Recreation when she writes that the planning response has tended to adjust to, rather than anticipate, public demand. This is not necessarily a problem, because much countryside recreation is axiomatically about spontaneity and freedom, and so it is entirely appropriate that many forms of countryside management are neither vigorous nor visible, and do not rely on high-profile management or deterrent policies. Instead much more success has been achieved by modest, subtle, participative and positive means. However, although there has been a good deal of success in recreational planning, access to countryside recreation resources remains unbalanced, both geographically and socially. In particular, barriers of access, mobility, awareness and confidence continue to stifle the recreational aspirations of many people. Pro-active encouragement of participation, backed with the practical removal of barriers, is all too rare. Fears of hordes of urban invaders wreaking havoc across a fragile countryside remain firmly entrenched in the mythology of countryside management, but do little justice, according to Glyptis, to either the motives of the visitors or the ingenuity of planners and managers to cope with them. Thus although Glyptis uses a logical positivist approach, she comes to a political economy and behavioural conclusion. These two approaches are now considered.

Structural analyses focused on political economy and regulation theory

Harrison (1991) in her book *Countryside Recreation in a Changing Society* employs such an approach as well as cultural factors. She argues that policy makers only decided to intervene in the 1960s, because the main policy areas of town and country planning and the National Parks Acts had 'failed' to protect and provide access to the kind of countryside desired by various recreational and environmental interests. Her analysis argues that three main policy responses were forthcoming.

The first response has been the provision of site access, notably Country Parks, which she concludes in general have been a success both for the public, and as cost-effective public facilities. Second, management projects have been provided which centre on small sums of money being spent across a wide area, mainly by farmers to defuse conflicts between recreation and conservation. This may be done by upgrading footpaths and their signage, notably in the urban fringe and in the uplands, for example, Groundwork in the urban fringe. According to Harrison, these are essentially conciliatory in style and operate outside the statutory planning system and thus raise important questions about the role of the public sector. Although they are innovatory and are often more effective than traditional site-based approaches, they are not politically neutral and raise the same issues of bi-polar standards raised by Country Parks.

Third, access to the wider countryside has continued to be denied not just because of the continued centrality of property rights in organising society, but also because a third element has been added to the dual alliance of landowners and conservationists, outlined earlier in this chapter: that is, the service class identified in Chapter 1, who have consciously acquired good taste, and thus the middle-class countryside aesthetic. However, this new class pay scant regard to the traditional community of rural areas and wish to impose their acquired version of rural recreation, thus 'urbanising' or 'Disneyfying' traditional pursuits, and inventing new pursuits based on some romanticised view of the rural idyll. The new service class thus has no interest in opening up general access to what they perceive as the 'real' countryside which they have just attempted to reinvent. These eclectic activities which transcend cultural norms will, according to Harrison, be less easy to predict since they are part of post-modernism. Thus attention is now turned to behavioural and post-modern studies.

Human agency, behaviouralist and post-modern perspectives

These have focused on how people use sites. For example, Cloke and Park (1982) in their study of the use of two contrasting sites, an informal open moorland site along a road and a formal Country Park site, found significant differences in behaviour between the two types of site. There is thus a clear behavioural difference between those who like the security of managed sites, and those who prefer the freedom of doing their own thing. Although this can also be related to class and notably education, these differences cut across class as well.

Behaviour can also be strongly influenced by management. This was first really demonstrated by the Goyt Valley experiment in the 1970s. This closed off a popular round-trip route, and only allowed access to a popular beauty spot by foot or minibus. This manifestly altered behaviour, with a significant number turning away, but an equally significant number abandoning their cars and enjoying the relative peace and quiet engendered by the scheme (Countryside Commission, 1972). Similar improvements were found in a longer project designed to improve the recreational and conservation environment of Tarn Hows in the Lake District. This project used a combination of redesigned and/or relocated car parks, paths and signs to significantly change behaviour (Countryside Commision, 1976).

The lessons from these projects have been widely employed since, and much has been learnt about how to manage people, both explicitly and implicitly. However, little is known about why we participate in and if we actually enjoy recreation. Harrison (1991) argues that we take part in recreation for several reasons, for example, as an antithesis to work, as a life-enhancing activity, as a post-material activity, or because it is fashionable. Appleton (1975) has, however, provided the best argument, and one that links the activity to landscape and conservation. His theory is that human beings have only recently evolved from being hunter-gatherers, and that our genes are still dominated by the need to survive in the wild. We thus see the landscape as a set of features which can provide life, but also death, since there are other predators too. Appleton thus produced his famous 'prospect-refuge' theory.

Prospect-refuge theory postulates that we seek sites for recreation on two basic counts. First, a place from which we can see over the landscape, and thus not only food sources, but also possible predators. Second, a place in which we can hide if these predators come too close. According to this theory this is why we seek out viewpoints (prospects) and why we stay so close to our cars (refuge). If the viewpoint provides a rich and varied habitat, full of potential food, to our Neanderthal lens, then so much the better. If the landscape also contains, flowing water, and snowy white peaks, then this appeals to our Freudian or sexual need to reproduce. Thus it is possible to hypothesise that we see landscapes, not aesthetically, as the romantic school would argue, nor as symbolic class-dominated visions, as Cosgrove (1990) would aver, but as primitive species. This may be disturbing, but has a deep ring of truth, confirmed only too recently in the Balkans. In addition, Appleton backed up his theory with a lifelong study of landscape depiction in the arts over the centuries, and his theory offers the most likely explanation about why we enjoy recreation, but more importantly why we need to conserve the natural environment, for without it we are in deep psychological trouble. A synthesis of psychological studies in this area has been provided by Kaplan and Kaplan (1989). Mainly from American experience, they do not attempt a major unifying theory, but do provide a theoretical classification of factors affecting the restorative power of nature. The four basic properties of a restorative experience are seen as: a sense of being away; a feeling of extent; fascination; and compatibility – by which is meant the resonance between one's purposes and the demands imposed by the environment. Accordingly, attention is now turned to studies of environmental conservation.

Conservation studies (including studies by the Countryside Commission)

Traditional meta-narratives based on empirical observation

These approaches have centred on two issues. First, the loss of wildlife and landscape, and second, the implementation of the voluntary/compensatory payment approach. The loss of wildlife and landscape has already been dealt with in Chapters 3, 4 and 5 to a large extent, so here it is only necessary to discuss some recent overview studies, notably the Countryside Survey 1990 carried out by the Institute of Terrestrial Ecology (ITE).

The Countryside Survey 1990 was set up to provide information on the changing countryside of Britain (Barr, 1994). Its main objectives were to:

record the stock of countryside features in 1990; determine change with reference to earlier surveys in 1978 and 1984; and provide a firm baseline for assessing future changes (Scott, 1994), which were due to be assessed by another survey in 2000. Two other things set this apart from earlier surveys. First, its breadth of coverage, and second the fact that it was the first national survey to combine satellite mapping and detailed field surveys of vegetation.

Although it is hard to generalise the findings, three general conclusions can be made. First, the proportion of the main semi-natural vegetation types in the countryside has remained constant but the quality has declined. Second, although there has been a loss of habitat and species diversity in linear features, these features are still important reservoirs of plant species. Third, losses of species and decreases in quality of vegetation were greater in the lowlands than the uplands. In more detail, hedgerows had decreased by 23 per cent between 1984 and 1990, with a decrease in species diversity, but broadleaved woodland had increased by 3 per cent.

Wider overviews of environmental change have been aggregated by the Environmental Challenge Group (1994a) which includes CPRE, RSPB, and the Friends of the Earth as well as several other groups. For example, farmland bird populations fell from around a 100 species in 1980 to only 45 in 1993; the length of hedgerow fell from 783,000 kms in 1950 to 366,000 kms in 1992; and more than 4,000 kms of rivers are graded as poor or bad with low levels of oxygen. The arithmetic of woe could go on and on, but what is important is that the Government in its annual environment White Paper *This Common Inheritance* and in its response to Rio has committed itself to the use of environmental indicators of this sort, as a measure of progress towards improving environmental quality. The Environmental Challenge Group (1994b) has welcomed this numerical approach and published its own criteria by which data to measure any progress should be collected and analysed, since at present data collection in this area is inadequate in terms of its relevance, reliability, frequency of collection and accessibility. Therefore much remains to be done in linking policy with data that can evaluate its impact.

The implementation of the voluntary/compensatory payment approach has been subjected to a good deal of attention, in particular, the percentage of land-use changes going through the system, and its costs. In terms of the percentage of potentially damaging operations, Brotherton (1988), in a study of objections to the 14,531 applications made for farm grants in his study area between 1983 and 1986, found that 90 per cent attracted no objection. Of the 10 per cent that were objected to only 0.15 per cent were found to be difficult. Brotherton concluded that the system was OK, a view endorsed by Lomas (1994), who found that 4,625 notifications for land-use change in the Peak District National Park were agreed to without recourse to the management agreement system, leaving only 76 of the total of 4,501 notifications to be negotiated, between 1980 and 1992.

Other commentators have pointed out that management agreements are short-term and expensive and Colman (1989) has shown that ownership is in fact cheaper over the longer term. However, Lomas, an officer with the Peak National Park Board, has argued from experience in practice (1994) that management agreements based on the concept of reward for conservation achievement offer the best partnership between agencies and land managers. In contrast, agreements based on compensation are expensive and promote negative attitudes, while purchase is also expensive and discourages the notion of environmental guardianship by private landowners.

Perhaps not surprisingly, therefore, Splash and Simpson (1994) have pointed out that expenditure on management agreements by NCC rose from only £1 million in 1985–6 to £7 million in 1991, or from 0.5 per cent of expenditure in the early 1980s to over 15 per cent by 1991. In spite of this expenditure, damage to SSSIs has continued (Splash and Simpson, 1994), and Friends of the Earth (1994) has estimated that at least 637 or more than 10 per cent of SSSIs are under threat from, for example, acid rain, eutrophication, road proposals or mineral extraction. Brotherton (1994) has also shown that the propensity to apply for Section 29 compulsory purchase orders, fell markedly after the mid 1980s, although the proportion of SSSIs threatened in any one year remained much the same at around 0.8 per cent. In addition, a number of very expensive agreements costing up to £1 million have created bad publicity for the system. Fraser (1995) also points out that the bargaining system between English Nature and farmers is inefficient and may not be sustainable in the long term, especially if a more positive and pro-active approach is adopted. Brotherton (1994) has nonetheless argued that Section 29 of the 1981 Act, the power to compulsorily purchase a threatened site, provides a winning hand for conservation that needs to be played more often, especially in the future over SACs under the Habitats Directive. In the wider arena offered by 'planning gain' under the Town and Country Planning Acts, Boucher and Whatmore (1993) have argued that the relatively low numbers of such agreements that involve a nature conservation gain could be increased.

To turn to studies of those affected by SSSI designation, Mather (1993) in a study of farmers in Scotland found that about two-thirds were positive, neutral, or had no comment about designation, but that one-third were negative about it. Similarly, two-thirds said that designation had not affected their use and management of land, but one-third said that it had. However, when respondents were asked to specify an actual obstruction only one-fifth could do so, indicating that the effect of SSSI designation was more in the mind than in practice. These behavioural attitudes have been reported already in Chapter 3, and work by MacGregor and Stockdale (1995) again in Scotland has confirmed that conservation objectives are supported when they fortuitously coincide with private enjoyment, which remains the main motivation for conserving the environment. These findings therefore confirm one of the key findings of this book, namely, that planners in a free-market economy can only guide and shape, and that if they are to succeed they must lead people gently towards more acceptable environmental practices by employing the full range of powers in the Gilg/Selman spectrum.

Structural analyses focused on political economy and regulation theory

Most of the approaches under this umbrella, namely, how farmers react to conservation policies, have already been discussed in Chapter 3. In this chapter discussion is focused on environmental policy in general and the organisations involved in conservation policy in particular. McCormick (1991) has provided a very useful overview of how politics relates to the environment. The key point he makes is that Britain does not in fact have a coherent and recognisable environmental policy. He then provides several possible explanations.

First, the environment as a policy area has not only been misunderstood by British governments, but it has also been particularly prone to the kind of ad hoc improvisational and piecemeal responses that characterise the policy process generally. This process of disjointed incrementalism has resulted in a confused and confusing medley of institutions and laws. Second, public policy is worked out by consensus and consultation, but certain information is kept secret, and the consultation is selective. Third, in spite of Britain being a centralised state, much detailed power in it is devolved, and local government has traditionally been responsible for environmental regulation, thus allowing considerable local interpretation of the statutes.

Fourth, the environment is difficult to compartmentalise, and overlaps with virtually every other aspect of public policy. Fifth, the environment is a relative newcomer to the policy agenda and has had to compete for funds and power as a junior partner. Sixth, the causes, effects and cures of environmental problems are still poorly understood, and even if they were understood, they represent bad news, which politicians do not like. The environment is like a pain in the chest, it is preferable to think it away, rather than begin the long course of treatment that may cure it. But what if the pain does not go away? In this sense, McCormick argues that in the absence of a coherent response to the environment by the British Government, the environmental lobby will continue to be the only effective force for positive and rational environmental protection.

This is a theme taken up by O'Riordan (1992), who uses Habermas's concept of a *legitimation crisis* by which the state seeks to parry the onset of political or economic embarrassment by driving inconsistencies into other realms of policy. For example, the crisis in agriculture in the 1980s, caused by over-supportive farm policies, and which was manifested in lower farm incomes and environmental damage, was responded to by apparently transferring spending from production to conservation, but not by dealing with the root causes. O'Riordan uses other examples of how the old way of looking at things is no longer tenable in the modern world of environmental crisis, and argues that the terrific flux of recent events allows Habermas's Critical Theory to show its full glory. He thus uses Critical Theory to examine three forms of greenism: dry; shallow; and deep. Dry greenism places great reliance on scientific and voluntary solutions. Shallow greenism attempts to be more pro-active by designing with nature and in combination with dry greenism represents the current position, in which three shallow green measures – national regulatory bubbles; best practice and strict liability; and public trust and equivalent compensation – are the main means of implementation.

O'Riordan, however, believes that both dry and shallow greenism will not survive beyond 2010, and that by then the largely cosmetic changes outlined above will have been shown to fail. Until then however, the industrialist, the politician and the citizen will be persuaded that modest levels of reform will somehow do the trick, that capitalism will essentially survive, and that wealth creation and an element of sharing can continue in a shadowy state of sustainable development. In contrast, deep greenism demands radical changes which place the Earth or Gaia at the centre of thought, and replace national governments with de-federated individual and communal action which stresses equality and rejects the tyranny of the market and its invisible hand which cannot value environmental goods.

Turning to the particular, the restructuring of the NCC in the early 1990s has attracted a good deal of attention (Box, 1994). Most commentators agree that the breakup of the NCC was a Government response to its growing stridency, pointing out that the Chairman at the time was not re-appointed, a key indicator of Government unhappiness (Lawson, 1993). The policy response was thus one of divide and rule, and not extending the tenure of over–critical people, thereby sending a powerful message in the form of *pour encourager les autres*. The policy seems to be only to appoint people with a 'safe pair of hands', and to make sure that these agencies remain friendly critics and do not become *agents provocateurs*, let alone quislings. As with the Countryside Commission, the 1980s and 1990s has seen the replacement of lively representatives from academia and the pressure groups with entrepreneurs, and career committee people, sometimes known as 'the great and the good.'

Brotherton (1994) has also argued that the lack of will to implement Section 29 Orders under the 1981 Act is not due to a lack of funds or any perception that the threats to SSSIs are unimportant, but because a growing pattern of rejection by ministers has led to a marked loss of enthusiasm by officials who wish not to incur further disapproval from Ministers. Chatters (1992) in a review of legal test cases has also found that the Government has been lukewarm in responding to cases which have shown how easily English Nature's ability to prosecute under Section 28 of the 1981 Act, when damage has occurrred, can be undermined. For example, an occupier need only wait long enough to escape the Act, or distance himself from ownership, since a third party also escapes from the Act's provisions. According to decisions by the Law Lords quoted by Chatters, the statutory scheme for compulsorily purchasing threatened sites or prosecuting for damage is flawed.

In the same way the committees of the National Park Authorities have been filled with people who, unlike the MacEwens in the 1970s, will not disrupt National Park business, and may not even rock the boat. This new generation have ensured that National Park Plans, on the surface at least, are very logically based on conserving the core and managing the farmed framework in which most recreational pressure is absorbed, but this can be interpreted as a plot by the cognoscenti to keep the 'riff-raff' out and to keep the core environment for themselves on the principle of ignorance.

Human agency, behaviouralist and post-modern perspectives

These have concentrated on the impact of personalities, and the meanings that we attach to conservation and the environment. These can be combined by examining the changing role of the Countryside Commission between its foundation in 1968 and its modest transformation in 1992, and in particular the great dilemma that the Commission has had to face, whether its overall priority was conservation or recreation, given that it had a rival in the conservation field in the form of the NCC and could thus be tempted to carve a niche for itself, as a recreational body above all, especially when its budgets and staff were threatened. The twists and turns of the Commission in this period have been graphically portrayed by two key actors, a Chairman, and a Director of the Commission, who have emphasised the importance of one or two powerful personalities (behaviouralism) and the importance of one or two key chance events (post-modernism).

The Chairman of the Countryside Commission between 1970 and 1978, Sir John Cripps, in a review of the Commission's first decade (1980) makes two key points. First, incrementalism was a crucial factor because, instead of setting up a *de novo* body as in Scotland, the Government was tempted to graft the Commission onto the National Parks Commission. This meant that the Commission was not a grant-in-aid body like its Scottish counterpart, but was part of the civil service. This meant that the free-thinking and flexible attitude to staffing needed by an agency with few executive powers was frustrated. In addition, the type of work carried out by the Commission, based on new ideas and exhortation, required skills which were difficult to find and were sometimes lacking altogether in the civil service. Second, with duties far greater than resources, a key personality at the time,

Reg Hookway, promoted the concept of management which became the flagship idea of the Government from the 1980s onwards. Thus the twin concept of seeking cooperation was born not just from expediency but also from conviction.

Cooperation was also the order of the day for much of the period between 1979–1992 considered by Phillips (1993), who was the Director of the Commission during most of this time. At the outset the Commission was threatened with abolition, by the then Secretary of State for the DOE, but after considering the need for primary legislation and the criticisms it would engender, he decided to cut the staff from 124 to 93, and to peg the budget. Yet when Heseltine returned to the DOE in 1991 he found a staff of 250 and a rising budget. Phillips argues that this transformation was the result of six key factors.

First, the *modus operandi* of the Commission, based on working through others, flexibility, and positive financial inducements rather than negative restraint, gave it considerable freedom of manoeuvre to respond to changing circumstances. In addition, the change to a grant-in-aid body in 1981 was a key factor in giving the Commission all sorts of freedoms it had not had before by instigating a managerial revolution. Second, by developing ideas and demonstrating their practicality the Commission acted as a testbed for subsequent Government policies. Third, good housekeeping, which ensured that no summons was received to appear in front of the Public Accounts Committee, and thus by implication an endorsement that the Commission provided value for money. Fourth, public opinion swung markedly towards conservation values in the 1980s. Ministers thus found it useful to have an agency like the Commission to back up their claims to be 'green', especially at election time, when the Commission found it was useful to step up their policy proposals. Fifth, ministers found the Commission useful when they were in tight spots, notably after the great gale of 1987, when the Commission came up with the Task Force Trees programme in two weeks, and was thus able to allow the then Secretary of State to demonstrate that he was doing something. This chance event changed his formerly negative attitude to the Commission. Sixth, the Commission had a patrician, but skilful Chairman, Sir Derek Barber, who throughout most of the period was able to build bridges, and sought compromise, not confrontation. However, his appointment was to some extent a chance event, since he came to the attention of Michael Heseltine, a bird enthusiast, largely because he was Chairman of the RSPB.

In spite of the overall progress recorded by Phillips, he also regrets the 1974 move from London to Cheltenham, because it removed the crucial ability to do business over lunch and an after-work drink, thus providing a useful insight into how opinions and decisions are really made. In addition, although the Commission itself strengthened itself and its influence, the environment as a whole still deteriorated.

Phillips concludes by making six observations about how environmental politics works. First, he notes how ministers exercise power over bodies like the Commission, by the appointments they make, through the legislation they promote, and through the resources they make available. Second, he argues that policies often change when, by chance, a number of factors coincide, for example, at Halvergate (see Chapter 3) where the need to curb farm production and appease the green lobby fortuitously coincided with the realisation by farmers that farm subsidies would be withdrawn, and thus a new source of public subsidy was needed.

Third, he notes that the influence of exceptional individuals is important, and that compromisers are far more successful than conflict seekers and those who antagonise. Fourth, although Whitehall has apparently been 'greened', Phillips concludes that many departments are still not sympathetic to conservation and that the 'greening' of Whitehall has a long way to go. Fifth, a maverick element is at work, and events elsewhere often have unforseen knock-on effects, for example, inner city riots in the early 1980s convinced Heseltine of the need for Groundwork, and the 1987 gales created new impetus for lowland tree planting. Taken together, these two unrelated events led to the groundswell of opinion in favour of the Community Forestry programme. Sixth, and the most enduring lesson, is that a body like the Commission must have clear principles, but be subtle, opportunistic and flexible in its approach, and cultivate allies and links with both the Government and pressure groups. In other words pragmatic incrementalism will be the most effective strategy.

In addition to these two important inside studies of the Countryside Commission, some people have examined how the behavioural culture of the NCC changed when it became English Nature. For example, Box (1994) has described how the first need was to turn a bureaucratic, risk-averse organisation with complex internal procedures and processes into one that was pro-active, progressive and customer-orientated. Accordingly, staff were asked to identify common beliefs from which a corporate vision emerged. Restructuring also focused on moving from top-down to bottom-up controls. Much progress was made in the first two to three years, but Box expected that it would take 5 to 10 years for the culture of the organisation to complete its transformation.

Sidaway (1990) has also argued that there is a major gap between research and policy, and that policy and professional practice are more often dependent on ideology, intuition and tradition than experimentation and critical assessment. For example, the selection of scenic areas for protection has been based on two key selection factors – national significance and consensus as to landscape value – but these have been assumed by decision makers, rather than based on popular opinion, which would probably have selected different areas, since many popularly valued scenic areas are not represented in the present system.

E: COMMENTARY AND THE FUTURE

The postwar period has witnessed a continuing loss of environmental quality, not by a few dramatic activities as highlighted by the media, but by the insiduous loss of a hedgerow here and a pond there, and by the slow but steady spread of new developments. The favoured policy response has been one of negative controls over endangered species and protected landscapes, site protection for special areas, and cosmetic schemes elsewhere. The mid 1990s has witnessed the first serious questioning of the agricultural, forestry and development policies that brought this steady erosion about. The mid 1990s thus provide an opportunity to first stop the erosion, and then to reverse it by creating new types of countryside based not on production, but on consumption. If this is to happen a vision of what could be created is needed, as well as a will for it be done. The rest of this chapter discusses whether these can be found.

The Countryside Commission (1995c) has provided a lead with a prospectus for the next century. This identified five challenges. First, the quality and diversity of the countryside could be further enhanced in general by developing the New Map of England concept by preparing countryside policies and programmes for Countryside Character areas at the regional level, and in particular by encouraging more cooperation between land-management bodies. Second, multi-purpose farming and forestry could be further encouraged by a combination of modifying agricultural policy as it affects the conservation and enjoyment of the countryside, and by further identifying and pioneering schemes to help farmers and land managers to take practical action on the ground. Third, sustainable development could be further developed by assessment measures like environmental indicators, State of the Environment Reports, and environmental appraisals of development, while building the concept into all aspects of policy and practice affecting the countryside. Fourth, to build on the current priority of getting the rights of way network legally defined by the year 2000, by promoting new forms of access, like greenways. Fifth, to involve the public more fully in issues of countryside conservation and recreation.

Such statements of intention ignore two harsh realities. First, that much of the problem stems from the unsustainable activities of the developed world, and second, that this implies a major change in lifestyles if real progress is to be made. Therefore the real environmental problem is one of imbalance between different regions of the Earth, and the need for there to be a levelling down of living standards in the 'developed' world, to allow a levelling up in the 'undeveloped' world. This does not necessarily mean a lower quality of life, but it does mean a massive transformation in attitude, and it is hard to see how this can be achieved. This is especially so when it is pointed out that some of the richest nations can also afford to be the greenest, in that they can afford hi-tech solutions to pollution. This is one of the disturbing paradoxes of environmentalism, namely, that poverty brings in train perhaps more environmental damage than affluence, since there are no resources to prevent basic pollution. Somewhere, a balance has to be found between affluence and effluence.

Owens (1994) has provided an exploration of the opportunities and contradictions in applying the concepts of sustainable development to land-use policy. She begins by setting out a framework for planning and sustainability which contains three types of capital – critical natural capital that is not substitutable; natural capital; and human-made capital – and different dimensions of sustainability from the luxurious to the necessary. From this framework, three increasingly demanding interpretations of sustainable development can be distinguished:

1 Development that passes on at least the same total capital stock to future generations;
2 Development that achieves 1) but passes on at least the current stock of critical natural capital intact, otherwise different types of capital may be traded off against each other to maintain at least the same total stock; and
3 Development that keeps critical natural capital intact as well as handing down no less natural capital than current generations enjoy.

Whichever of the interpretations are employed, implementing them will be difficult, and although the concepts of sustainability are gaining ground in planning and related disciplines, political commitments to sustainability, at first made before the extent of the challenge they posed to a demand-led economy was fully grasped, are now being toned down. This is because the principles of sustainability challenge the presumption in favour of development and sit uneasily with the utilitarian notion of balance. Instead they require an alternative ethical basis and, especially in the post-material realm, are inherently bound up with value theory.

At the moment, only hierarchical scales, as for instance in nature conservation sites, are used rather than value systems which might tell us how to weigh different kinds of interests against each other, although the Habitats Directive does suggest how the preservation of habitats may be weighed against human life and safety *vis-à-vis* economic interests. Therefore, Owens concludes that sustainability principles will not be widely espoused outside planning circles, nor will they be much use in reducing the conflict between conservation and development, but they will bring that conflict forward to an earlier stage of the planning process.

Rydin (1995), however, doubts whether any of the three models of planning can deal with sustainability. In more detail, the technocratic model based on urban form and controlling land use cannot do much, even in the medium term, since so much of the infrastructure is already in place. The mediation/bargaining model cannot be effective, since the most powerful groups have little interest in sustainability and the move towards a more equitable world that it implies. The advocacy model based on the empowerment of the disadvantaged and the silent also cannot work since ironically the poor in the short term aspire to material goods just as much as the already advantaged. More prosaically, Rowan-Robinson *et al.* (1995) argue that there are four difficulties relating to development control. First, planners are advised by PPGs to turn down applications only on the tough grounds of 'demonstrable harm', rather than the weaker 'precautionary principle' of sustainability which would probably be defeated at appeal. Second, the interpretation of 'other material considerations' leaves much room for manouevering by developers. Third, development control needs to have much more freedom to rescind or modify approvals if they have been shown to be harmful later. Fourth, Rown-Robinson *et al.* agree with Owens (1994) that defining critical natural capital is difficult and therefore that development control could only ensure weak rather than strong sustainable development.

Selman (1995), however, is more optimistic and argues that the three key concepts of sustainability – inter-generational equity (principle of futurity); intra-generational equity (principle of social justice); and the principle of trans-frontier responsibility – and the two concepts of natural capital – critical natural capital – and substitutable natural capital – have had four effects on planning. First, they have given planning renewed vigour and justification, and revived the concept of wise stewardship. Second, governments are now committed to sustainability via public and unavoidable international agreements. Third, sustainability has allowed a counterattack to the rhetoric of lifting the burden of planning control. And fourth, new techniques of ecological planning are beginning to emerge that will allow planning to deliver sustainability.

In addition to the problems caused by unequal standards of living, there are also problems relating to unsustainable activities, based on the division of labour and space. The modern world is characterised by increasing divisions and distances between one type of land use and another. Food and drink are now carted from one end of the globe to the other, people live further and further away from their place of work, leisure, and shopping, and tourism now treats the globe as its oyster. There is little that is sustainable in recreation or tourism that depends on people driving in cars or sitting in airplanes to travel to far-distant locations. Not only do cars and planes pollute the environment in their use, but their manufacture and disposal also consume large quantities of energy, non-renewable raw materials, and huge amounts of water. Those advocating so called 'green tourism' need to be reminded of these harsh facts about travel and its wider resource implications. A final major problem will continue to be the division of responsibility between agencies, in terms of their sectoral responsibilities, but also in their spatial territories.

Gare (1995) has also juxtaposed the paradox of growing concern for the environment with the paltry achievements of environmental movements. Not surprisingly some commentators, for example, Burbridge (1994) have recognised a new breed of environmentalists who are angry, impolite, disrespectful of authority and have a healthy contempt for the law and the property that it protects. These are *radical* environmentalists, for instance Earth First, a group founded in Britain in 1991. They believe that mass society and its reliance on hierarchical structures is inherently unsustainable and must be replaced. Their main weapon is non-violent direct action, and activities in the mid 1990s have included halting the destruction of rare habitats, acting in solidarity with activists overseas, and educating the public. For the future Earth First intends to take on the issue of concentrated land ownership, the root cause of the enslavement of the British landscape.

At the other end of the political spectrum Newbould (1994) has provided examples of how the 1994 inception of the National Lottery could help fund nature conservation. The private sector has also been investing heavily in outdoor recreation for those who can afford it, especially in golf courses, leisure centres, Center Parcs, farm-based recreation and tourism, and in some cases in partnership with the public sector, for example supermarkets linked to football clubs/leisure centres at edge-of-town sites.

Turning specifically to recreation, Glyptis (1991) argues that the informality and spontaneity of much countryside recreation, axiomatically, implies little need for positive planning and management. Nonetheless, quite small measures can have disproportionate impacts in particular sites, and thus she advocates the continuation of this approach. More problematically, Glyptis argues that conservation and recreation are not incompatible, and that the 'them and us' attitudes portrayed throughout this chapter can, and should, be replaced with greater cooperation. Glyptis advocates a four-pronged approach to combine them effectively.

The first is to undertake comprehensive resource appraisals to identify sites and areas of crucial importance for conservation or recreation reasons. The second need is to evaluate demand, addressing not merely the volume of demand for particular activities, but also the needs of particular minority groups and minority activities. The third need is for partnerships between those who share an interest in, or use, the same resources. Separation leads to extreme claims, entrenched views, and a polarisation of interests. In contrast, joint working can usually reach honest appraisals of need, minor concessions, mutual respect, and a sensible sharing of resources. The fourth need is to understand public interests and facilitate access. Even among frequent countryside users, patterns of activity are habitual and unadventurous. Among non users there is a lack of awareness of opportunities and lack of confidence in using them. Lack of knowledge is no less a discriminating factor than lack of transport or finance.

Harrison (1991) is less optimistic and concludes that certain privileged groups who accept the 'countryside aesthetic' ideology and the correctness of market mechanisms will continue to gain preferential access both through public and private provision. Nonetheless, Harrison believes that this ideology is losing ground, and the belief that countryside recreation should be 'fun' is gaining ground. Thus she argues that the public sector is uniquely positioned to develop new entrepreneurial and innovative approaches which could mesh with the views of contemporary society. More general access, which would mean wresting the bundle of rights that landownership confers from those who own them, is likely to require nothing less than a social revolution. Finally, Harrison argues that, given the plurality of views held by the public about what the countryside is, what it should be, and how they could use it, a general countryside policy combining conservation and recreation would be very hard to achieve intellectually, let alone practically.

Curry (1994) argues instead for recreation planning to be given greater priority. He puts forward five propositions which have caused our current difficulties, and some possible solutions as follows. First, the fragmented nature of the organisational structure for countryside recreation has inhibited the implementation of comprehensive policies and plans, and so Curry proposes the realignment of the Countryside Commission, into a Countryside Recreation Commission, and the expansion of farm-based recreation. Second, there has been a confusion between the responsibilities of the public and the private sector, and so Curry proposes an increased market-orientation for public sector facilities, with more emphasis on charging for access; treating information as promotion; and a greater use of powers to increase access, so that access rights are considered as commodities. Third, recreation has been dominated by middle-class groups, but other groups seem reticent to partake, even when offered the chance, and so Curry cautions against the misplaced philanthropy of current Reithian-style policies, and their reliance on a predetermined profile of participation. Fourth, recreation policies have seen recreation as an invasion to be controlled and channelled, but in reality it does little damage compared with agriculture and forestry, and so Curry proposes more promotional policies encouraging recreation as a potentially friendly land use. Fifth, recreation policies have been residual to other rural policies, but with land surpluses a reality, a golden opportunity now exists to promote recreation facilities as market commodities, and to provide people with what they want, measured through the market, rather than any philanthropic notions of what they ought to have.

This theme has also been developed by Clark *et al.* (1994), who highlight the recent growth of new very intrusive forms of recreation, for example, war games. They argue that a major public debate is needed to place these desires alongside the prevailing white middle-class aesthetic values based on quiet enjoyment, but warn that one view is not likely to emerge, since a survey of 1,000 people (McNaghten, 1995) found that attitudes to recreation in the countryside varied by the introductory comments made by the interviewer. Although some general trends emerged, a number of ambivalent, ambiguous and contradictory views also emerged, and this research suggests that we may have to abandon attempts to seek concensus about rural recreation and allocate different spaces for different activities. In contrast, research into the future landscape preferences for the Yorkshire Dales (O'Riordan *et al.*, 1993) found that about half wanted the status quo to continue, with another quarter wanting a loosely related ESA-type landscape.

Finally, and perhaps most importantly of all, not just attitudes but motivations for visiting the countryside are still poorly understood. They should be systematically explored, since it is all too easy for providers to impose on others their own tastes, rather than what people really want. Developing new styles of rural recreation in a re-created natural environment thus presents countryside planners with their greatest challenge for the twenty-first century. All this, of course, may be mere fiddling while Rome burns, or rearranging the deck chairs on the *Titanic*, since, as Porritt and Winner (1988) have argued:

> Sooner or later the realisation will dawn on more people that solving the planet's problems is going to require breathtaking radical action and international cooperation on a scale not seen since the Second World War.
>
> (263)

Policy evaluation checklist derived from Chapter 1.

1 *Were the problems understood and which value system dominated?* The unsustainability of the global over-use of resources and grotesque imbalances between nations and classes have been understood by only a few, and the value system has been one of selfish complacency and cosmetic tinkering.
2 *Which goals were to be achieved and were the policy objectives clearly spelt out?* The main goals have been: maintaining modern farming, and conceding conservation and recreation goals at the margin, mainly on small selected sites. The policies have not been very clear due to sectoral fragmentation.
3 *Were the policy responses accretionary, pragmatic and incremental?* Extremely so, as most of the authors referred to in the chapter have pointed out. Nonetheless, progress has been made and the policies have become stronger if not altogether coordinated or strategically coherent.
4 *Which powers in the Gilg/Selman spectrum were mainly used?*

Exhortation and financial mechanisms have been the main weapon, allied to the use of regulatory controls in site-specific policies like SSSIs. Public land ownership has been rejected, but private groups have increasingly used this route.

5 *Were planners given sufficient powers/resources and sufficient flexibility for implementation?* The lack of powers and resources to do much has been one of the main weaknesses, especially the lack of money. In response planners have reacted by inventing often non-statutory mechanisms.
6 *Were performance indicators built into the policies?* Not at the outset, but especially in pollution control, emission standards, etc. have become the main mechanism. The arithmetic of woe, in contrast, has been widely used to portray environmental decline, for example, the loss of wildlife and hedgerows.
7 *Have the policies been underresearched?* Yes, except for the voluntary compensation approach which has been heavily researched in terms of its processes and costs. Recreational activity patterns are well known, but why people do things and what they get out of recreation are less well understood.
8 *Have the policies been effective overall?* Partially, to the extent that the countryside would have been worse without them, and people would have had less access. The least effective area has been developing holistic conservation, while site-specific policies have been quite successful.
9 *Have there been uneven policy impacts, by group or by area?* Yes, middle-class groups and landowners have benefited, while other groups have been excluded. Areas selected for special conservation attention have benefitted but areas selected for recreational use have often been over-used.
10 *Have the policies been cost-effective and are they environmentally sustainable?* Yes, in that the budgets are small, but no, in that they are only temporary bribes. Site selection in a sea of decay is also unsustainable as is the huge (in comparison) CAP budget.
11 *What have been the side-effects?* The development of complacency among politicians and the public that something is being done, and thus postponing the evil day when something really will have to be done. So-called 'green tourism' is a symptom of this unfortunate disease.
12 *Were the policies modified by unforeseen events/policy changes elsewhere?* Yes, as two commentaries on the Countryside Commission show in the Analysis section D on pages 231–3. In addition, the EU has prompted several UK changes. Nonetheless, the underlying concepts have been unaffected.
13 *Were the policies overtly sectoral and were they compatible with*

other policy areas? The sectoral division between agriculture, nature conservation, recreation, and the wider environment, has been a fundamental weakness with four bodies involved in England. Scotland may be developing some coherence.

14 *What have been the effects of 16 years of Tory rule?* The development of the management approach pioneered in the 1970s has been their flagship, allied to a belief in exhortation and private landownership, thus releasing large areas of former utility land into potentially exclusive facilities.

15 *Should the policies be modified?* Yes, either radically, or incrementally. Incrementally, the emphasis should be shifted from short-term financial measures to long-term ownership by trusts.

16 *What alternatives are there?* Radically, the whole ethos needs to shift from site-specific policies to holistic environmental stewardship, and then to a whole shift in our culture away from unsustainable lifestyles, with a return to self-contained communities and recreation based on muscles not on fossil fuels.

17 *Overall verdict.* Limited progress has been made and key sites/species have been conserved, and many more people enjoy rural recreation than ever before. Much of Britain still looks very attractive, but how long it can last is the $64,000 question.

7

AN EVALUATION AND THE WAY FORWARD

All the chapters in the book have concluded with the checklist produced in Chapter 1, and so it seems perfectly logical to use the same checklist as the basis for the concluding chapter.

1 Were the problems understood and which value system dominated?

In the 1940s the problems were simple and focused on the twin-track approach of encouraging agricultural production and safeguarding agricultural land from development. The value system was thus one of 'agricultural fundamentalism' and was expressed very clearly in the Scott Report (Cmd 6378 (1942)). Even then, however, one member of the Committee dissented and with each decade that passed new problems emerged and agriculture gradually lost its position as the focus. The need to conserve the countryside has by and large replaced it, and we can now recognise an almost sacred cow based on Harrison's (1991) conception of a 'countryside aesthetic', in which all countryside images are good – a belief that is vividly used by advertisers. The countryside has become wholesome and untouchable, a self-reinforcing icon. The question is how to reconcile the starker reality of, say, broiler chicken production, with the cosy images of a 'rural idyll'. Thus the real problem now is What role should the countryside play? Should we continue down the road to 'The Rustic Theme Park' or return to reality?

2 Which goals were to be achieved and were the policy objectives clearly spelt out?

The original goals were food and timber production and preventing urban sprawl, but in 1949, recreation, landscape and nature conservation goals were added. From the 1960s onwards these goals have become increasingly interlinked and the policy objectives in turn have become less clearly spelt out. The problem has been exacerbated by the evolution of planning organisations which not only cut across each other sectorally, but are inconsistently organised spatially. It has thus been very difficult to find comprehensive statements of countryside policy, but the 1990s series of White Papers on 'This Common Inheritance' and the 1995 White Paper on 'Rural England' have to some extent filled the gap.

3 Were the policy responses accretionary, pragmatic and incremental?

Extremely so, except in one or two areas. Almost every page of this book confirms one of its themes, namely, that incremental 'muddling through' has been the order of the day, in both policy development and implementation. In more detail, policies have been accretionary, notably in agriculture and forestry, and overall Hill and Young (1989) found 178 rural support systems even before the further explosion of schemes in the period after the 1992 CAP reforms. Policies have also been pragmatic, and rooted in political reality, a point often ignored by some of the critics who criticise policies, not the society that constrains them. There is little point in criticising planning for not delivering social policies when it was not intended to.

4 Which powers in the Gilg/Selman spectrum were mainly used?

Working through most of the spectrum, from exhortation to regulation, has been the main policy response (Table 7.1). Thus voluntary methods have been the favoured first option, notably in environmental issues. Financial incentives have been the main tool in agriculture and forestry, and these have been extended to environmental schemes as well. Monetary disincentives have been used widely in general environmental issues, and their use is growing in the field of farm pollution control, and curbs on production. Regulatory controls provide town and country planning with its teeth, but the system also relies heavily on voluntary methods centred around bargaining. Finally, public ownership has hardly been used.

5 Were planners given sufficient powers/resources and sufficient flexibility for implementation?

The answer to this varies widely across the field. Agricultural production was given very useful powers initially, but these have become much more complex and less useful. In contrast, nature and landscape conservation were given only limited powers and resources. Gilg (1995) has shown for example that agricultural support costs over £3,000 million per year, compared to the £300 million or so spent on *all* forms of countryside conservation and recreation. Most policies are flexible enough, notably in town and country planning, where flexibility is the great virtue of British planning.

Table 7.1 The differential use of planning powers in the Gilg/Selman spectrum

Activity/power	*Public ownership or long term lease*	*Regulatory controls*	*Monetary disincentive*	*Financial incentive*	*Voluntary methods*	*Site + area designation/ special bodies*
Agriculture as a productive activity	*	***	***	*****	****	**/****
Agriculture as a land manager	*	***	**	*****	****	****/**
Public Forestry	*****	*	*	**	*	***/**
Private Forestry	**	***	**	*****	***	***/***
Planning the built environment	**	*****	**	*	****	*****/***
Nature conservation	****	**	*	***	****	*****/***
Landscape conservation	**	***	*	***	***	****/***
Recreation	**	***	*	***	*****	****/***
Total	19	23	13	27	28	30/23

Key:

*	Used hardly at all or not at all
**	Not used very much
***	A significant power in certain areas
****	Used extensively
*****	The dominant power

Source: Author

6 Were performance indicators built into the policies?

Not at the outset, but in the 1980s and 1990s, the 'bottom-line' accountancy-driven *Zeitgeist* has become all-pervasive. Apart from production-based policies, however, these indicators are not very useful, since they either measure the wrong indicators, for example, speed of decision making, or attempt to measure the unmeasurable, for example, environmental quality. Nonetheless, the approach has led to much more focused annual reports from the main countryside planning organisations, and to a set of useful reports from the National Audit Office and the various Committees of Parliament.

7 Have the policies been underresearched?

In general, yes. For example, the impact of agricultural and town and country planning policies has hardly been examined in terms of both the overall impact, and the variable impact between areas and groups of people. In contrast, some areas, such as the efficacy of environmental schemes and the recreational use of certain sites, have been relatively overresearched. There is also a research gap between contract research for planning organisations, with a limited agenda, and academic research which has tended to be too theoretical and lost touch with how real people think. There thus needs to be a revival of applied research which combines sensible theoretical approaches with pragmatic systems for measuring policy impacts.

8 Have the policies been effective overall?

Yes, in that Britain now enjoys plentiful food which is assured and relatively cheap, and has preserved its countryside from the depradations that have so disfigured other countries. However, the food that is produced has become standardised and bland, and is not as cheap as it could be because of the CAP. In addition, self-sufficiency has fallen in response to the need to cut European surpluses. The countryside has been preserved, but many villages have been swamped by characterless housing estates. The wider environment has also suffered, with landscapes, habitats, and peace and quiet suffering reverses. Forestry has been spectacularly successful in doubling the forested area, but there have been teething problems with the transition. The overall verdict is thus one of overall success, but with considerable reservations.

9 Have there been uneven policy impacts, by group or by area?

Yes. In agriculture, regional distinctions have been increased and arable farmers have in general done better than livestock farmers. In addition, large farms have fared best from the CAP, while smaller farms and farm workers have lost out. The impact of afforestation has been felt almost entirely in upland Scotland, but from the 1990s urban fringe forestry in England could become more important. Large parts of rural England, however, remain very poor in tree cover. Town and country planning has had its biggest impact in southern England, where it has been able to guide development pressures to certain areas and settlements, but it has been less sucessful elsewhere in stimulating growth in areas of decline. The middle classes have done best out of town and country planning, in that they have been able to afford the higher house prices encouraged by restrictive planning policies, which they have supported out of social and financial self-interest. Similarly, the middle classes have imposed the concept of quiet countryside recreation, in an unholy alliance with landowners and conservationists. Conservation policies have been quite effective in protecting key sites, but far less so in the wider countryside where serious damage has been done. Overall, the policies have favoured the wealthier, more articulate sections of society, but this is not surprising, since planning is a reflection of society, and not an agent of social change, however much some academics would like it to be or erroneously think it is.

10 Have the policies been cost-effective and are they environmentally sustainable?

Again, the answer varies across the field. Agricultural planning was initially cost-effective, but the CAP did not have the checks and balances of the 1947 system, and in the 1980s and 1990s the policy has become bloated and subject to fraud. Nonetheless, the policy only costs a few per cent of GDP, and it is axiomatic that full food shelves are worth such a small expenditure. Environmental policies, in contrast, are very cheap and they have been very cost-effective given their limited resources. Town and country planning represents remarkable value for money, which is more than could be said for forestry, which has its own peculiar economics but is unlikely ever to be cost-effective in strict economic terms. Forestry needs to be judged on wider environmental grounds. Very few of the policies are sustainable, notably agriculture with its emphasis on non-renewable fossil fuels and high technology which keeps one step ahead of the power of pests and diseases to mutate in self-defence. Town and country planning is little better, with its feeble attitude towards the car and lorry culture and the insiduous movement towards a 'spread city'. However, both of these policy areas cannot ignore global competition, and sustainability issues can only be treated at the global level.

11 What have been the side-effects?

The main side-effect has been the destruction of habitats and the loss of wildlife and landscapes as agriculture has embarked on a policy-supported and technology-induced tide of destruction. This has slowed down and even been reversed in some areas, but a time-bomb of long-term deterioration will be inherited as the existing landscape loses the habitats that were developed in the long agricultural depression of 1870 to 1939. The main difficulty is that modern agriculture no longer needs the matrix of buildings and landscape features that it once did and that have produced characteristic countryside areas. The short-term response has been to subsidise the continuation or re-creation of redundant systems. In the long term, side-effects will lead to a continuing erosion, unless policy makers wake up to the stark reality that farming no longer creates good countryside, any more than chemical engineering produces good townscape. Modern agriculture is an industry with all the side-effects that this implies.

Elsewhere, town and country planning has aided and abetted the transformation of villages from self-contained, socially stratified communities to increasingly middle-class commuter or retired/weekender places with little soul or character. This has been particularly marked in green belts and protected landscape areas. Although frequently criticised, it may be no bad thing, in that the good side-effect has been an increase in care and maintenance of the built and natural environment. In addition, the accident of birth gives no more right to live in a place than it would to follow one's parents as a brain surgeon or an airline pilot. The continued evolution of the village thus adds one more layer to its rich patina. The bad side-effect has been the suburban styles that have been allowed, and the fact that too many large estates have been allowed, intruding poor environments and car-borne pollution and noise into the countryside.

12 Were the policies modified by unforeseen events/policy changes elsewhere?

The main changes have been brought about by political forces which can be traced back to random rather than logical events. In particular, the advent of Mrs Thatcher as Prime Minister in 1979 was the result of a long.sequence of rather bizarre circumstances. This began in 1970, when the Labour Party lost the general election in the last two days of the election campaign, as poor trade figures and the loss of a football match in the World Cup changed the mood of the country and confounded the opinion polls. The incoming Prime Minister, Ted Heath, unadvisedly took on the coal miners in two strikes, during a worldwide energy crisis precipitated by war in the Middle East. He lost the subsequent election and Mrs Thatcher defeated him to become leader. The Labour Government, although in disarray, might have been able to mount a sensible campaign in 1979, but were forced to resign overnight by their own MPs when they failed to obtain a decisive referendum result on devolution to Scotland. Mrs Thatcher was then kept in power by a perverse swing to the left by the Labour Party in an over-reaction to Mrs Thatcher's 'New Right' policies. In the 1990s, the return of the Labour Party to the centre ground, under the leadership of two Scotsmen and a Scottish-dominated front bench, ushered in opinion poll ratings of nearly 60 per cent. In contrast, Mrs Thatcher never gained much more than 40 per cent of the vote, and only 30 per cent of the electorate. Thus the major changes brought about by Mrs Thatcher were not the result of deep-seated beliefs in the electorate, but by the failure of the opposition parties to mount a sensible or coordinated alternative. Many more examples could be provided of policies being developed in such haphazard ways and then being blown off course by unexpected weather, but in spite of these cautionary comments, policies have kept their underlying stability, notably in town and country planning.

13 Were the policies overtly sectoral and were they compatible with other policy areas?

Yes, largely because they have been produced by government departments which are organised on sectoral lines, as graphically shown in Figure 7.1, and thus have a vested interest in keeping policy areas to themselves. This has been manifested most strongly in the battle between MAFF and DOE over which should be the leading authority, and in the clash between the Countryside Commission and the former NCC. However, consultation between authorities is built into the system at all levels and Mather (1994) has argued that 'constructive tension' between rival agencies may not be a bad thing. Nonetheless, the tension in MAFF, between its responsibilities to food consumers and food producers, has been highlighted vividly in the BSE fiasco of the 1990s. When responsibilities for the environment are also added to the equation, the danger of internal contradictions probably becomes too great. A favoured solution has been a Ministry of Rural Affairs, but, like a Ministry for Men, this would be an over-artificial device merely adding another layer of bureaucracy. The solution is to build coordination into decision making, possibly by setting up some sort of sustainability test for all policies with a checklist of their side-effects and which agency could best advise on these, all set in the context of the 'integrated strategy for the countryside' advocated by the Countryside Commission (1991d).

14 What have been the effects of 16 years of Tory rule?

These are often exaggerated, especially in the area of planning, where in most cases attempts at radical reform in the early 1980s were rebuffed by the

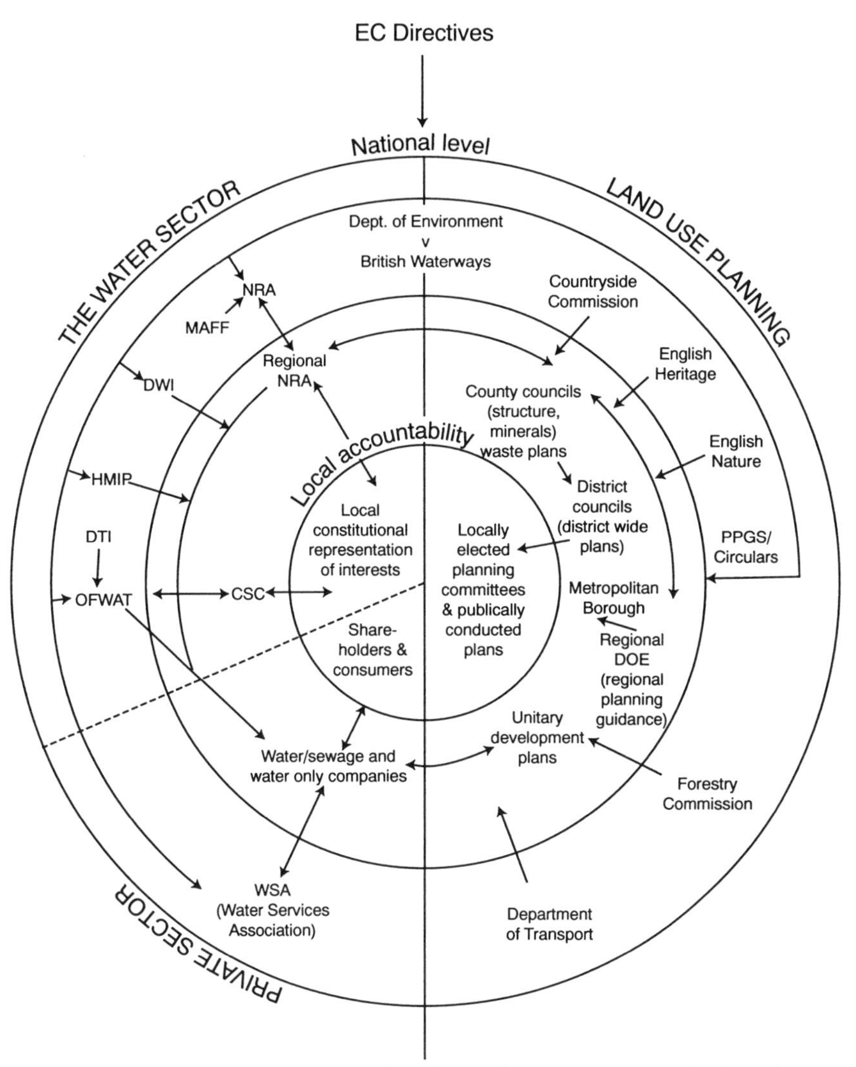

Figure 7.1 The sectoral system of land-use planning: an example from the perspective of the water sector

Source: Slater *et al.*, 1994

not unsubstantial conservation lobby in the Conservative Party. From the late 1980s the Tories were also partially converted to the environment as a middle-class vote winner, and by Secretaries of State who had strong personal convictions. In addition, the CAP remains a bête noir of the Party which they have been unable radically to reform in spite of being ideologically opposed to it.

15 Should the policies be modified?

This depends on the vision of the countryside and of politics that is adopted. If muddling through incrementalism based on concensus and the status quo is acceptable, then the policies only need to be modified in the eclectic way they have been so far. If, however, a vision of the countryside based on agriculture as an unfriendly activity and a return to urbane lifestyles is predicated, then only a radical reform will do.

16 What alternatives are there?

Gilg (1991a and 1992a) has produced analyses which have summarised the policy options both sectorally and by the Gilg/Selman spectrum. He was also able to subdivide them into evolutionary or revolutionary proposals, which reinforces the two choices outlined above.

Evolutionary changes are predicted by the 1995 'Rural White Paper' produced jointly by MAFF and the DOE (Cm 3016, 1995). In particular this addressed the problem of sectorality by expanding the remit of the Cabinet Committee dealing with the environment so that it would in future consider rural affairs. Elsewhere, apart from a continuing commitment to reform the CAP radically , it only proposed nine new measures, all of which involved relatively minor amendments to existing powers. Similar evolutionary proposals have been put forward by groups of professionals including the Royal Town Planning Institute *et al.* (1995) and the Countryside Commission (1991d) and by academics (Gilg, 1991a) and Bryden, 1994). Most notably, the Countryside Commission (1990) in their agenda for the 1990s has called, in general, for: 1) conservation to mean creation as well as preservation; 2) but nonetheless for the wild places to be cherished; and then, in detail, for 3) conserving through farming; 4) creating a new face for forestry; 5) making the countryside more accessible; but 6) averting damage from visitor pressure; 7) setting a new direction for transport; 8) establishing a new role for planning; 9) building on popular support, while 10) calling on the government to give the policy lead.

Revolutionary changes of varying degrees of change have been put forward mainly by academics. Many are based on sustainability, as for example, the Town and Country Planning Association's manifesto for a 'sustainable environment' (Blowers, 1993) and idealistic models of how planning could once claim to be the master allocator of environmental resources, as shown in Figure 7.2, or Green's (1995) call for a more sustainable agricultural

Figure 7.2 (opposite) Schematic of the 'planning for sustainability' framework

Source: Amundson, 1993

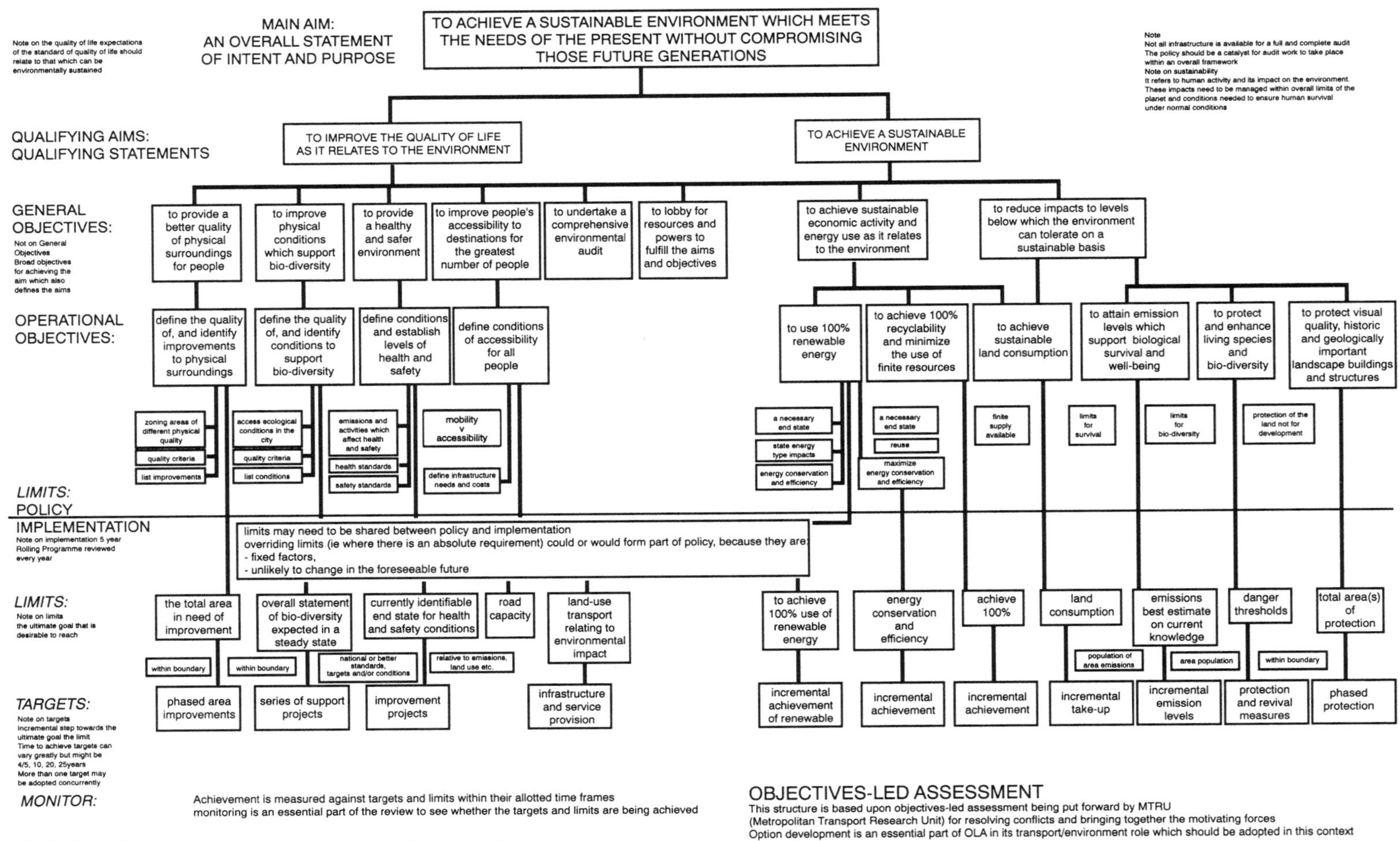
ENVIRONMENTAL AUDIT
An audit of the state of the environment and its impacts on it, needs to be undertaken to support the policy position and provide a strategy position for the future generations
Note on the quality of life expectations of the standard of quality of life should relate to that which can be environmentally sustained
MAIN AIM: AN OVERALL STATEMENT OF INTENT AND PURPOSE
TO ACHIEVE A SUSTAINABLE ENVIRONMENT WHICH MEETS THE NEEDS OF THE PRESENT WITHOUT COMPROMISING THOSE FUTURE GENERATIONS
Note
Not all infrastructure is available for a full and complete audit
The policy should be a catalyst for audit work to take place within an overall framework
Note on sustainability
It refers to human activity and its impact on the environment.
These impacts need to be managed within overall limits of the planet and conditions needed to ensure human survival under normal conditions
QUALIFYING AIMS: QUALIFYING STATEMENTS
TO IMPROVE THE QUALITY OF LIFE AS IT RELATES TO THE ENVIRONMENT
TO ACHIEVE A SUSTAINABLE ENVIRONMENT
GENERAL OBJECTIVES:
Not on General Objectives
Broad objectives for achieving the aim which also defines the aims
to provide a better quality of physical surroundings for people
to improve physical conditions which support bio-diversity
to provide a healthy and safer environment
to improve people's accessibility to destinations for the greatest number of people
to undertake a comprehensive environmental audit
to lobby for resources and powers to fulfill the aims and objectives
to achieve sustainable economic activity and energy use as it relates to the environment
to reduce impacts to levels below which the environment can tolerate on a sustainable basis
OPERATIONAL OBJECTIVES:
define the quality of, and identify improvements to physical surroundings
define the quality of, and identify conditions to support bio-diversity
define conditions and establish levels of health and safety
define conditions of accessibility for all people
to use 100% renewable energy
to achieve 100% recyclability and minimize the use of finite resources
to achieve sustainable land consumption
to attain emission levels which support biological survival and well-being
to protect and enhance living species and bio-diversity
to protect visual quality, historic and geologically important landscape buildings and structures
zoning areas of different physical quality
quality criteria
list improvements
access ecological conditions in the city
quality criteria
list conditions
emissions and activities which affect health and safety
health standards
safety standards
mobility v accessibility
define infrastructure needs and costs
a necessary end state
state energy type impacts
energy conservation and efficiency
a necessary end state
reuse
maximize energy conservation and efficiency
finite supply available
limits for survival
limits for bio-diversity
protection of the land not for development
LIMITS: POLICY
IMPLEMENTATION
Note on implementation 5 year Rolling Programme reviewed every year
limits may need to be shared between policy and implementation
overriding limits (ie where there is an absolute requirement) could or would form part of policy, because they are:
- fixed factors,
- unlikely to change in the foreseeable future
LIMITS:
Note on limits the ultimate goal that is desirable to reach
the total area in need of improvement
overall statement of bio-diversity expected in a steady state
currently identifiable end state for health and safety conditions
road capacity
land-use transport relating to environmental impact
to achieve 100% use of renewable energy
energy conservation and efficiency
achieve 100%
land consumption
emissions best estimate on current knowledge
danger thresholds
total area(s) of protection
within boundary
within boundary
national or better standards, targets and/or conditions
relative to emissions, land use etc.
population of area emissions
area population
within boundary
TARGETS:
Note on targets
Incremental step towards the ultimate goal the limit
Time to achieve targets can vary greatly but might be 4/5, 10, 20, 25years
More than one target may be adopted concurrently
phased area improvements
series of support projects
improvement projects
infrastructure and service provision
incremental achievement of renewable
incremental achievement
incremental achievement
incremental take-up
incremental emission levels
protection and revival measures
phased protection
MONITOR:
Achievement is measured against targets and limits within their allotted time frames
monitoring is an essential part of the review to see whether the targets and limits are being achieved
OPERATIONAL IMPLEMENTATION:
This is the end point at which people try to implement the policies
Results will vary according to resources (finance knowledge, technology)
The degree of success in implementation will help determine the need for resources
OBJECTIVES-LED ASSESSMENT
This structure is based upon objectives-led assessment being put forward by MTRU
(Metropolitan Transport Research Unit) for resolving conflicts and bringing together the motivating forces
Option development is an essential part of OLA in its transport/environment role which should be adopted in this context

policy. Others consider the options posed by different scenarios. For example, Masser *et al.* (1992) examine three scenarios based on growth, equity, and the environment, and Meeus *et al.* (1990) examine four scenarios centred around: optimising agricultural production; promoting regional differences; optimising landscape preservation; and the sustainable use of natural resources.

Two themes stand out in both these choices, first, the issue of sustainability, and second, whether the countryside should move from being a place of production to a place of consumption. In some ways this helps to polarise the choice further. The first choice would appear to be a move towards a devolved, more self-contained environment in which globalisation trends were reversed and people produced and consumed local goods, and thus reduced the need for energy to move goods from place to place. The second choice would appear to be a reinforcement of the division of space, with consumption moving back towards the town and the re-creation of urban lifestyles, thus reducing the enormous waste of energy of the putative 'spread city'. A third choice of letting existing trends continue is also an option but is ultimately unsustainable in that it is based on non-renewable and highly polluting fossil fuels.

The most likely option is the theme of this book, the development of policy from insufficient information and the incremental random accretion of policies in response to crises as they occur. The future ahead of us is, therefore, unready, and the only certainty about it, is uncertainty.

17 Overall verdict

It would be wrong, however, to end on too pessimistic a note, since planning, like religion, offers hope that there may be a better tomorrow, and there is little doubt that countryside planning for all its sectoral fragmentation and frailties has provided a better environment than would otherwise have been the case. Indeed the British countryside still contains many beautiful places and therefore provides a vision that in due course we can build a society that will be able not only to conserve the best of our heritage but also to create new countrysides for our successors.

GUIDE TO FURTHER READING AND INFORMATION

There are four main sources for developing your interest in Countryside Planning: first, text and research books; second; official publications; third, journals and periodicals; and fourth, electronic media.

TEXT AND RESEARCH BOOKS

Good texts on environmental planning and countryside issues in general are provided by:

Blowers, A. (ed.) (1993) *Planning for a Sustainable Environment*, London: Earthscan.

Bowler, I., Bryant, C., and Nellis, D. (1992) *Contemporary Rural Systems in Transition*, Wallingford: CAB International, 2 vols.

Cloke, P. (ed.) (1987) *Rural Planning Policy into Action*, London: Harper and Row.

Cloke, P. and Little, J. (1990) *The Rural State: Limits to Planning in Rural Society*, Oxford: Oxford University Press.

Gilg, A. (1991) *Countryside Planning Policies for the 1990s*, Wallingford: CAB International.

Gilg, A. (ed.) (1992) *Restructuring the Countryside: Environmental Policy in Practice*, Andover: Avebury.

Marsden, T., Murdoch, J., Lowe, P., Munton, R. and Flynn, A. (1993) *Constructing the Countryside*, London: University College London Press.

Murdoch, J. and Marsden, T. (1994) *Reconstituting Rurality*, London: University College London Press.

Robinson, G. (1990) *Conflict and Change in the Countryside*, London: Belhaven.

Selman, P. (1992) *Environmental Planning*, London: Paul Chapman.

Winter, A. (1996) *Rural Politics: Policies for Agriculture, Forestry and the Environment*, London: Routledge.

Good texts on specific aspects of countryside planning are provided by:

Bowler, I. (ed.) (1992) *The Geography of Agriculture in Developed Market Economies*, Harlow: Longman.

Champion, A. and Watkins, C. (eds) (1991) *People in the Countryside*, London: Paul Chapman.

Cullingworth, J. and Nadin, V. (1994) *Town and Country Planning in Britain*, London: Routledge.

Curry, N. (1994) *Countryside Recreation: Access and Land Use Planning*, Andover: E. & F.N. Spon.

Glasson, J., Therivel. R. and Chadwick, A. (1994) *Introduction to Environmental Impact Assessment, Principles and Procedures, Process, Practice and Prospects*, London: University College London Press.

Glyptis, S. (1991) *Countryside Recreation*, Harlow: Longman.

Goodman, D. and Redclift, M. (1991) *Refashioning Nature: Food, Ecology and Culture*, London: Routledge.

Groome, D. (1993) *Planning and Rural Recreation in Britain*, Aldershot: Avebury.

Harrison, C. (1991) *Countryside Recreation in a Changing Society*, London: TMS Partnership, University College London.

Hill, B. and Ray, D. (1987) *Economics for Agriculture: Food Farming and the Rural Economy*, Basingstoke: Macmillan.

Ilbery, B. (1992) *Agricultural Change in Great Britain*, Oxford: Oxford University Press.

Le Heron, R. (1993) *Globalized Agriculture: Political Choice*, Oxford: Pergamon.

Pierce, J. (1990) *The Food Resource*, Harlow: Longman.

Rydin, Y. (1993) *The British Planning System: An introduction*, Basingstoke: Macmillan.

Tracy, M. (1989) *Government and Agriculture in Western Europe 1880–1988*, London: Harvester.

Whitby, M. (ed.) (1994) *Incentives for Countryside Management: The Case of Environmentally Sensitive Areas*, Wallingford: CAB International.

OFFICIAL PUBLICATIONS

Cullingworth and Nadin (1994) provide an excellent summary of most of the official publications on pp. 316–31, in particular, of Command Papers, DOE advice, parliamentary enquiries in the form of House of Commons and House of Lords Committee reports, and Acts of Parliament. These are restricted to the broad area of Town and Country Planning however.

For a more comprehensive review see the 'Annual Reviews of Rural Planning', produced by Gilg, A. in *The Countryside Planning Yearbook*, 1980–86, *The International Yearbook of Rural Planning* 1987 and 1988, and *Progress in Rural Policy and Planning* 1990–95. In addition to the publications listed above, this series also includes: statutory instruments; consultation papers; departmental research publications; and departmental annual reports.

The annual reports of each organisation also make extremely useful reading. In the 1990s most of these are now published by the organisation involved except the major organisations such as MAFF, DOE and the Forestry Commission. However, the work of all publicly funded organisations is covered in the excellent series of Command Paper Annual Reports under their umbrella organisation. For more detail see the 'Evolutionary Events' sections of chapters 2 to 6.

In addition, most of the main organisations publish statistical series, notably MAFF and the DOE. For example, *Agriculture in the UK* (MAFF); *Farm Incomes in the UK* (MAFF); and *Land Use Change* (DOE). For more detail on agricultural data see Clark (1992).

The 1995 rural white paper Cmd 3016 (1995) *Rural England: A Nation Committed to a Living Countryside*, London: HMSO provides a useful if bland summary of rural issues and policies. It is a classical example of incremental policy evolution. White Papers for Scotland Cm 3041 (1995) and for Wales Cm 3180 (1996) were published after the manuscript was completed.

Another fairly self-congratulatory, but very useful, series is the annual summary of Government environmental policy which reports on progress since the first white paper was published under the generic title: *This Common Inheritance*. The series began with an overview in 1990 (Cmd 1200, 1990) and since then there have been four progress reports (Cmd 1655, 1991; Cmd 2068, 1992; Cmd 2549, 1994; and Cmd 2822, 1995).

JOURNALS AND PERIODICALS

Ecos provides a series of lively polemics.
Environment and Planning in its various forms (A–D) provides a mix of theoretical and empirical papers.
Farmers Weekly provides contemporary and alternative views. If planners think their powers are too weak, a week or two reading this publication will change their mind.
Journal of Agricultural Economics provides a mix of mathematical models and empiricism.
Journal of Environmental Planning and Management provides sound impact studies.
Journal of Rural Studies is the premier journal and provides a mix of theoretical and empirical papers.
Land Use Policy provides a series of pragmatic interpretations.
Planning provides news, short articles, and job adverts.
Planning Week provides contemporary news and views from the Royal Town Planning Institute.
Progress in Rural Policy and Planning and its predecssors *The Countryside Planning Yearbook* and *The International Yearbook of Rural Planning* provide a mix of legislative and literature reviews and empirical articles.
From 1997 *Perspectives on UK Rural Policy Planning* will provide a triannual review.
Town and Country Planning provides mainly exhortation and polemic.
Town Planning Review provides mainly empirical material, on both Town and Country Planning and Countryside Planning.

ELECTRONIC MEDIA

The Internet and the World Wide Web provide a fast evolving service. First, the Internet offers bulletin boards and the group exchange of information between researchers. Second, the World Wide Web offer pages from the various planning organisations, and in 1995 MAFF, the DOE, and the Countryside Commission all had pages on the Web, although some were better than others. Potentially, these pages offer a real alternative to the annual white papers.

In addition, bibliographical services such as BIDS (Bath Information and Data Services) offer the student access to a mass of literature in the journals. These services are as yet, however, less good at other publications, notably from the presure groups.

Finally, television offers a series of programmes ranging from the dramatic to the journalistic which report on current affairs and which, in combination with reading planning-related novels and quality newspapers, should enable the good student to assemble a large portfolio of case studies with which to put flesh on the bones of this book. However, the student is reminded that reading and viewing are no substitute for the real thing, and that though cyberspace may be fun, the countryside has so many more sensual pleasures.

BIBLIOGRAPHY

Abercrombie, N., Warde, A., Soothill, K., Urry, J. and Walby, S. (1988) *Contemporary British Society*, Oxford: Basil Blackwell.

Adams, D. (1994) *Urban Planning and the Development Process*, London: University College London Press.

Adams, W. (1984) *Implementing the Act: A Study of Habitat Protection under Part II of the Wildlife and Countryside Act 1981*, Godalming: World Wildlife Fund.

Adams, W., Hodge, I. and Bourne, N. (1994) 'Nature conservation and the management of the wider countryside in Eastern England', *Journal of Rural Studies*, 10: 147–57.

Agriculture, Fisheries and Food, Ministry of (1972) *Forestry Policy*, London: HMSO.

—— (1993) *Agriculture and England's Environment: Consultation Documents*, London: The Ministry.

—— (1995a) *European Agriculture: The Case for Radical Reform*, London: The Ministry.

—— (1995b) *Environmental Land Management Schemes: A Consultation Document*, London: The Ministry.

Almas, R. (1994) 'The rise and fall of agricultural policy cycles: from planned economy to green liberalism', *Journal of Rural Studies*, 10: 15–25.

Amundson, C. (1993) 'Sustainable aims and objectives: a planning framework', *Town and Country Planning*, 62 (1/2): 20–22.

Anderson, M. (1981) 'Planning policies and development control in the Sussex Downs AONB', *Town Planning Review*, 52: 5–25.

Anon (1995) 'CAP claims are exaggerated', *Farmers Weekly*, 10 November 1995: 15.

Appleton, J. (1975) *The Experience of Landscape*, London: John Wiley.

Audit Commission (1992) *Building in Quality: A Study of Development Control*, London: HMSO.

Ball, S.N. (1985) 'Sites of Special Scientific Interest', *Journal of Planning and Environment Law*, November, 767–77.

Bandarra, N. (1993) The Community Structural Funds and post-Maastricht cohesion, *Progress in Rural Policy and Planning*, 3: 221–42.

Barr, C. (1994) 'Countryside Survey 1990: an overview', *Ecos*, 15(3/4): 9–18.

Barr, C., Benefield, C., Bunce, B., Rinsdale, H. and Whitaker, M. (1986) *Landscape Changes in Britain*, Grange-over-Sands: Institute of Terrestrial Ecology.

Barrass, R. (1979) 'The first ten years of English Structure Planning: current progress and future directions', *Planning Outlook*, 22: 19–23.

Bateman, I. (1995) 'Environmental and economic appraisal', in O'Riordan, T. (ed) *Environmental Science for Environmental Management*, Harlow: Longman, 45–66.

Bell, S. (1995) 'From politics to tree planting', *Countryside*, March/April: 4–5.
Best, R. (1977) 'Agricultural land loss: myth or reality?', *The Planner*, 63: 15–16.
Bishop, K. (1992) Britain's new forests: Public dependence on private interest? in Gilg A. (ed.) *Restructuring the Countryside: Environmental Policy in Practice*, Aldershot: Avebury, 138–58.
Bishop, K. and Phillips, A. (1993a) 'Seven steps to the market – the development of the market-led approach to countryside conservation and recreation, *Journal of Rural Studies*, 9: 315–38.
—— (1993b) 'Integrating conservation, recreation and agriculture through the market place', *Ecos*, 14(2): 36–46.
Blacksell, M. and Gilg, A. (1981) *The Countryside: Planning and Change*, London: Allen & Unwin.
Blowers, A. (ed.) (1993) *Planning for a Sustainable Environment: A Report by the Town and Country Planning Association*, London: Earthscan.
Boucher, S. and Whatmore, S. (1993) 'Green gains? Planning by agreement and nature conservation', *Journal of Environmental Planning and Management*, 36: 33–49.
Bowers, J. (1990) *Economics of the Environment: The Conservationist's Response to the Pearce Report*, Telford: British Association of Nature Conservationists.
—— (1995) 'Sustainability, agriculture and agricultural policy', *Environment and Planning A*, 27: 1231–43.
Bowers, J. and Cheshire, P. (1983) *Agriculture, the Countryside and Land Use*, London: Methuen.
Bowler, I. (1976a) 'Regional agricultural policies: experience in the United Kingdom', *Economic Geography*, 52: 267–80.
—— (1976b) 'Spatial responses to agricultural subsidies in England and Wales', *Area*, 8: 225–9.
—— (1976c) 'The adoption of grant aid in agriculture', *New Transactions of the Institute of British Geographers*, 1: 143–58.
—— (1979) *Government and Agriculture*, Harlow: Longman.
—— (1981) 'Regional specialisation in the agricultural industry', *Journal of Agricultural Economics*, 32: 43–55.
—— (1983a) 'The agricultural pattern', in Johnston, R. and Doornkamp, J. *The Changing Geography of the United Kingdom*, London: Methuen: 75–104.
—— (1983b) 'Structural change in agriculture', in Pacione, M., *Rural Geography*, London: Harper & Row: 46–73.
—— (1985) *Agriculture under the Common Agricultural Policy*, Manchester: Manchester University Press.
—— (1992a) 'The Industrialisation of agriculture', In I. Bowler (ed.) *The Geography of Agriculture in Developed Market Economies*, Harlow: Longman, 7–31.
—— (1992b) 'Sustainable Agriculture as an alternative path of farm business development', in Bowler, I., Bryant, C. and Nellis, D. (eds) *Contemporary Rural Systems in Transition*, vol. 1, *Agriculture and Environment*, Wallingford, CAB International, 237–53.
Box, J. (1994) 'Changing the conservation culture', *Ecos*, 15(2): 16–22.
Bracken, I. and Hume, D. (1981) 'Problems and issues in structure planning', *Planning Outlook*, 23: 85–89.
Breheny, M. (1993) 'Planning the sustainable city region', *Town and Country Planning*, 62 (4): 71–5.
—— (1995a) 'The housing numbers game – again', *Town and Country Planning*, 64: 170–2.
—— (1995b) 'The compact city and transport energy consumption', *New Transactions of the Institute of British Geographers*, 20: 81–101.

Briggs, D. (1991) 'Environmental issues and policies in the European Community', *The Planner*, 70 (40): 7–10.

Briggs, D. and Kerrell, E. (1992) 'Patterns and implications of policy-induced agricultural adjustments in the European Community', in Gilg, A. (ed.) *Restructuring the Countryside: Environmental Policy in Practice*, Aldershot: Avebury, 85–102.

Brindley, T., Rydin, Y. and Stoker, G. (1989) *Remaking Planning: The Politics of Urban Change in the Thatcher Years*, London: Unwin Hyman.

Brotherton, I. (1973) 'The concept of carrying capacity of countryside recreation areas', *Recreation News Supplement*, 9: 6–10.

—— (1982) 'Development pressures and control in the National Parks', *Town Planning Review*, 53: 439–59.

—— (1985) 'Issues in National Park administration', *Environment and Planning A*, 17: 47–58.

—— (1986) 'Agriculture and afforestation controls: conservation and ideology', *Land Use Policy*, 3: 21–30.

—— (1987) 'The case for consultation', *Ecos*, 8(4): 18–23.

—— (1988) 'Grant-aided agricultural activity in National Parks', *Journal of Agricultural Economics*, 39: 376–81.

—— (1992) *Towards a Theory of Planning Control*, Sheffield: Department of Landscape, Sheffield University.

—— (1993a) 'The interpretation of planning appeals', *Journal of Environmental Planning and Management*, 36: 179–86.

—— (1993b) 'The interpretation of planning refusals', *Journal of Environmental Planning and Management*, 36: 167–78.

—— (1994) 'SSSIs: betting on a busted flush? *Ecos*, 15(2): 33–7.

Brotherton, I. and Devall, N. (1988) 'Forestry conflicts in National Parks', *Journal of Environmental Management*, 26: 229–38.

Brotherton, I. and Lowe, P. (1984) 'Statutory bodies and rural conservation: Agency or instrument', *Land Use Policy*, 1: 147–53.

Bruton, M. (1983) 'Local plans, local planning and development schemes in England 1974–82', *Town Planning Review*, 54: 4–23.

Bruton, M. and Fisher, E. (1980) 'The future of development plans', *Town Planning Review*, 51: 131–51.

Bruton, M. and Nicholson, D. (1983) 'Non-statutory local plans and supplementary planning guidance', *Journal of Planning and Environment Law*, July: 432–44.

—— (1985) 'Strategic land use planning and the British development plan system', *Town Planning Review*, 56: 21–41.

—— (1987a) 'Planners alter role to manage change', *Town and Country Planning*, 56: 22–3.

—— (1987b) 'A future for development plans', *Journal of Planning and Environment Law*, October: 687–703.

—— (1987c) *Local Planning in Practice*, London: Hutchinson.

Bryden, J. (1994) 'Prospects for rural areas in an enlarged Europe', *Journal of Rural Studies*, 10: 387–94.

Buckwell, A., Harvey, D., Thompson, M. and Parton, K. (1982) *The Costs of the Common Agricultural Policy*, London: Croom Helm.

Buller, H. and Hoggart, K. (1986) 'Non-decision making and community power: residential development control in rural areas', *Progress in Planning*, 25: 135–203.

Bunnell, G. (1995) 'Planning gain in theory and practice: Negotiation of agreements in Cambridgeshire', *Progress in Planning*, 44: 1–113.

Burbridge, J. (1994) 'Radical action and the evolution of consistency', *Ecos*, 15(2): 7–11.

Burton, R. (1974) *The Recreational Carrying Capacity of the Countryside*, Stoke on Trent: Keele University.

Centre for Agricultural Strategy (1980) *Strategy for the UK Forestry Industry*, Reading: The Centre.

Champion, A. (1993) 'A decade of regional and local population change: Census 91', *Town and Country Planning*, 62: 42–45.

Champion, A. and Watkins, C. (eds) (1991) *People in the Countryside*, London: Paul Chapman.

Champion, A., Wong, C., Rooke, A., Dorling, D., Coombes, M., Charlton, M. and Brunsdon, C. (1995) *The Population of Britain in the 1990s*, Oxford: Oxford University Press.

Chatters, C. (1992) 'Wildlife Act Flawed – It's Official', *Ecos*, 13(2): 49.

Cherry, G. (1979) 'The town planning movement and the late Victorian city', *New Transactions of the Institute of British Geographers*, 4: 306–19.

—— (1985) 'Scenic heritage and national park lobbies and legislation in England and Wales', *Leisure Studies*, 4: 127–39.

Cherry, G., Minett, J., Ward, S., Schaffer, F., and Bor, W. (1994) 'The Housing Town Planning Act 1909; the Housing Town Planning Act 1919; The Town and Country Planning Act 1932; The Town and Country Planning Act 1947; and the Town and Country Planning Act 1968', *The Planner*, 60: 75–703.

Christensen, K. (1993) 'Teaching Savvy', *Journal of Planning Education and Research*, 12: 202–12.

Clark, D. and Dunmore, K. (1990) *Involving the Private Sector in Rural Housing*, Cirencester: Action with Communities in Rural England (ACRE).

Clark, G. (1982) 'Housing policy in the Lake District', *New Transactions of the Institute of British Geographers*, 7: 59–70.

—— (1992) 'Data sources for studying agriculture', in I. Bowler (ed.) *The Geography of Agriculture in Developed Market Economies*, Harlow: Longman: 32–55.

Clark, G., Darrall, J., Grove-White, R., MacNaghten, P. and Urry, J. (1994) *Leisure, Culture and the English Countryside: Challenges and Conflicts*, London: Council for the Protection of Rural England.

Clarke, J. (ed.) (1993) *Nature in Question: An Anthology of Ideas and Arguments*, London: Earthscan.

Clarke, P., Jackman, B. and Mercer, D. (1980) *The Sunday Times Book of the Countryside*, London: Macdonald.

Cloke, P. (1977) 'An index of rurality in England and Wales', *Regional Studies*, 11: 31–46.

—— (1979) *Key Settlements in Rural Areas*, London: Methuen.

—— (1983) *An Introduction to Rural Settlement Planning*, London: Methuen.

—— (1987) 'Policy and implementation decisions', in P. Cloke (ed.) *Rural Planning Policy into Action*, London: Harper & Row: 19–34.

—— (1994) '(En)culturing political economy: A life in the day of a 'rural geographer', in Cloke, P., Doel, M., Matless, D., Phillips, M. and Thrift, N., *Writing the Rural*, London: Paul Chapman, 149–90.

Cloke, P., Doel, M., Matless, D., Phillips, M. and Thrift, N. (1994) *Writing the Rural: Five Cultural Geographies*, London: Paul Chapman.

Cloke, P. and Edwards, G. (1985) *Rurality in England and Wales 1991*, Lampeter: Department of Geography, St David's University College.

Cloke, P., Lapping, M. and Phillips, M. (1995) *Myth and Rural Culture*, London: Edward Arnold.

Cloke, P. and Little, J. (1987) 'The impact of decision making on rural communities: an example from Gloucestershire', *Applied Geography*, 7: 55–77.

—— (1990) *The Rural State: Limits to Planning in Rural Society*, Oxford: Oxford University Press.

Cloke, P. and McLaughlin, B. (1989) 'Politics of the alternative land use and rural economy (ALURE) proposals in the UK: crossroads or blind alley?', *Land Use Policy*, 6: 235–48.

Cloke, P. and Milbourne, P. (1992) 'Deprivation and lifestyles in rural Wales: rurality and the cultural dimension', *Journal of Rural Studies*, 8: 359–71.

Cloke, P. and Park, C. (1982) 'Country Parks in National Parks: a case study of Craig-y-Nos in the Brecon Beacons', *Journal of Environmental Management*, 12: 173–85.

Cloke, P. and Shaw, D. (1983) 'Rural settlement policies in structure plans', *Town Planning Review*, 54: 338–54.

Cloke, P. and Thrift, N. (1987) 'Intra-class conflict in rural areas', *Journal of Rural Studies*, 3: 321–34.

Clout, H. (1972) *Rural Geography*, Oxford: Pergamon Press.

—— (1984) *A Rural Policy for the EEC?*, London: Methuen.

Cm 56–II (1987) *The Government's Expenditure Plans 1987–88 to 1989–90*, London: HMSO.

Cm 569 (1989) *The Future of Development Plans*, London: HMSO.

Cm 1503 (1991) *Departmental Report by the Ministry of Agriculture, Fisheries and Food and the Intervention Board: The Government's Expenditure Plans 1991/2 to 1993/94*, London: HMSO.

Cm 1903 (1992) *Departmental Report by the Ministry of Agriculture, Fisheries and Food and the Intervention Board: The Government's Expenditure Plans 1992/93 to 1994/95*, London: HMSO.

Cm 2155 (1993) *Local Government in Wales: A Charter for the Future*, London: HMSO.

Cm 2267 (1993) *Shaping the Future: The New Councils*, London: HMSO.

Cm 2426 (1994) *Sustainable Development: The UK Strategy*, London: HMSO.

Cm 2427 (1994) *Climate Change: The UK Programme*, London: HMSO.

Cm 2428 (1994) *Biodiversity: The UK Action Plan*, London: HMSO.

Cm 2429 (1994) *Sustainable Forestry: The UK Programme*, London: HMSO.

Cm 2644 (1994) *Our Forests – The Way Ahead: Enterprise, Environment and Access: Conclusions from the Forestry Review*, London: HMSO.

Cm 2645 (1994) *The Government's Response to the Welsh Affairs Committee Report on 'Forestry and Woodlands'*, London: HMSO.

Cm 2803 (1995) *Departmental Report by the Ministry of Agriculture, Fisheries and Food and the Intervention Board, The Government's Expenditure Plans 1995/96 to 1997/98*, London: HMSO.

Cm 2807 (1995) *Department of the Environment: Annual Report 1995: The Government's Expenditure Plans 1995–96 to 1997–98*, London: HMSO.

Cm 2811 (1995) *The Department of National Heritage: The Government's Expenditure Plans 1995–96 to 1997–98*, London: HMSO.

Cm 2814 (1995) *The Government's Expenditure Plans 1995–96 to 1997–98: Departments of the Secretary of State for Scotland and the Forestry Commission*, London: HMSO.

Cm 2815 (1995) *The Government's Expenditure Plans 1995–96 to 1997–98: Departmental Report of the Welsh Office*, London: HMSO.

Cm 2822 (1995) *This Common Inheritance: UK Annual Report 1995*, London: HMSO.

Cm 3016 (1995) *Rural England: A Nation Committed to a Living Countryside*, London: HMSO.

Cmd 6153 (1940) *Report of the Royal Commission on the Distribution of the Industrial Population*, London: HMSO.

Cmd 6378 (1942) *Report of the Committee on Land Utilisation in Rural Areas*, London: HMSO.
Cmd 6386 (1942) *Final Report of the Expert Committee on Compensation and Betterment*, London: HMSO.
Cmd 6628 (1945) *National Parks in England and Wales: A Report by John Dower for the Ministry of Town and Country Planning*, London: HMSO.
Cmd 6631 (1945) *A Report by Sir J.D. Ramsay on National Parks for Scotland*, London: HMSO.
Cmd 7121 (1947) *National Parks (England and Wales): Report of the Committee: Chairman Sir Arthur Hobhouse*, London: HMSO.
Cmd 7122 (1947) *Conservation of Nature in England and Wales*, London: HMSO.
Cmd 7207 (1947) *Footpaths and Access to the Countryside*, London: HMSO.
Cmd 7814 (1947) *Final Report of the Scottish National Parks Committee and the Scottish Wildlife Convention*, London: HMSO.
Cmnd 2928 (1966) *Leisure in the Countryside*, London: HMSO.
Cmnd 6020 (1975) *Food from our Own Resources*, London: HMSO.
Cmnd 6447 (1943) *Postwar Forestry Policy*, London: HMSO.
Cmnd 7458 (1979) *Farming and the Nation*, London: HMSO.
Cmnd 9522 (1985) *Operation and Effectiveness of Part II of the Wildlife and Countryside Act, 1981*, London: HMSO.
Cmnd 9734 (1986) *Privatisation of the Water Authorities in England and Wales*, London: HMSO.
Coates, D. (1992) 'Affordable housing: the private sector perspective', *The Planner*, 78 (21): 57–9.
Cole, J. and Cole, F. (1993) *The Geography of the European Community*, London: Routledge.
Coleman, A. (1976) 'Is planning really necessary?', *Geographical Journal*, 142: 411–37.
—— (1977) 'Land use planning: success or failure?, *Architects Journal*, 165: 93–134.
Colman, D. (1989) 'Economic issues from the Broads Grazing Marshes Conservation Scheme', *Journal of Agricultural Economics*, 40: 336–44.
—— (1994) 'Comparative evaluation of environmental policies', in Whitby, M. (ed.) *Incentives for Countryside Management: The Case of Environmentally Sensitive Areas*, Wallingford: CAB International, 219–52.
Committee on the Management of Local Government (1967) *Report of the Committee Chaired by Sir John Redcliffe-Maud to the Ministry of Housing and Local Government*, London: HMSO.
Committee on Public Participation in Planning (1969) *People and planning: Report of the Committee chaired by A. Skeffington to the Ministry of Housing and Local Government*, London: HMSO.
Connell, B. (1995) 'Development pressures, environmental limits', *Town and Country Planning*, 64: 177–9.
Cook, H. (1993) 'Progress in water management in the lowlands', *Progress in Rural Policy and Planning*, 3: 91–103.
Coppock, T. (1961) 'The parish as a geographical statistical unit', *Tijdscrift voor Economische en Sociale Geografie*, 51: 317–26.
—— (1964) *An Agricultural Atlas of England and Wales*, London: Faber.
—— (1971) *An Agricultural Geography of Great Britain*, London: Bell.
Coppock, T. and Duffield, B. (1975) *Recreation in the Countryside*, London: Macmillan.
Copus, A. and Thomson, K. (1993) The budgetary effects of the CAP reforms in the UK, *Progress in Rural Policy and Planning*, 3: 208–18.

Cosgrove, D. (1990) 'Environmental thought and action: pre-modern and post-modern, *New Transactions of the Institute of British Geographers*, 15: 344–58.

Council for the Protection of Rural England (1994) 'Regional planning guidance', *Countryside Campaigner*, Spring: 8.

—— (1995) *Index of Planning Policy Guidance Notes*, London: The Council. (Produced in association with *Planning*.)

Countryside Commission (1970) *Countryside Recreation Glossary*, London: The Commission.

—— (1972) *The Goyt Valley Traffic Experiment*, London: The Commission.

—— (1976) *Tarn Hows: An Approach to the Management of a Popular Beauty Spot: CCP106*, Cheltenham: The Commission.

—— (1977) *New Agricultural Landscapes: Issues, Objectives and Action: CCP102*, Cheltenham: The Commission.

—— (1983) *The Changing Uplands*, Cheltenham: The Commission.

—— (1984a) *Agricultural Landscapes: Demonstration Farms: CCP170*, Cheltenham: The Commission.

—— (1984b) *Agricultural Landscapes: A Second Look: CCP168 and also CCP 169, 170, 171, 172 and 173*, Cheltenham: The Commission.

—— (1984c) *A Better Future for the Uplands: CCP162*, Cheltenham: The Commission.

—— (1984d) *The Broads: A Review: Conclusions and Recommendations: CCP163*, Cheltenham: The Commission.

—— (1984e) *National Countryside Recreation Survey: CCP201*, Cheltenham: The Commission.

—— (1985) *Cannock Chase 1979–84: A Country Park Plan on Trial: CCP181*, Cheltenham: The Commission.

—— (1986a) *Capital Tax Relief for Outstanding Scenic Land: CCP204*, Cheltenham: The Commission.

—— (1986b) *Grants for Landscape Conservation: CCP207*, Cheltenham: The Commission.

—— (1986c) *Grants for Countryside Conservation and Recreation*, Cheltenham: The Commission.

—— (1987a) *Recreation 2000. Policies for Enjoying the Countryside: CCP234*, Cheltenham: The Commission.

—— (1987b) *Recreation 2000. Enjoying the Countryside. Priorities for Action: CCP235*, Cheltenham: The Commission.

—— (1989a) *Forests for the Community CCP270; The Community Forest: Planning Design and Implementation: CCP271*; and *Your Community Forest: CCP272*, Cheltenham: The Commission.

—— (1989b) *Planning for a Greener Countryside: CCP264*, Cheltenham: The Commission.

—— (1990) *Ten Critical Years: An Agenda for the 1990s: CCP282*, Cheltenham: The Commission.

—— (1991a) *Landscape Change in the National Parks: CCP359*, Cheltenham: The Commission.

—— (1991b) *Fit for the Future: Report of the National Parks Review Panel: CCP 334*, Cheltenham: The Commission.

—— (1991c) *Fit for the Future: The Countryside Commission's Response: CCP337*, Cheltenham: The Commission.

—— (1991d) *Caring for the Countryside: A Policy Agenda for the Nineties: CCP351*, Cheltenham: The Commission.

—— (1992) *Implementing the Edwards Report: Report of the Seventh National Parks Workshop: CCP368*, Cheltenham: The Commission.
—— (1993b) *Hedgerow Incentive Scheme: CCP393*, Cheltenham: The Commission.
—— (1993a) *England's Trees and Woods: CCP408*, Cheltenham: The Commission.
—— (1993c) *It's Your Forest: Consultation Paper on the National Forest: CCP 410*, Cheltenham: The Commission.
—— (1993d) *The National Forest Strategy: Draft for Consultation: CPP411*, Cheltenham: The Commission.
—— (1994a) *The National Forest: The Strategy: CCP468*, Cheltenham: The Commission.
—— (1994b) *Working for the Countryside: CCP467*, Cheltenham: The Commission.
—— (1995a) *Annual Report: The 28th report of the Countryside Commission 1994–1995: CCP480*, Cheltenham: The Commission.
—— (1995b) *Grants and Payments Schemes: CCP422*, Cheltenham: The Commission.
—— (1995c) *Quality of Countryside: Quality of Life – The Countryside Commission's Prospectus into the Next Century: CCP470*, Cheltenham: The Commission.
Countryside Review Committee (1976) *The Countryside – Problems and Policies: A Discussion Paper*, London: HMSO.
Cowell, R. and Jehlicka, P. (1995) 'Backyard and biosphere: the spatial distribution of support for English and Welsh environmental organisations', *Area*, 27: 110–17.
Cox, G. and Lowe, P. (1983) 'A battle not the war: The politics of the Wildlife and Countryside Act', *Countryside Planning Yearbook*, 4: 48–76.
Cox, G., Lowe. P and Winter, M. (1990) *The Voluntary Principle in Conservation: The Farming and Wildlife Advisory Group*, Packard Publishing, Chichester.
Crabtree, J. and Appleton, Z. (1992) 'Economic evaluation of the Farm Woodland Scheme in Scotland', *Journal of Agricultural Economics*, 43: 355–67.
Crabtree, J. and Chalmers, N. (1994) 'Economic evaluation of policy instruments for conservation: standard payments and capital grants', *Land Use Policy*, 11: 94–106.
Crabtree, J and Macmillan, D. (1989) 'UK fiscal policy and new forestry planting', *Journal of Agricultural Economics*, 40: 314–22.
Craighill, A. and Goldsmith, E. (1994) 'A future for Set-Aside?', *Ecos*, 15 (3/4): 58–62.
Cramon-Taubadel, S. von (1993) The reform of the CAP from a German perspective, *Journal of Agricultural Economics*, 44: 394–409.
Cripps, J. (1980) 'The Countryside Commission: its first decade', *Countryside Planning Yearbook*, 1: 38–48.
Cross, D. and Bristow, M. (eds) (1983) *English Structure Planning*, London: Pion.
Cullingworth, J. and Nadin, V. (1994) *Town and Country Planning in Britain*, London: Routledge.
Curry, N. (1994) *Countryside Recreation: Access and Land Use Planning*, Andover: E. & F.N. Spon.
Curry, N. and McNab, A. (1986) 'Development control and landscape protection', *Countryside Planning Yearbook*, 7: 89–110.
Curtis, L. (1984) 'Reflections on management agreements for conservation of Exmoor moorland', *Journal of Agricultural Economics*, 34: 397–406.
de Gorter, H. and Swinnen, J. (1994) 'The economic polity of farm policy', *Journal of Agricultural Economics*, 45: 312–26.
de Loe, R.C. (1995) 'Exploring complex policy questions using the policy Delphi', *Applied Geography*, 15, 1: 53–68.
Dear, M. (1986) 'Postmodernism and planning', *Environment and Planning D*, 4: 367–84.
Delafons, J. (1995) 'Policy forum: Planning research and the policy process', *Town Planning Review*, 66: 83–109.

Denyer-Green, B. (1983) *Wildlife and Countryside Act: The Practitioner's Companion*, London: Surveyors Publications.

Derrida, J. (1995) *Spectres of Marx: The State of the Debt, the Work of Mourning, and the New International*, Routledge: London.

Diamond, D. (1995) 'Geography and planning in the information age', *New Transactions of the Institute of British Geographers*, 20: 131–8.

Diamond, J. (1975) 'The island dilemna: lessons of modern biogeographic studies for the design of natural reserves', *Biological Conservation*, 7: 129–46.

Dower, M. (1965) *The Challenge of Leisure*, London: Civic Trust.

Dudley, N. (1994) 'Setting good criteria for forestry', *Ecos*, 15(3/4): 68–71.

Edwards, A. (1986) *An Agricultural Land Budget for the United Kingdom*, Ashord: Wye College.

Eldon, J. (1988) 'Agricultural change, conservation and the role of advisors', *Ecos*, 9: 14–20.

Elson, M. (1981) 'Structure plan policies for pressured rural areas', *Countryside Planning Yearbook*, 2: 49–70.

—— (1986) *Green Belts: Conflict Mediation in the Urban Fringe*, London: Heinemann.

—— (1993) *The Effectiveness of Green Belts: Report for the Department of the Environment*, London: HMSO.

English Nature (1993) *Strategy for the 1990s*, Peterborough: Nature.

Environment Committee of the House of Commons (1985) *Operation and Effectiveness of Part II of the Wildlife and Countryside Act 1981: HC 6 (84–85)*, London: HMSO.

Environment, Department of (1972) *Sinews for Survival*, London: HMSO.

—— (1974) *Report of the National Park Policy Review Committee: Chairman Lord Sandford*, London: HMSO.

—— (1975) *Review of the Development Control System: Final Report by George Dobry QC*, London: HMSO.

—— (1976a) *Circular 9/76: The Dobry Report: Action by Local Authorities*, London: HMSO.

—— (1976b) *Circular 4/76: Report of the National Park Policy Review Committee*, London: HMSO.

—— (1977a) *Circular 108/77: Nature Conservation and Planning*, London: HMSO.

—— (1977b) *A Study of Exmoor*, London: HMSO.

—— (1978) *Land Availability: A Study of Land with Residential Planning Permission*, London: The Department.

—— (1980) *Circular 22/80: Development Control – Policy and Practice*, London: HMSO.

—— (1983) *Circular 4/83: Wildlife and Countryside Act 1981: Financial Guidelines for Management Agreements*, London: HMSO.

—— (1984a) *Memorandum on Structure and Local Plans*, London: HMSO.

—— (1984b) *Circular 14/84: Green Belts*, London: HMSO.

—— (1985) *Circular 1/85: The Use of Conditions in Planning Permission*, London: HMSO.

—— (1986) *Conservation and Development: The British Approach*, London: The Department.

—— (1987a) *Circular 16/87: Development Involving Agricultural Land*, London: HMSO.

—— (1987b) *Circular 27/87: Nature Conservation*, London: HMSO.

—— (1988a) *Circular 15/88: Environmental Assessment*, London: HMSO.

—— (1988b) *PPG2: Green Belts*, London: HMSO.

—— (1988c) *PPG12: Local Plans*. London: HMSO.
—— (1988d) *Our Common Future: A Perspective by the United Kingdom on the Report of the World Commission on Environment and Development*, London: The Department.
—— (1991) Circular 16/91: *Planning Obligations*, London: HMSO.
—— (1992a) *Circular 29/92: Indicative Forestry Strategies*, London: HMSO.
—— (1992b) *Development Plans and Regional Planning Guidance*, London: HMSO.
—— (1992c) *Development Plans: A Good Practice Guide*, London: HMSO.
—— (1992d) *PPG1: General Policies and Principles*, London: HMSO.
—— (1992e) *PPG3: Housing*, London: HMSO.
—— (1992f) *PPG7: The Countryside and the Rural Economy*, London: HMSO.
—— (1992g) *PPG2: Green Belts*, London: HMSO.
—— (1992h) *The Relationship between House Prices and Land Supply: A Report by G. Eve in Association with the Department of Land Economy at the University of Cambridge*, London: HMSO.
—— (1993) *Environmental Appraisal of Development Plans: Good Practice Guide*, London: HMSO.
—— (1994a) *PPG15: Planning and the Historic Environment*, London: HMSO.
—— (1994b) *PPG9: Nature Conservation*, London: HMSO.
—— (1995a) *PPG2: Green Belts*, London: HMSO.
—— (1995b) *Land Use Change in England: Issue 10*, London, The Department.
—— (1995c) *Towards Sustainability – Government Action in the United Kingdom*, London: HMSO.
Environmental Challenge Group (1994a) *Green Gauge*, Godalming: Worldwide Fund for Nature.
—— (1994b) *Environmental Measures: Indicators for the Environment*, Sandy: Royal Society for the Protection of Birds.
Errington, A. (1994) 'The peri-urban fringe: Europe's forgotten rural areas', *Journal of Rural Studies*, 10: 367–75.
Ervin, D. (1988) Set-aside programmes: using US experience to evaluate UK proposals, *Journal of Rural Studies*, 4: 181–91.
Etzioni, A. (1993) *The Spirit of Community: Rights, Responsibilities, and the Communitarian Agenda*, New York: Crown Publishers.
European Environmental Agency (1995) *European Environmental Agency: Putting Information to Work*, Copenhagen: The Agency.
Eyre, S. (1963) *Vegetation and Soils*, London: Edward Arnold.
Fairgreive, R. (1979) *A Policy for Forestry*, London: Conservative Political Centre.
Fearne, A. (1989) 'A "satisficing" model of CAP decision making', *Journal of Agricultural Economics*, 40: 71–81.
Fischer, F. and Forester, J. (eds) (1993) *The Argumentative Turn in Policy Analysis and Planning*, Durham NC: Duke University Press.
Flynn, A. (ed.) (1993) 'Special issue on costing the countryside', *Journal of Environmental Planning and Management*, 36: 3–116.
Flynn, A. and Murdoch, J. (1995) 'Rural change, regulation and sustainability: Guest editorial', *Environment and Planning A*, 27: 1180–92.
Flynn, A. and Pratt, A. (1993) 'Costing the countryside', *Journal of Environmental Planning and Management*, 36: 3–14.
Forester, J. (1993) *Critical Theory, Public Policy and Planning Practice*, New York: State University of New York Press.
Forestry Authority (1994) *Forest Landscape Design Guidelines*, London: HMSO.
Forestry Authority for England (1993) *Landscape Assessment for Indicative Forestry Strategies*, Cambridge: The Authority.

Forestry Commission (1978) *The Wood Production Outlook in Britain*, London: HMSO.
—— (1991) *Forestry Policy for Great Britain*, Edinburgh: The Commission.
—— (1994) *Annual Report 1993–94 HC 661(93–94)*, London: HMSO.
Foster, I. and Ilbery, B. (1992) 'Water protection zones: a valid management stategy?', in *Restructuring the Countryside: Environmental Policy in Practice*, A.Gilg (ed.), Aldershot: Avebury, 178–202.
Fraser, I. (1995) 'An analysis of management agreement bargaining under asymmetric information', *Journal of Agricultural Economics*, 46: 20–32.
Friedmann, J. (1973) *Retracking America: A Theory of Transactive Planning*, New York: Doubleday Anchor.
Friends of the Earth (1994) *Losing Interest – A Survey of Threats to Sites of Special Scientific Interest in England and Wales*, London: Friends of the Earth.
Froud, J. (1994) 'The impact of ESAs on lowland farming', *Land Use Policy*, 11: 107–18.
Gardner, R. and Hay, A. (1992) 'Geography in the United Kingdom 1988–92', *Geographical Journal*, 158: 13–30.
Gare, A. (1995) *Postmodernism and the Environmental Crisis*, London: Routledge.
Garner, J. (1993) *Garner's Rights of Way*, London: Longman.
Garrod, G., Willis, K. and Saunders, C. (1994) 'The benefits and costs of the Somerset Levels and Moors ESA', *Journal of Rural Studies*, 10: 131–45.
Garrod, G. and Willis, K. (1995) 'Valuing the benefits of the South Downs Environmentally Sensitive Area', *Journal of Agricultural Economics*, 46: 160–73.
Gasson, R. (1973) 'Goals and values of farmers', *Journal of Agricultural Economics*, 24: 521–37.
Gilder, I. (1979) 'Rural planning policies: an economic appraisal', *Progress in Planning*, 11: 213–71.
Gilg, A. (1978) *Countryside Planning*, Newton Abbot: David & Charles.
—— (1981) 'Planning for nature conservation: a struggle for survival and political respectability', in Kain, R. (ed.) *Planning for Conservation*, London: Mansell, 97–116.
—— (1985) *An Introduction to Rural Geography*, London: Edward Arnold.
—— (1986a) 'Annual Review', *Countryside Planning Yearbook*, 7: 1–42.
—— (1986b) 'The regional development of agriculture', *Geographica Polonica*, 51: 301–12.
—— (1987) 'Legislative review 1985–86', *International Yearbook of Rural Planning*, 1: 65–92.
—— (1991a) *Countryside Planning Policies for the 1990s*, Wallingford: CAB International.
—— (1991b) 'Planning for agriculture: the growing case for a conservation component', *Geoforum*, 22: 75–9.
—— (1992a) 'Policy options for the British countryside', in Bowler, I., Bryant, C. and Nellis, D. (eds) *Contemporary Rural Systems in Transition*, vol. 1, *Agriculture and Environment*, Wallingford: CAB International, 206–18.
—— (1992b) 'Restructuring the countryside: an introductory essay', in A. Gilg (ed.) *Restructuring the Countryside: Environmental Policy in Practice*, Aldershot: Avebury, 3–18.
—— (1993) 'Annual review of rural planning in the United Kingdom', *Progress in Rural Policy and Planning*, 3: 104–92.
—— (1994) 'Planning from first principles: introductory concepts: a voyage of discovery and rediscovery', completed chapter 1 of *Regulating Rural Environments*, in preparation for Routledge, available on request from the author for legitimate scholarly purposes before publication.

—— (1995) 'Annual review of rural planning in the United Kingdom', *Progress in Rural Policy and Planning*, 5: 31–88.

Gilg, A. and Battershill, M. (1996) 'Environmentally friendly farming in southwest England: an exploration and analysis', in *Changing Rural Policy in Britain*, (eds) N. Curry and S. Owen, Cheltenham: Countryside and Community Press, 200–24.

Gilg, A. and Blacksell, M. (1977) 'Planning control in an Area of Outstanding Natural Beauty', *Social and Economic Administration*, 11: 206–15.

Gilg, A. and Kelly, M., (1996a) 'The analysis of development control decisions: a position statement and some new insights from recent research in south west England, *Town Planning Review*, 67: 203–28.

Gilg, A. and Kelly, M. (1996b) 'Farmers, planners and councillors: an insider view of their interaction', in *Changing Rural Policy in Britain*, (eds) N. Curry and S. Owen, Cheltenham: Countryside and Community Press, 145–64.

—— (1996c) 'The implementation of planning policies in practice: a study of agricultural dwellings by an academic and a planning practitioner', *Progress in Planning*, in press.

Glasson, J., Therivel, R. and Chadwick, A. (1994) *Introduction to Environmental Impact Assessment, Principles and Procedures, Process, Practice and Prospects*, London: University College London Press.

Glyptis, S. (1991) *Countryside Recreation*, Harlow: Longman.

Goodman, D. and Redclift, M. (1985) 'Capitalism, petty commodity production and the farm enterprise', *Sociologia Ruralis*, 25: 231–47.

—— (1991) *Refashioning Nature: Food, Ecology and Culture*, London: Routledge.

Goodwin, M., Cloke, P. and Millbourne, P. (1995) 'Regulation theory and rural research: theorising contemporary rural change', *Environment and Planning A*, 27: 1245–60.

Gordon, D. and Forrest, R. (1995) *People and Places 2: Social and Economic Distinctions in England*, Bristol: School for Advanced Urban Studies.

Grainger, A. (1995) 'The forest transition: an alternative approach', *Area*, 27: 242–51.

Green, B. (1989) 'Agricultural impacts on the rural environment', *Journal of Applied Ecology*, 26: 793–802.

—— (1995) 'Plenty and wilderness? Creating a new countryside', *Ecos*, 16: 3–9.

Greenaway, D. (1991) 'The Uruguay round of multilateral trade negotiations: last chance for GATT?', *Journal of Agricultural Economics*, 42: 365–79.

Gregory, D. (1971) *Green Belts and Development Control: A Case Study in the West Midlands*, Birmingham: Centre for Urban and Regional Studies.

Groome, D. (1993) *Planning and Rural Recreation in Britain*, Aldershot: Avebury.

Groome, D. and Tarrant, C. (1985) 'Countryside recreation: achieving access for all', *Countryside Planning Yearbook*, 6: 72–100.

Guyomard, H., Mahe, L., Munk, K. and Roe, T. (1993) 'Agriculture in the Uruguay rounds: ambitions and realities', *Journal of Agricultural Economics*, 44: 245–63.

Hagerstrand, T. (1967) *Innovation Diffusion as a Spatial Process*, Chicago: Chicago University Press.

Haigh, N. (1992) *Manual of Environmental Policy: The EC and Britain*, Harlow: Longman.

Halfacree, K. (1993) 'Locality and social representation: space discourse and alternative definitions of the rural', *Journal of Rural Studies*, 9: 1–15.

—— (1995) 'Talking about rurality: social representations of the rural as expressed by residents of six English parishes', *Journal of Rural Studies*, 11: 1–20.

Hall, J. (1982) *The Geography of Planning Decisions*, Oxford: Oxford University Press.

Hall, P. (1974) 'The containment of urban England', *Geographical Journal*, 140: 386–418.

—— (1988) 'The industrial revolution in reverse?', *The Planner*, 74 (1): 15–20.
—— (1992) *Urban and Regional Planning*, 3rd edn, London: Routledge.
Hall, P., Drewett, R., Gracey, H. and Thomas, R. (1973) *The Containment of Urban England*, London: Allen & Unwin.
Ham, C. and Hill, M. (1984) *The Policy Process in the Modern Capitalist State*, Brighton: Harvester Wheatsheaf.
Harper, S. (1987) 'The rural-urban interface in England: a frmework for analysis', *New Transactions of the Institute of British Geographers*, 12: 284–302.
Harrison, C. (1991) *Countryside Recreation in a Changing Society*, London: TMS Partnership, Univerity College London.
Harvey, D. (1989) *The Condition of Postmodernity: An Inquiry into the Origins of Social Change*, Oxford: Blackwell.
Healey, P. (1983) *Local Plans in British Land Use Planning*, Oxford: Pergamon.
—— (1988) 'The British planning system and managing the urban environment', *Town Planning Review*, 59: 397–417.
—— (1989) 'Directions for change in the British planning system', *Town Planning Review*, 60: 125–49 and 319–32.
Healey, P. and Shaw, T. (1994) 'Changing meanings of "environment" in the British planning system', *New Transactions of the Institute of British Geographers*, 19: 425–38.
Herington, J. (1984) *The Outer City*, London: Harper Row.
—— (1985) 'Small settlements planning and residential patterns in outer Leicester', in J. Herington (ed.) *Planning and Residential Change in Outer Metropolitan Areas*, Loughborough: Department of Geography, Loughborough University, 22–34.
—— (1989) *Planning Processes: A Introduction for Geographers*, Cambridge: Cambridge University Press.
Hill, B. and Ray, D. (1987) *Economics for Agriculture: Food Farming and the Rural Economy*, Basingstoke: Macmillan.
Hill, B. and Young, N. (1989) *Alternative Support Systems for Rural Areas*, Ashford: Wye College.
Hill, H. (1980) *Freedom to Roam: The Struggle for Access to Britain's Moors and Mountains*, Ashbourne: Moorland Publishing.
Hobbs, P. (1992) 'Economic determinants of post-war British town planning', *Progress in Planning*, 38: 179–300.
Hooper, A., Pinch, P. and Rogers, S. (1988) 'Housing land availability: circular advice, circular arguments and circular methods, *Journal of Planning and Environment Law*, April: 225–39.
Hoskins, W. (1955) *The Making of the English Landscape*, London: Hodder & Stoughton.
House of Commons Committee of Public Accounts (1987) *Forestry Commission: Review of Objectives and Achievements: HC185(86–87)*, London: HMSO.
—— (1993) *Timber Harvesting and Marketing: HC597(92–93)*, London: HMSO.
House of Commons Select Committee on Agriculture (1990) *Land Use and Forestry: HC16(89–90)*, London: HMSO.
House of Commons Select Committee on the Environment (1993) *Forestry and the Environment: HC257(92–93)*, London: HMSO.
—— (1995) *The Environmental Impact of Leisure Activities: HC246(94–95)*, London: HMSO.
House of Commons Select Committee on European Legislation (1991) *Development and Future of the Common Agricultural Policy: HC29-xxix(90–91), London: HMSO.*
House of Lords Select Committeee on the European Communities (1989) *Nitrates in Water: HL73(88–89)*, London: HMSO.

—— (1991) *The Development and Future of the CAP: HL79(90–91), London: HMSO.*

—— (1992) *Implementation and Enforcement of Environmental Legislation: HL53(91–92)*, London: HMSO.

Housing and Local Government, Ministry of (1955) *Circular 42/55: Green Belts*, London: HMSO.

Housing and Local Government, Ministry of (1969) *The South Atcham Scheme: A Report by J. Warford*, London: HMSO.

—— (1970) *Development Plans – A Manual on Form and Content*, London: HMSO.

Hutchinson, W., Chilton, S. and Davis, J. (1995) 'Measuring non-use value of environmental goods: problems of information and cognition of cognitive questionnaire design methods', *Journal of Agricultural Economics*, 46: 97–112.

Ilbery, B. (1977) 'Point score analysis: a methodological framework for analysing the decision making process in agriculture', *Tijdschrift voor Economische en Sociale Geografie*, 68: 66–71.

—— (1978) 'Agricultural decision-making: a behavioural perspective', *Progress in Human Geography*, 2: 448–66.

—— (1992a) *Agricultural Change in Great Britain*, Oxford: Oxford University Press.

—— (1992c) 'State-assisted farm diversification in the United Kingdom', in Bowler, I., Bryant, C. and Nellis, D. (eds) *Contemporary Rural Systems in Transition*, vol. 1, *Agriculture and Environment*, Wallingford: CAB International, 100–16.

—— (1992b) 'Agricultural policy and land diversion in the European Community', *Progress in Rural Policy and Planning*, 2: 153–66.

Ilbery, B., Healey, M. and Higginbottom, J. (1995) 'On and off-farm business diversification by farm households in England', paper presented to second Anglo/Canadian/USA symposium on rural geography, University of North Carolina, August 1995.

Institute of Terrestrial Ecology (1993) *Countryside Survey 1990*, London: Department of the Environment.

Jessop, B. (1990) 'Regulation theories in retrospect and prospect', *Economy and Society*, 19: 153–216.

Johnson, A. (1994) 'What's super about big quarries?', *Ecos*, 15 (3/4): 35–42.

Johnson, J. and Price, C. (1987) 'Afforestation, employment and depopulation in Snowdonia', *Journal of Rural Studies*, 3: 195–205.

Jones, O., (1995) 'Lay discourses of the rural: developments and implications for rural studies', *Journal of Rural Studies*, 11: 35–49.

Jones, R. (1994) 'National forestry strategy required', *Land Use Policy*, 11: 124–7.

Kaplan, R. and Kaplan, S. (1989) *The Experience of Nature: A Psychological Perspective*, Cambridge: Cambridge University Press.

Kaydier, R. (1977) 'The pursuit of perfection', *The Planner*, 63: 84–5.

Kelly, M. and Gilg, A. (1996) 'The delivery of planning policy: an account of decision-making in a rural planning authority', *Environment and Planning C*, forthcoming.

Keyes, J. (1986) 'Controlling residential development in the Green Belt: a case study', *The Planner*, 72 (11): 18–20.

Kinniburgh, S. and Marshall, R. (1988) 'Management agreements in National Parks', *The Planner*, 74: 29–32.

Korbey, A. (ed.) (1984) *Investing in Rural Harmony: A Critique*, Reading: Centre for Agricultural Strategy.

Kula, E. (1988) *The Economics of Forestry: Modern Theory and Practice*, London: Croom Helm.

Larkham, P. (1992) 'Conservation and the changing urban landscape', *Progress in Planning*, 37: 84–181.

Lassey, W. (1977) *Planning in Rural Environments*, New York: McGraw-Hill.
Lavery, P. (1971) *Recreation Geography*, Newton Abbot: David & Charles.
Lawson, T. (1993) 'The implications of staff reforms at English Nature', *Ecos*, 14(1): 55–8.
—— (1995) 'Green hopes for the millenium', *Ecos*, 16(2): 42–7.
Le Heron, R. (1993) *Globalized Agriculture: Political Choice*, Oxford: Pergamon.
Leonard, P. (1982) 'Management agreements: a tool for conservation', *Journal of Agricultural Economics*, 33: 351–60.
Lloyd, T., Watkins, C. and Williams, D. (1995) 'Turning farmers into foresters via market liberalisation', *Journal of Agricultural Economics*, 46: 361–70.
Local Government Management Board (1995) *Sustainable Settlements – A Guide for Planners, Designers and Developers*, London: The Board and University of the West of England.
Lomas, J. (1994) 'The role of management agreements in rural environmental conservation', *Land Use Policy*, 11: 119–23.
Lowe, P. (1982) 'A question of bias', *Town and Country Planning*, 52: 132–4.
Lowe, P., Murdoch, J., Marsden, T., Munton, R. and Flynn, A. (1993) 'Regulating the new rural spaces: the uneven development of land', *Journal of Rural Studies*, 9: 205–22.
Lucas, R. and Sargent, T. (1981) 'After Keynesian macroeconomics', in *Rational Expectations and Economic Practice*, (eds) R. Lucas and T. Sargent, London: George Allen & Unwin: 295–320.
Lukes, S. (1974) *Power: A Radical View*, London: Macmillan.
Lund, P. and Price, R. (1995) 'UK overseas trade statistics: food, feed and drink', *Journal of Agricultural Economics*, 46: 252–60.
MacEwen, A. and MacEwen, M. (1982a) *National Parks: Conservation or Cosmetics?*, London: George Allen & Unwin.
—— (1982b) 'The Wildlife and Countryside Act 1981: an unprincipled Act?', *The Planner*, 68: 69–71.
MacGregor, B. and Ross, A. (1995) 'Master or servant? The changing role of the development plan in the British planning system', *Town Planning Review*, 66: 41–59.
MacGregor, B. and Stockdale, A. (1995) 'Land use change on Scottish Highland Estates', *Journal of Rural Studies*, 10: 301–9.
MacLaren, D. (1991) 'Agricultural trade policy analysis and international trade theory: a review of recent developments', *Journal of Agricultural Economics*, 42: 250–97.
Macmillan, D. (1993) 'Commercial forests in Scotland: an economic appraisal of replanting', *Journal of Agricultural Economics*, 44: 51–66.
McNaghten, P. (1995) 'Public attitudes to countryside leisure: a case study on ambivalence', *Journal of Rural Studies*, 11: 135–47.
Marsden, T. and Arce, A. (1995) 'Constructing quality: emerging food networks in the rural transition', *Environment and Planning A*, 27: 1261–79.
Marsden, T., Murdoch, J., Lowe, P., Munton, R. and Flynn, A. (1993) *Constructing the Countryside*, London: University College London Press.
Marsden, T., Whatmore, S., Munton, R. and Little, J. (1986a) 'The restructuring process and economic centrality in capitalist agriculture', *Journal of Rural Studies*, 2: 271–80.
—— (1986b) 'Towards a political economy of capitalist agriculture: a British perspective', *International Journal of Urban and Regional Research*, 10: 498–521.
Martin and Voorhees Associates (1980) *Review of Rural Settlement Policies*, London: Department of the Environment.
Masser, I., Sviden, O. and Wegener, M. (1992) *The Geography of Europe's Future*, London: Belhaven, .

Massey, D. (1994) *Space, Place and Gender*, London: Polity Press.
Mather, A. (1991a) 'The changing role of planning in rural land use: the example of afforestation in Scotland', *Journal of Rural Studies*, 7: 299–309.
—— (1991b) 'Pressure on Bristish forest policy: prologue to the post-industrial forest, *Area*, 23: 245–53.
—— (1993) 'Protected areas in the periphery: conservation and controversy in Northern Scotland', *Journal of Rural Studies*, 9: 371–84.
—— (1994) 'Policy reform and institutional restructuring: the case of the Forestry Commission', *Progress in Rural Policy and Planning*, 4: 69–78.
Mather, A. and Murray, N. (1987) 'Employment and private-sector afforestation in Scotland', *Journal of Rural Studies*, 3: 207–18.
Mather, A. and Thomson, K. (1995) 'The effects of afforestation on agriculture in Scotland', *Journal of Rural Studies*, 11: 187–202.
McClenaghan, J. and Blatchford, C. (1993) 'Development plan slippage', *The Planner*, 79(2): 29–30.
McCormick, J. (1991) *British Politics and the Environment*, London: Earthscan.
McDonald, G.J. (1989) 'Rural land use planning decisions by bargaining', *Journal of Rural Studies*, 5: 325–35.
McInerney, J. (1986) 'Agricultural policy at the crossroads', *Countryside Planning Yearbook*, 7: 44–75.
McLaughlin, B. (1986a) 'Rural policy in the 1980s: the revival of the rural idyll', *Journal of Rural Studies*, 2: 81–90.
—— (1986b) 'The rhetoric and reality of rural deprivation', *Journal of Rural Studies*, 2: 291–308.
—— (1987) 'Rural policy into the 1990s: self-help or self-deception?', *Journal of Rural Studies*, 3: 361–4.
McLoughlin, J. (1973) *Control and Urban Planning*, London: Faber.
Meeus, J., Wijermans, M. and Vroom, M. (1990) 'Agricultural landscapes in Europe and their transformation', *Landscape and Urban Planning*, 18: 289–352.
Meikle, J. (1991) 'Costs of residential development on greenfield sites', *The Planner*, 77 (13) 5–7.
Miles, I. (1995) 'IT makes itself at home?', *Town and Country Planning*, 64: 13–15.
Miller, R. (1981) *State Forestry for the Axe*, London: Institute for Economic Affairs.
Millward, H. (1993) 'Public access in the West European countryside: a comparative survey, *Journal of Rural Studies*, 9: 39–51.
Morgan, G. (1991) *A Strategic Approach to the Planning and Management of Parks and Open Spaces*, Basildon: Institute of Leisure Management.
MORI (1987) *Public Attitudes Towards Nature Conservation*, London: MORI.
Morphet, J. (1992) 'Subsidiarity and EC environmental futures', *Town and Country Planning*, July/August: 192–4.
Morris, C. and Potter, C. (1995) 'Recruiting the new conservationists: farmers', adoption of agri-environment schemes in the UK', *Journal of Rural Studies*, 11: 51–63.
Morris, P. and Therivel, R. (eds) (1995) *Methods of Environmental Impact Assessment*, London: University College London Press.
Moseley, M. (1979) *Accessibility: The Rural Challenge*, London: Methuen.
—— (1980) 'Is rural deprivation really rural?', *The Planner*, 66: 97.
Moss, G. (1978) 'The village – a matter of life and death', *Architects Journal*, 167 (3): 100–39.
Moxey, A., White, B., Sanderson, R. and Rushton, S. (1995) 'An approach to linking an ecological vegetation model to an agricultural economic model', *Journal of Agricultural Economics*, 46: 381–97.

Munton, R. (1977) 'Financial institutions: their ownership of agricultural land in Great Britain', *Area*, 9: 29–37.

—— (1983) *London's Green Belt: Containment in Practice*, London, George Allen & Unwin.

—— (1992) 'Factors of production in modern agriculture', in Bowler, I. (ed) *The Geography of Agriculture in Developed Market Economies*, Harlow: Longman, 56–84.

Munton, R., Marsden, T. and Ward, N. (1992) 'Uneven agrarian development and the social relations of farm households', in Bowler, I., Bryant, C. and Nellis, D. (eds) *Contemporary Rural Systems in Transition*, vol. 1, *Agriculture and Environment*, Wallingford: CAB International, 61–73.

Murdoch, J. (1995) 'Middle-class territory? Some remarks on the use of class analysis in rural studies', *Environment and Planning A*, 27: 1213–30.

Murdoch, J. and Marsden, T. (1992) 'A fair way to plan? Assessing the golf course boom', *The Planner*, 78(16): 8–10.

—— (1994) *Reconstituting Rurality*, London: University College London Press.

Mynors, C. (1995) *Listed Buildings and Conservation Areas*, London: Longman.

National Audit Office (1986) *Review of Forestry Commission Objectives and Achievements: HC75(86–87)*, London: HMSO.

—— (1989) *Grants to Aid the Structure of Agriculture in Great Britain*, HC105(89–90), London: HMSO.

—— (1993) *Timber Harvesting and Marketing: HC526(92–93)*, London: HMSO.

Nature Conservancy Council (1984) *Nature Conservation in Great Britain: Summary of Objectives and Strategy*, Peterborough: The Council.

Newbould, P. (1994) 'Nature conservation and the National Lottery', *Ecos*, 15(3/4): 63–5.

Newby, H., Bell, C., Rose, D. and Saunders, P. (1978) *Property, Paternalism and Power: Class and Control in Rural England*, London: Hutchinson.

North, J. (1989) 'How much land will be available for development in 2015?', *Development and Planning*, 1: 29–34.

Norton-Taylor, R. (1982) *Whose Land Is It Anyway?*, Wellingborough: Turnstone Press.

O'Riordan, T. (1985b) 'Halvergate: the politics of policy change', *Countryside Planning Yearbook*, 6: 101–6.

—— (1985a) 'Future directions for environmental policy', *Environment and Planning A*, 17: 1431–46.

—— (1990) 'Environmental assessment and a future strategy for the Broads', *The Planner*, 76: 7–10.

—— (1991) 'The new environmentalism and sustainable development', *The Science of the Total Environment*, 108: 5–15.

—— (1992) 'The Environment', in *Policy and Change in Thatcher's Britain*, Oxford: Pergamon Press.

—— (1995) 'Environmental science on the move', in O'Riordan, T. (ed.) *Environmental Science for Environmental Management*, Harlow: Longman, 1–16.

O'Riordan, T., Wood, C. and Shadrake, A. (1993) 'Landscapes for tomorrow', *Journal of Environmental Planning and Management*, 36: 123–47.

Office of Population Censuses and Surveys (1995a) 'The new OPCS area classifications', *Population Trends*, 79 Spring 1995: 15–30.

—— (1995b) *A 1991 Socio-economic Classification of Local and Health Authorities of Great Britain*, London: HMSO.

Oglethorpe, D. (1995) 'Sensitivity of farm plans under risk-averse behaviour: a note on the environmental implications', *Journal of Agricultural Economics*, 46: 227–32.

Organisation for Economic Cooperation and Development (1989) *Agricultural and Environmental Policies: Opportunities for Integration*, London: HMSO.
Orwin, C. (1944) *Country Planning*, Oxford: Oxford University Press.
Owen, D. (1995) 'The spatial and socio-economic patterns of minority ethnic groups in Great Britain', *Scottish Geographical Magazine*, 11: 27–35.
Owen, S. (1995) 'Local distinctiveness in villages: overcoming some impediments to clear thinking about village planning', *Town Planning Review*, 66: 143–61.
Owens, P. (1984) 'Rural leisure and recreation research: a retrospective evaluation', *Progress in Human Geography*, 8: 157–88.
Owens, S. (1994) 'Land, limits and sustainability: a conceptual framework and some dilemnas for the planning system', *New Transactions of the Institute of British Geographers*, 19: 439–56.
Pacione, M. (1980) 'The quality of life in a metropolitan village', *New Transactions of the Institute of British Geographers*, 5: 185–206.
—— (1990) 'Private profit and public interest in the residential development process: a case study of conflict in the urban fringe', *Journal of Rural Studies*, 6: 103–16.
—— (1991) 'Development pressure and the production of the built environment in the urban fringe', *Scottish Geographical Magazine*, 107: 162–9.
Pahl, R. (1965) *Urbs in Rure*, London: London School of Economics.
—— (1966) 'The rural/urban continuum', *Sociologia Ruralis*, 6: 299–329.
—— (1967) 'The rural/urban variable reconsidered: the cross cultural perspective', *Sociologia Ruralis*, 7: 21–30.
Paice, C. (1994) 'Market strength is false', *Farmers Weekly*, 23 December: 65.
Parry, M. (1982) *Surveys of Moorland and Roughland Change*, Birmingham: Department of Geography, University of Birmingham.
Passmore, W. (1974) *Man's Responsibility for Nature*, London: Duckworth.
Pattie, C.J., Russell, A.T. and Johnston, R.J. (1991) 'Going green in Britain? Votes for the Green Party and attitudes to green issues in the late 1980s', *Journal of Rural Studies*, 7, 3: 285–97.
Pearce, M. (1992) 'The effectiveness of the British land use planning system', *Town Planning Review*, 63: 13–28.
Petit, M., de Benedictus, M., Britton, M., de Groot, M., Henrichsmeyer, W. and Lechi, F. (1987) *Agricultural Policy Formulation in the European Community*, Amsterdam: Elsevier.
Phillips, A. (1993) 'The Countryside Commission and the Thatcher years', *Progress in Rural Policy and Planning*, 3: 63–89.
Phillips, D. and Williams, A. (1983) 'The social implications of rural housing policy', *Countryside Planning Yearbook*, 4: 77–102.
Pickering, M. (1987) 'Development control topics', *The Planner*, 73: 94–98.
PIEDA plc (Planning, Economic and Development Consultants) in asociation with CUDEM Leeds Polytechnic and Diamond, D. (1992) *Evaluating the Effectiveness of Land Use Planning*, London: Department of the Environment.
Pierce, J. (1990) *The Food Resource*, Harlow: Longman.
—— (1992) 'The policy agenda for sustainable agriculture', in Bowler, I., Bryant, C. and Nellis, D. (eds) *Contemporary Rural Systems in Transition* vol. 1, *Agriculture and the Environment*, Wallingford: CAB International.
Pigram, J. (1983) *Outdoor Recreation and Resource Management*, Beckenham: Croom Helm.
Pile, S. (1992) 'All winds and weathers: uncertainty, debt and the subsumption of the family farm', in Gilg, A. (ed.) *Restructuring the Countryside: Environmental Policy in Practice*, Aldershot: Avebury: 69–84.
Pitman, J. (1992) 'Changes in crop productivity and water quality in the United

Kingdom', in *Agriculture Change, Environment and Economy*, K. Hoggart (ed.), London: Mansell, 89–122.

Planning Advisory Group (1965) *The Future of Development Plans: Report to the Ministry of Housing and Local Government*, London: HMSO.

Porritt, J. and Winner, D. (1988) *The Coming of the Greens*, London: Collins Fontana.

Potter, C. (1986) 'Processes of countryside change in lowland England', *Journal of Rural Studies*, 2: 187–95.

—— (1987) 'Set-aside friend or foe?', *Ecos*, 8, 1: 36–8.

—— (1988) 'Making waves: farm extensification and conservation', *Ecos*, 9(3): 32–7.

—— (1990) 'Conservation under a European Farm Survival Policy', *Journal of Rural Studies*, 6: 1–7.

Potter, C., Burnham, P., Green, B. and Shinn, A. (1987) *Targetting for Conservation Set-aside*, Ashford: Wye College.

Potter, C., Cook, H. and Norman, C. (1993) 'The targeting of rural environmental policies: an assessment of Agri-Environment schemes in the UK', *Journal of Environmental Planning and Management*, 36: 199–216.

Potter, C. and Gasson, R. (1988) 'Farmer participation in voluntary land diversion schemes: some predictions from a survey', *Journal of Rural Studies*, 4: 365–75.

Potter, C. and Lobley, M. (1992) 'Elderly farmers as countryside managers', in Gilg, A. (ed.) *Restructuring the Countryside: Environmental Policy in Practice*, Aldershot: Avebury, 54–68.

Pountney, M. and Kingsbury, P. (1983) 'Aspects of development control: part 1: the relationship with local plans', *Town Planning Review*, 54: 139–54.

Poustie, M. (1993) 'The Environmental Information Regulations', *Scottish Planning and Environment Law*, 39: 43–44.

Powe, N. and Whitby, M. (1994) 'Economies of settlement size in rural settlement planning', *Town Planning Review*, 65: 415–34.

Price, C. (1987) 'The economics of forestry – any change?', *Ecos*, 8: 31–5.

—— (1989) *The Theory and Application of Forest Economics*, Oxford: Basil Blackwell.

Putnam, R.D., Leonardi, R. and Nanetti, R.Y. (1993) *Making Democracy Work: Civic Traditions in Modern Italy*, Princeton: Princeton University Press.

Pye-Smith, C. and Rose, C. (1984) *Crisis and Conservation*, London: Penguin.

Rackham, O. (1986) *The History of the Countryside*, London: J.M. Dent.

Ramblers Association (1980) *The Case Against Afforestation*, London: The Association.

Randolph, W. and Robert, S. (1981) 'Population redistribution in Great Britain 1971–81', *Town and Country Planning*, 50: 227–30.

Rawls, J. (1971) *A Theory of Justice*, Oxford: Clarendon Press.

—— (1993) *Political Liberalism*, New York: Columbia University Press.

Raymond, W. (1985) 'Options for reducing inputs to agriculture: a non-economist's view', *Journal of Agricultural Economics*, xxxvi: 345–54.

Redclift, M. and Benton, T. (eds) (1994) *Social Theory and the Global Environment*, London: Routledge.

Richardson, R., Gillespie, A. and Cornford, J. (1995) 'Low marks for rural home work', *Town and Country Planning*, 64: 82–4.

Robinson, G. (1991) 'EC agricultural policy and the environment: land use implications in the UK', *Land Use Policy*, 8: 95–107.

—— (1993) 'Trading strategies for New Zealand: The GATT, CER and trade liberalisation', *New Zealand Geographer*, 49 (2): 13–22.

—— (1994) 'The greening of agricultural policy: Scotland's ESAs', *Journal of Environmental Planning and Management*, 37: 215–25.

Robinson, G. and Ilbery, B. (1993) 'Reforming the CAP: beyond MacSharry', *Progress in Rural Policy and Planning*, 3: 197–208.

Rowan-Robinson, J., Ross, A. and Walton, W. (1995) 'Sustainable development and the development control process', *Town Planning Review*, 66: 269–86.

Rowswell, A. (1989) 'Rural depopulation and counterurbanisation: a paradox', *Area*, 21: 93–94.

Royal Town Planning Institute (1976) 'Memorandum', of evidence to the Environmental Sub-Committee of the House of Commons, *National Parks and the Countryside*, London: HMSO.

Royal Town Planning Institute, Royal Institution of Chartered Surveyors, the County Planning Officers Society, and the District Planning Officers Society (1995) *Tomorrow's Countryside: A Rural Strategy*, London: The Institute.

Rudlin, D. and Falk, N. (1995) *21st Century Homes: Building to Last*, London: Urban and Economic Development Group and the Joseph Rowntree Foundation.

Russell, N. (1994) 'Issues and options for agri-environmental policy', *Land Use Policy*, 11: 83–7.

Russell, N. and Fraser, I (1995) 'The potential impact of environmental cross-compliance on arable farming', *Journal of Agricultural Economics*, 46: 70–79.

Rydin, Y. (1985) 'Residential development and the planning system: a study of the housing land system at the local level', *Progress in Planning*, 24 (1): 1–69.

—— (1993) *The British Planning System: An Introduction*, Basingstoke: Macmillan.

—— (1995) 'Sustainable development and the role of land use planning', *Area*, 27: 369–77.

Rydin, Y. and Myerson, G. (1989) 'Exploring and interpreting ideological effects: a rhetorical approach to green belts', *Environment and Planning D: Society and Space*, 7: 463–79.

Schrader, H. (1994) 'Impact assessment of the EU Structural Funds to support regional economic development in rural areas of Germany', *Journal of Rural Studies*, 10: 357–65.

Scottish Office (1989) *Central Scotland Woodlands*, Edinburgh: Scottish Office.

Selman, P. (1988a) 'Potential planning implications of changing agricultural priorities', *Scottish Planning Law and Practice*, 23: 8–9.

—— (1988b) 'Rural land-use planning – resolving the British paradox', *Journal of Rural Studies*, 4: 277–94.

—— (1993) 'Landscape ecology and countryside planning: vision, theory and practice', *Journal of Rural Studies*, 9: 1–21.

—— (1995) 'Local sustainability: can the planning system help get us from here to there?', *Town Planning Review*, 66: 287–302.

Selznick, P. (1992) *The Moral Commonwealth: Social Theory and the Promise of Community*, Berkeley: University of California Press.

Seymour, S. and Cox, G. (1992) 'Nitrates in water: the politics of pollution regulation', in *Restructuring the Countryside: Environmental Policy in Practice*, A. Gilg (ed.), Aldershot: Avebury, 203–20.

Shaw, M. (ed.) (1979) *Rural Deprivation and Planning*, Norwich: Geo Books.

Sheail, J. (1995) 'Nature protection, ecologists and the farming context: a UK historical context', *Journal of Rural Studies*, 11: 79–88.

Short, J., Fleming, S. and Witt, S. (1986) *Housebuilding, Planning and Community Action*, London: Routledge & Kegan Paul.

Short, J., Witt, S. and Fleming, S. (1987) 'Conflict and compromise in the built environment: housebuilding in central Berkshire', *New Transactions of the Institute of British Geographers*, 12: 29–42.

Shucksmith, M., Chapman, P., Clark, G. and Black, S. (1994) 'Social welfare in rural Europe', *Journal of Rural Studies*, 10: 343–56.

Sidaway, R. (1990) 'Contemporary attitudes to landscape and implications for policy: a research agenda', *Landscape Research*, 15(2): 2–6.

Sidaway, R. and Thompson, D. (1991) 'Upland recreation: the limits of acceptable change', *Ecos*, 12(1): 31–7.

Sillince, J. (1986) 'Why did Warwickshire key settlement policy change in 1982? An assessment of the political implications of cuts in social services', *Geographical Journal*, 152: 176–92.

Sinclair, G. (1985) *How to Help Farmers and Keep England Beautiful*, London: Council for the Protection of Rural England.

—— (1992) *The Lost Land: Land Use Change in England 1945–90*, London: Council for the Protection of Rural England.

Slater, S., Marvin, S. and Newson, M. (1994) 'Land use planning and the water sector', *Town Planning Review*, 65: 375–97.

Smith, M. (1990) *The Politics of Agricultural Support*, Aldershot: Dartmouth Publishing.

Smith, S. (1983) *Recreation Geography*, London: Longman.

Southgate, M. (1995) 'Nature conservation and planning: implications of recent guidance and the Habitats Directive for Planners', *Report for the Natural and Built Environment Professions*, 5: 34–7.

Splash, C. and Simpson, I. (1994) 'Utilitarian and rights-based alternatives for protecting Sites of Special Scientific Interest', *Journal of Agricultural Economics*, 45: 15–26.

Stewart, P. (1985) 'British forestry policy: time for change?', *Land Use Policy*, 2: 16–29.

Stott, A. (1995) 'Monitoring land-use change – the context for the Countryside Survey 1990', *Ecos*, 15 (3/4): 3–8.

Sundseth, K. (1994) 'LIFE: a new European fund to help nature', *Ecos*, 15(1): 23–7.

Swales, V. (1994) 'Incentives for countryside management', *Ecos*, 15(3/4): 52–7.

Tarrant, J. (1992) 'Agriculture and the state', in I. Bowler (ed.) *The Geography of Agriculture in Developed Market Economies*, Harlow: Longman, 239–74.

Tarrant, J. and Cobb, R. (1991) 'The convergence of agricultural and environmental policies: the case of extensification in Eastern England', in Bowler, I., Bryant, C. and Nellis, D. (eds) *Contemporary Rural Systems in Transition*, vol. 1, *Agriculture and Environment*, Wallingford: CAB International, 153–65.

Tewdwr-Jones, M. (1995) 'Development control and the legitimacy of planning decisions', *Town Planning Review*, 66: 163–181.

Thornley, A. (1990) *Urban Planning under Thatcherism*, London: Routledge.

Thorns, D. (1970) 'Participation in rural planning', *International Review of Community Development*, 24: 129–38.

Tickle, A. (1994) 'Protected areas in Europe – the threat from acid rain', *Ecos*, 15(1): 33–9.

Tickner, J. (1978) Personal communication. On reading the first edition, my future stepfather noted that there aren't any tigers in Africa, so to perpetuate at least one error, another has been added with thanks to the classic Monty Python sketch.

Tompkins, S. (1986) *The Theft of the Hills*, London: The Ramblers Asociation.

—— (1989) 'Forestry in crisis', *Town and Country Planning*, 58: 276–7.

—— (1993) 'Conifer conspiracy', *Ecos*, 14(2): 23–7.

Toogood, M. (1995) 'Representing ecology and Highland tradition', *Area*, 27: 102–9.

Town and Country Planning Association (1990) 'The people and the land: strategic planning for the future, *Town and Country Planning*, 59: 239–46.

Townsend, A. (1993) 'The urban-rural cycle in the Thatcher growth years', *New Transactions of the Institute of British Geographers*, 18: 207–21.

Towse, R. (1988) 'Industrial location and site provision in an area of planning constraint: part of South West London's metropolitan green belt', *Area*, 20: 323–32.

Tracy, M. (1989) *Government and Agriculture in Western Europe 1880–1988*, London: Harvester.

Treasury, Her Majesty's (1972) *Forestry in Great Britain: An Inter-departmental Cost/benefit Study*, London: HMSO.

Troughton, M., (1991) 'Canadian agriculture and the Canada-US Free Trade Agreement.: a critical appraisal', *Progress in Rural Policy and Planning*, 1: 176–96.

Turoff, M. (1975) 'The policy Delphi', in H.A. Linstone and M. Turoff (eds) *The Delphi Method: Techniques and Applications*, Reading MA: Addison Wesley: 84–100.

Underwood, J. (1982) 'Local plans: the state of play', *The Planner*, 68: 136–53.

Vickery, D. (1995) 'Southern councils go for short cut plans' (1995), *Planning*, 1146, 24 November: 30–1.

Walford, N. (1995) 'A new agricultural geography of England and Wales', paper presented to the joint Anglo–Canadian–US symposium on rural geography, University of North Carolina at Charlotte, August.

Walker, A. (ed.) (1978) *Rural Poverty: Poverty, Deprivation and Planning in Rural Areas*, London: Child Poverty Action Group.

Wall, G. (1972) 'Socio-economic variations in pleasure trip patterns: the case of Hull car-owners', *Transactions of the Institute of British Geographers*, 59: 45–58.

Ward, N and Lowe, P. (1994) 'Shifting values in agriculture: the farm family and pollution regulation', *Journal of Rural Studies*, 10: 173–84.

Ward, N., Lowe, P., Seymour, S. and Clark, J. (1995) 'Rural restructuring and the regulation of farm pollution', *Environment and Planning A*, 27: 1193–211.

Ward, N., Talbot, H. and Lowe, P. (1995) 'Environmental agencies: a case of missed opportunities', *Ecos*, 16(2): 47–53.

Waters, G. (1994) 'Government policies for the countryside', *Land Use Policy*, 11: 88–93.

Watkins, C. (1984) 'The public control of woodlands', *Town Planning Review*, 54: 437–59.

Watkins, C. (1992) 'Forestry as an alternative land use', in Bowler, I., Bryant, C. and Nellis, D. (eds) *Contemporary Rural Systems in Transition*, Wallingford: CAB International, 182–94.

—— (1994) *Farmers Not Foresters: Constraints on the Planting of New Farm Woodland*, Nottingham: Department of Geography, University of Notingham.

Webster, S. and Felton, M. (1993) 'Targeting for nature conservation in agricultural policy', *Land Use Policy*, 10: 67–82.

Weekley, I. (1988) 'Rural depopulation and counterurbanisation: a paradox', *Area*, 20: 127–34.

Westmacott, R. and Worthington, T. (1974) *New Agricultural Landscapes*, Cheltenham: Countryside Commission.

Whatmore, S. and Boucher, S. (1993) 'Bargaining with nature: the discourse and practice of "environmental planning gain"', *New Transactions of the Institute of British Geographers*, 18: 166–78.

Whatmore, S., Munton, R., Marsden, T. and Little, J. (1987a) 'Interpreting a regional typology of farm businesses in southern England', *Sociologia Ruralis*, 27: 103–22.

—— (1987b) 'Towards a typology of farm businesses in contemporary British agriculture', *Sociologia Ruralis*, 27: 21–37.

Whitby, M. (ed.) (1994) *Incentives for Countryside Management: The Case of Environmentally Sensitive Areas*, Wallingford: CAB International.

Wibberley, G.P. (1959) *Agriculture and Urban Growth: A Study of the Competition for Rural Land*, London, Michael Joseph.

—— (1972) 'Conflicts in the countryside', *Town and Country Planning*, 40: 259–64.
Wilcock, D. (1995) 'Top-down bottom-up approaches to nature conservation and countryside management in Northern Ireland', *Area*, 27: 252–60.
Willis, K. and Garrod, G. (1992) 'Amenity value of forests in Great Britain and its impact on the internal rate of return from forestry', *Forestry*, 65: 331–46.
Wilson, G. (1994) 'German agri-environment schemes – I: a preliminary review', *Journal of Rural Studies*, 10: 27–45.
Wilson, R. (1987) 'The political power of the forestry lobby', *Ecos*: 8(4): 11–17.
World Commission on Environment and Development (1987) *Our Common Future* (The Brundtland Report), Oxford: Oxford University Press.

INDEX